ACTIVE MATTER

Edited by Skylar Tibbits

This book was set in GT Haptik by Grilli Type.
Printed and bound in South Korea.

Library of Congress Cataloging-in-Publication
Data is available
ISBN: 978-0-262-03680-1

CONTENTS

ACKNOWLEDGMENTS

A special thank you to those who have contributed to this book and collaborated, inspired, and supported us throughout the process. First and foremost I would like to thank Leila Kinney, CAST executive director and our team: Anastasia Hiller, Athina Papadopoulou, Patsy Baudoin, and E Roon Kang. It was a pleasure working together to make *Active Matter* a reality. This book would not have been possible without generous support from the Andrew W. Mellon Foundation, which funds MIT's Center for Art, Science & Technology (CAST), and from the office of Associate Provost Philip S. Khoury; and without the help of so many others: MIT's Department of Architecture and School of Architecture and Planning, Dean Hashim Sarkis, Department Head Meejin Yoon, Andreea O'Connell, Melissa Vaughn, Tom Gearty, Terry Knight, Inala Locke, and Patricia Driscoll; CAST producers Meg Rotzel and Katherine Higgins; Arts at MIT staff Stacy DeBartolo, Heidi Erickson, Sharon Lacey, and Leah Talatinian; the MIT Press and our acquisitions editor, Roger Conover, as well as the production editor, copy editor, and indexer who make us look much better than we are; MIT's International Design Center, John Brisson, Jon Griffith, Deb Payson, and Lennon Rogers; Tal Danino for the guidance and sanity checks; E Roon Kang, Minsun Eo, and Math Practice; and editor Patsy Baudoin for all of the hours and dedication working with us to pull this together. Thank you!

Sincere gratitude to all of our amazing authors and contributors!
Petrus Aejmelaeus-Lindström, Abdikhalaq Bade, Nadia M. Benbernou, David Benjamin, Katia Bertoldi, Greg Blonder, P. T. Brun, Markus J. Buehler, Max Carlson, Daniel Cellucci, Marcelo Coelho, Grace Copplestone, Corentin Coulais, Tal Danino, Felecia Davis, Erik D. Demaine, Martin L. Demaine, Michael D. Dickey, Delia Dumitrescu, Behnaz Farahi, Merav Gazit, Jan Genzer, Neil Gershenfeld, Amanda Ghassaei, Kyle Gilpin, A. Sydney Gladman, Fabio Gramazio, Christophe Guberan, Candice Gurbatri, Tetsuhiro Harimoto, Volker Helm, Chuck Hoberman, Douglas P. Holmes, Eric Höweler, Henry Hwang, Mariana Ibañez, Lodovica Illari, Hiroshi Ishii, Heinrich M. Jaeger, Benjamin Jenett, Jessica Jiang, Randall Kamien, Paul Kassabian, Simon Kim, Leila Kinney, Matthias Kohler, Will Langford, Jared Laucks, Neil Leach, Suzanne Lee, Jennifer A. Lewis, Hannah Lienhard, Ying Liu, Jesse Louis-Rosenberg, Anna Lubiw, Dan Luo, L. Mahadevan, Elisabetta A. Matsumoto, Paul McEuen, Lucy McRae, Achim Menges, Martin Miller, Ammar Mirjan, Andrew Moorman, Kieran A. Murphy, Paul Noll, Fiorenzo G. Omenetto, Luvena Ong, Christine Ortiz, Giffen Ott, Johannes T. B. Overvelde, Neri Oxman, Athina Papadopoulou, Nadya Peek, Matteo Pezzulla, Zhao Qin, John A. Rogers, Jessica Rosenkrantz, David Rosenwasser, Leah K. Roth, Daniela Rus, Jenny E. Sabin, Tomás Saraceno, Jane Scott, Michael Short, Alex Slocum, Lotje Sodderland, Mark Steranka, Andreas Thoma, Sharon Tracy, Max Vanatta, Sally Van Gorder, Martin van Hecke, Eric VanWyk, James Weaver, Justin Werfel, Rob Wood, Shu Yang, Peng Yin, Meejin Yoon, Xuanhe Zhao, Katia Zolotovsky.

Thank you to the speakers and attendees who made the original Active Matter Summit an enormous success, which led directly to the possibility of this book: Alfredo Alexander-Katz, Paola Antonelli, David Benjamin, Markus Buehler, Marcelo Coelho, Christophe Cros, Tal Danino, Erik Demaine, Michael Dickey, Merton C. Flemings, Neil Gershenfeld, Christophe Guberan, Peko Hosoi, Mariana Ibañez, Lodovica Illari, Heinrich Jaeger, Paul Kassabian, Sheila Kennedy, Junus Khan, Philip S. Khoury, Simon Kim, Leila Kinney, Chris Lasch, Suzanne Lee, Jennifer Lewis, Hod Lipson, Ying Liu, Jesse Louis-Rosenberg, Rob MacCurdy, John Main, Paul McEuen, Achim Menges, Arthur Olson, Fiorenzo Omenetto, Neri Oxman, Athina Papadopoulou, Pedro Reis, John Romanishin, Jessica Rosenkrantz, Daniela Rus, Jenny Sabin, Tomás Saraceno, Rob Wood, and Peng Yin. Thank you especially to the Summit's organizers: Leila Kinney, CAST executive director; Athina Papadopoulou, Co-chair; Anastasia Hiller, Alexis Sablone, Sara Falcone, Dimitris Mairopoulos, Meg Rotzel, Katherine Higgins, Heidi Erickson, Sharon Lacey, Steelcase Inc., and Autodesk Inc.

A special thank you to Jared Laucks, Schendy Kernizan, and the amazing Self-Assembly Lab team for their ongoing work and support: Athina Papadopoulou, Christophe Guberan, Bjorn Sparrman, Cosima Du Pasquier, Kate Hajash, Jonah Ross-Marrs, Maggie Hughes, Kate Weishaar, Hannah Lienhard, Willy Wu.

Book credits:
Volume editor: Skylar Tibbits
MIT CAST: Leila Kinney
MIT Press: Roger Conover
Developmental editor: Patsy Baudoin
Editing collaborator: Anastasia Hiller
Book concept lead: Athina Papadopoulou
Design: E Roon Kang, Minsun Eo / Math Practice
Index: Tobiah Waldron

Thank you, all.
Skylar Tibbits

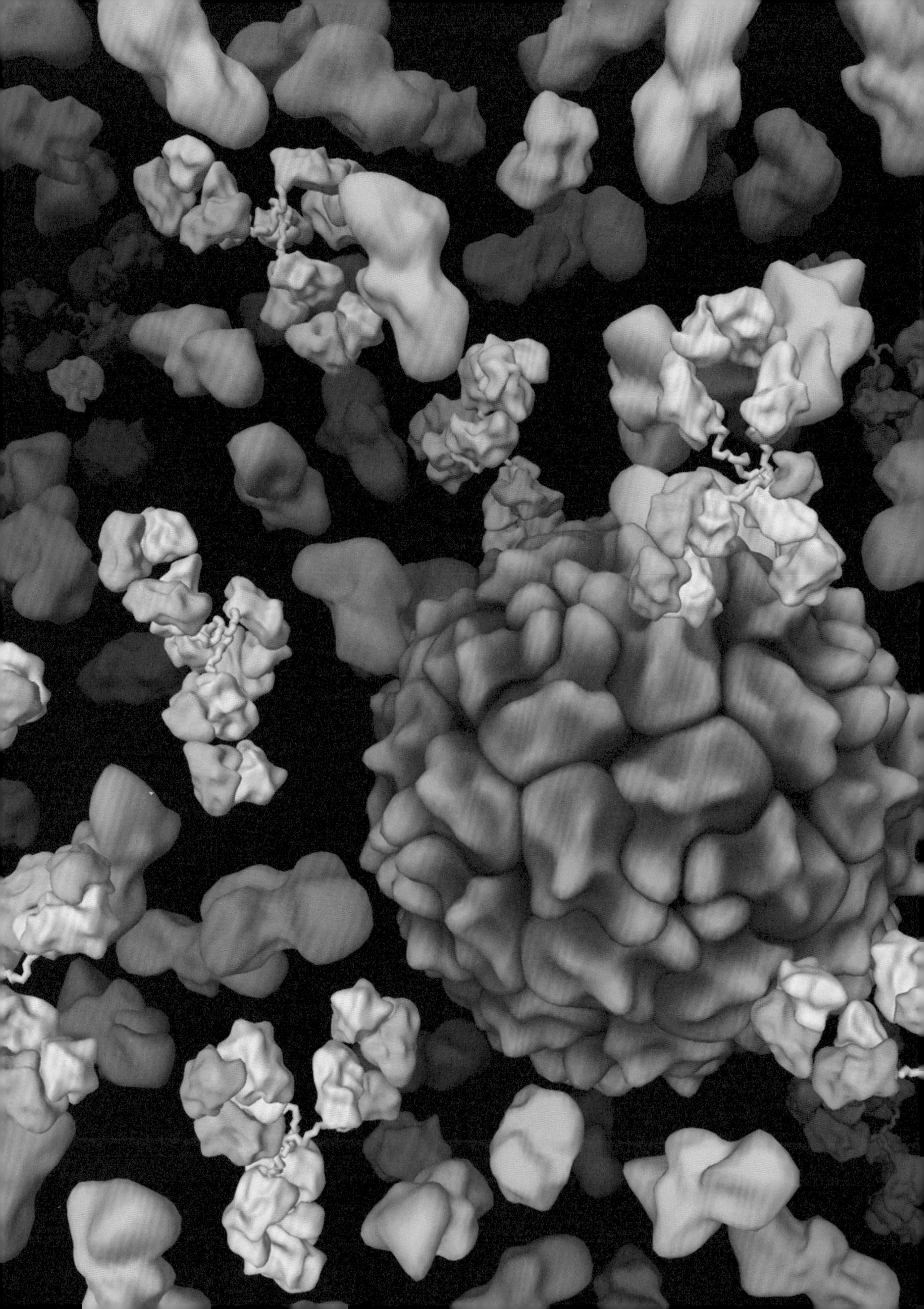

INTRODUCTION

Leila Kinney

Sitting in the audience during the Active Matter Summit in April 2014, I could not stop thinking about quorum sensing, a natural communications system that bacteria use to express themselves, by fluorescing when certain proteins reach a critical mass. The very first speaker, Neil Gershenfeld, thanked the Self-Assembly Lab for assembling, in this case, not materials but "such a fabulous group of people; it's a great reminder of the neighborhood that we sometimes take for granted." A sense of excitement began to pulse through the room as the conference continued, and speaker after speaker noted the variety of researchers working in different fields but on related problems, who may have known each other but not necessarily the details of what was going on in one another's labs, much of it unpublished. They began cross-referencing each other's presentations, finding connections and new possibilities among materials, miniature "machines," and fabrication processes as seemingly various as graphene kirigami, "Robobees" (inspired by insects and children's pop-up books), smorfs (smart morphable surfaces), "vegetable" leather (grown from green tea, sugar, yeast, and bacteria), mushroom bricks, and "HygroSkin" (meteorologically responsive) wood pavilions. It was as if the architects, engineers, designers, scientists, and industry representatives were themselves lighting up in unison in recognition of new possibilities, as organizer Skylar Tibbits, Co-Director of the Self-Assembly Lab, proclaimed, "to program nearly every material to assemble itself and transform in useful ways."

Many participants mentioned one defining new condition: the ability of scientists and engineers to see continuities across scales, from nano to micro, meso, and macro, in what Arthur Olson calls our current "bioatomic" age, where living things can be understood in a continuum from the atom to the organism. It is one reason why the *Active Matter* book itself is structured according to scale.

This conceptual framework is in striking contrast to the conditions that prevailed at MIT when Merton C. Flemings arrived in 1947. The Active Matter Summit and the accompanying design studio that Skylar Tibbits and Athina Papadopoulou taught (student projects from the class were exhibited during the Summit) were made possible by a gift from Ron Kurtz '54, '59, SM '60, who wished to honor the long, distinguished career of Flemings, his professor and mentor. When Mert came to MIT, the Institute had a department of Metallurgy where heavy industry was the focus and labs with big machinery, forges, and welding operations were the norm. He was an expert in applied properties of solidification and modern industrial foundry practices, but he went on to oversee a profound shift in the field toward engineering science and the transformation and expansion of the Metallurgy department into Material Sciences and Engineering. In a conversation with Philip Khoury, Associate Provost and Ford International Professor of History, Mert spoke of the transition from a focus on a "materials circle" (essentially a process of purification) for a single industry in the 1950s to a broader model in the 1960s, conceived as a "tetrahedron" (structure, properties, performance, and processing of materials) applicable to many areas.

In contrast, what drives the researchers who participated in the Active Matter Summit is a quest to miniaturize machines, on the one hand—think of proteins as "self-describing machines" (Arthur Olson) or

Illustration of the biomolecular self-assembly of a polio virus capsid. Credit Arthur Olson "wearable electronics for the individual cell" (Paul McEuen)—and on the other, an opportunity to scale up—think of studying the growth of slime mold to solve problems like the design of highway or railway networks (David Benjamin). What is propelling this emerging field and exciting its practitioners is the opportunity for architects, artists, and designers in all domains to leverage the properties, failures, and opportunities of these materials to do things that have never been imagined before. It is, as Neri Oxman observed, a special moment of confluence in four areas: (1) computational design—the ability to design complex shapes with very simple forms; (2) additive manufacturing—the availability of sophisticated tools to design by adding materials instead of carving them out; (3) materials science engineering—the availability of materials with very high spatial resolution for manufacturing; and (4) synthetic biology—the ability to design with the units of life.

The MIT Center for Art, Science & Technology (CAST) is very grateful to have been able to sponsor the Active Matter Summit and the companion studio, as a result of the enlightened philanthropy of Ron Kurtz, with additional support from the MIT Department of Architecture, Autodesk Inc., and Steelcase Inc. CAST was established in 2012 with a grant from the Andrew W. Mellon Foundation, and its mission is to be a catalyst for multidisciplinary creative experimentation and integration of the arts across all areas of MIT. A joint initiative of Philip S. Khoury, Associate Provost with responsibility for the arts and Ford International Professor of History, with the Deans of the School of Architecture and Planning and the School of Humanities, Arts, and Social Sciences,[1] the Center promotes research, teaching, and programming at the intersections of art, engineering, science, and the humanities. Led by faculty director Evan Ziporyn, Kenan Sahin Distinguished Professor of Music, since its inception, CAST has been the catalyst for more than 35 artist residencies and collaborative projects with MIT faculty and students, 20 cross-disciplinary courses and workshops, four concert series, and numerous multimedia projects, lectures, seminars, and symposia. The visiting artists program is a cornerstone of CAST's activities, which encourages cross-fertilization among disciplines and brings outstanding artists to campus to co-create with faculty and students. We were pleased to have the artist Tomás Saraceno as a participant in the Summit, who was CAST's inaugural Visiting Artist and has ongoing collaborations with two other speakers: with Markus Buehler, investigating the architecture of three-dimensional spiderwebs, and with Lodovica Illari, exploring atmospheric systems that can propel airborne sculptures without the use of helium, batteries, or solar panels.

Bringing together the disciplines and insights of artists, designers, scientists, and engineers may seem like a recent idea, and it is certainly a trending one, but those who know MIT's history will realize that creating a culture where the arts, science, and technology thrive as interrelated, mutually informing modes of exploration and discovery is deeply ingrained in the Institute's ethos. You can go back to the founding vision. In the 1861 pamphlet in which William Barton Rogers envisioned what was originally established as "Boston Tech" across the Charles River in Back Bay, he proposed not just a school of industrial science but also a museum and a society of arts.

However, it was not until the late 1960s and 1970s, in the era of György Kepes and the 13th President of MIT, Jerome B. Wiesner, that MIT made a serious commitment to the visual and performing arts. A Hungarian-born multimedia artist, designer, theorist, educator, and member of the Bauhaus, Kepes emigrated to the United States and, after a brief association with the New Bauhaus in Chicago, arrived at MIT in 1946. He immediately understood the distinctive nature of this institution. *The New Landscape in Art and Science*, published in 1956, and the seven-volume series *Vision and Value*, which appeared in 1965 and 1966, looked for connections among ways of thinking and viewing the world by artists and scientists. In 1967, he established the Center for Advanced Visual Studies (CAVS), one of the first programs to bring together artists, scientists, and engineers in a research environment. He articulated the ambition of creating "art on a civic scale" and "environmental art," using unconventional materials such as steam and "the fluid power of light in motion" as artistic media. Researchers at CAVS pioneered the use of technologies such as lasers, plasma sculptures, sky art, and holography as tools of expression in public and environmental art and established a legacy of creative practice at MIT that embraces the challenge of inventing new methods, media, and technologies for artistic production, alongside the goal of creating the most expressive artifacts, performances, and buildings.

As president, Jerry Wiesner fostered a multimedia program in the arts at MIT, with ambitions to firmly ground it in teaching and research at the Institute. Wiesner established the Council for the Arts at MIT in 1974, one of the first organizations of its kind in a US university, and his support led to the establishment of the List Visual Arts Center and the MIT Media Laboratory in 1985. As President L. Rafael Reif wrote recently in *Technology Review*,[2] his predecessor "saw vital connections among the communications he studied as an engineer, the means of communication between human beings (including media, music, and art), and what Norbert Wiener called cybernetics, or 'the scientific study of control and communication in the animal and the machine.'" CAVS, which merged with the Program in Art, Culture and Technology in 2009, the Media Lab, and the more recently established Self-Assembly Lab, all in the School of Architecture and Planning, along with many other labs in the Schools of Engineering and of Science that welcome artists and designers into their midst, have brought this vision forward and integrated it into the creative life of the campus far beyond what Weisner could have imagined.

Active Matter, in its multimodal expression of studio, summit, and book, is the brainchild of Skylar Tibbits, whose own ability to bridge a number of disciplines—architecture, computation, biology, materials science, and product design, to name the most pertinent—is a pioneering example of this promising, emerging field. Already the recipient of many awards for innovation in his young career, Skylar also has an extraordinary ability to collaborate and to convene leading thinkers around common concerns and fertile avenues of exploration. His colleague Athina Papadopoulou, PhD candidate and researcher at the Self-Assembly Lab, played a crucial role in the studio and took the lead in organizing the summit, with vital assistance from CAST producers Meg Rotzel and Katherine Higgins and important support from Anastasia Hiller and Alexis Sablone at the Self-Assembly Lab, and Arts at MIT staff Stacy DeBartolo, Heidi Erickson, Sharon Lacey, and Leah Talatinian. We deeply appreciate their tremendous effort and the excellent outcome. We are grateful to Patsy Baudoin, Matthew Christensen, and Anastasia Hiller for their invaluable help in preparing the manuscript and to E Roon Kang for his brilliant and flexible design "across scales" of a book containing a varied array of contributions and images. Our very special thanks to Roger Conover of the MIT Press, who readily embraced this project. It is the second publication documenting what CAST anticipates will be an ongoing series of exciting and unexpected revelations that emerge from bringing together such disparate thinkers; for, to paraphrase Arthur Olson, the best example of active matter is the contributors themselves.

Leila W. Kinney, Executive Director
MIT Center for Art, Science & Technology

NOTES

1 Adèle Naudé Santos was Dean of the School of Architecture and Planning and Deborah K. Fitzgerald was Dean of the School of Humanities, Arts, and Social Sciences when CAST was founded in 2012; Hashim Sarkis succeeded Santos in January 2015, and Melissa Nobles succeeded Fitzgerald in July 2015.

2 https://www.technologyreview.com/s/536471/from-the-president-mit-artist-on-a-global-stage/

AN INTRODUCTION TO ACTIVE MATTER

Skylar Tibbits

> *ACTIVE MATTER is a newly emerging field focused on physical materials that can assemble themselves, transform autonomously, and sense, react, or compute based on internal and external information.*

The impetus for this book grew out of the 2015 Active Matter Summit at MIT, when, for the first time, researchers from seemingly unrelated disciplines convened in one place to discuss their new efforts to design and create active, intelligent, and dynamic materials. With more than forty contributors, this book includes pioneering research and groundbreaking experiments in this emerging field as well as visionary perspectives on future materials, design products, and even entirely new industries. We hope that this book will serve as an introduction to active matter and the field guide for future generations of matter programmers.

The world of active matter spans disparate scales, from nanoscale to microscopic and macroscopic, as well as different disciplines, from synthetic biology to computer science, materials science, robotics, and even large-scale, active architectures. This book brings together leading academic researchers, industry practitioners, artists, designers, curators, and other thought leaders who are paving the way for a new, programmable, and highly active physical world that will transform design and manufacturing in the near future.

The initial insight for the Active Matter Summit began with people—all of our colleagues "in the field" who are not actually in the same field but from very different disciplines and departments working on completely different yet deeply related topics—and an interest in teasing out these connections. We often run into each other at seemingly unrelated conferences, cite publications from very different fields, stay up to date, and are inspired by one another, though we have no apparent disciplinary connection in the traditional sense. Yet beneath all of the schools, departments, and disciplinary problems that divide us, we are actually working on similar phenomena with similar goals. We aim to design and create active, intelligent, and dynamic materials with new emerging fabrication techniques, design tools, or material processes. This is the underlying thread tying all of these topics and individuals together, yet it hasn't been articulated. Thus, Active Matter emerged as an attempt at clarity and a unifying "field" that could tie all of the disconnected parts into a collective swarm. This book serves as a roadmap or momentary snapshot of the connections, relationships, and even differences highlighting the prominent figures and topics within this field.

Beyond being an introduction and map, we hope this book will inspire future generation of students, researchers, entrepreneurs, and policymakers. It truly feels like we are at a turning point in our relationship with matter. At some point we will look back on our world of programming computers and machines as we do on the mainframe computers of the last century. Why did we define academic degrees or entire fields (e.g., "computer science" or "computer programming") tied to a fleeting tool like today's "computer" rather than to the process and information or the fundamental principles of

Figure 0.1
The first transistor developed at Bell Laboratories in December 1947. Reprinted with permission of Nokia Corporation.

programming? Why did we make such efforts to create disciplinary borders rather than creating connections by focusing on principles, systems, and phenomena? We are more interested in a new way of thinking about matter and a unique perspective for interacting with the world. It will not just be computers that we program in the future, but everything will be computing and matter will be programmable. This book gathers the energy of the field in order to inspire the next generation.

> *If over the past half-century we have experienced a software and hardware revolution, we are now experiencing a true materials revolution. We can now sense, compute, and actuate with materials alone, just as one could previously with software and hardware platforms. It is becoming increasingly clear that materials are a platform for turning digital information into physical performance and functionality.*
>
> *If yesterday we programmed computers and machines, today we program matter itself.*

How did active matter emerge? We could go back to the history of computing with Ada Lovelace, Charles Babbage, or even the Jacquard loom as the implementation of mechanical computing for industrial production with a true material output. Or perhaps we could emphasize Turing, von Neumann, or any of the other incredibly important figures in computing. But it seems to make more sense to start at the beginning of the "digital" and how that relates to the physical. In 1937, Claude Shannon produced what has been described as one of the most influential master's theses of all time, in which he introduced the concept of Boolean logic for relays, digital logic eventually becoming the foundation of digital communication, electronics, and information theory. Since the subsequent invention of the transistor in

Figure 0.2
Ivan Sutherland in 1963 demonstrating his development of the first computer-aided design (CAD) tool called Sketchpad.

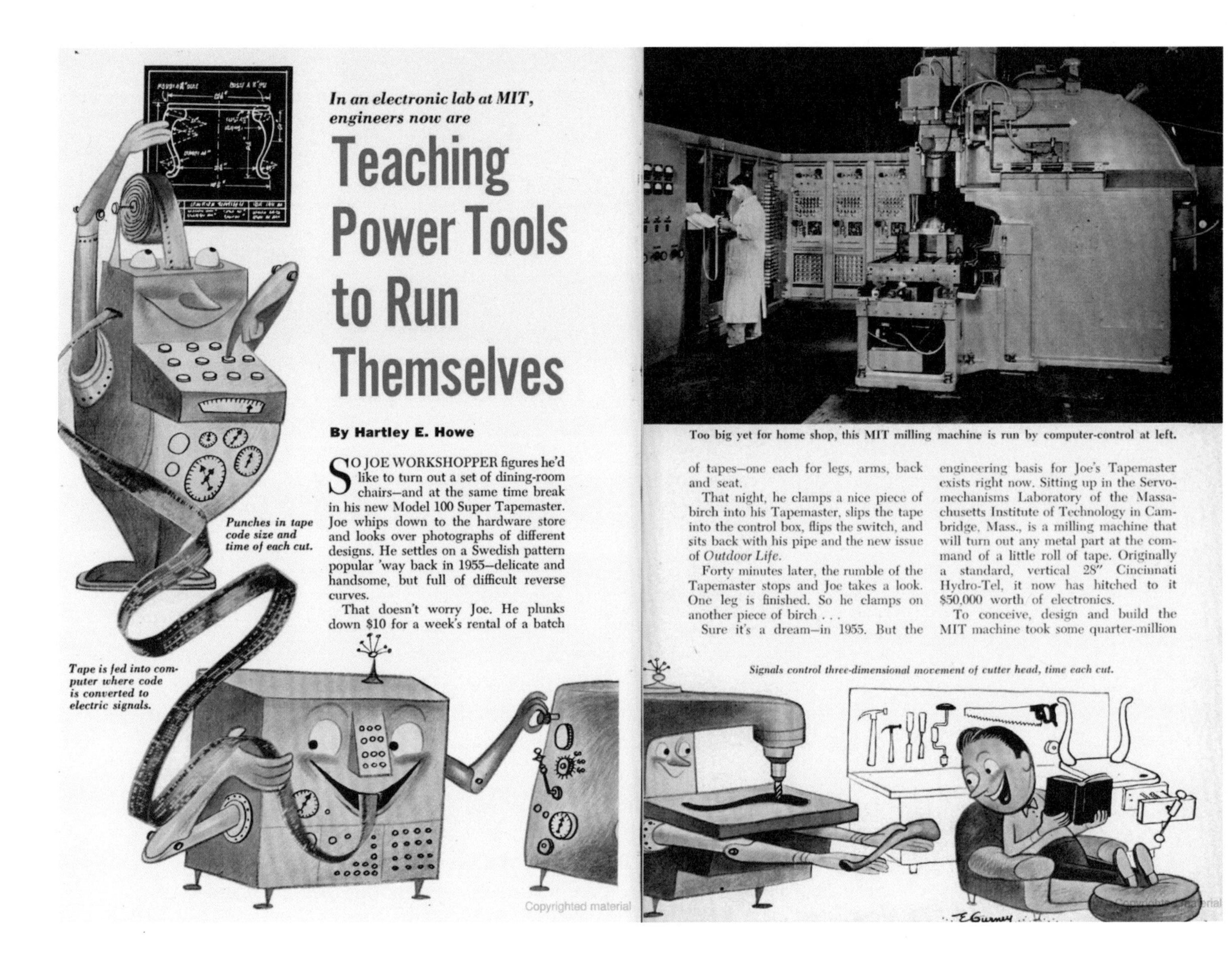

In an electronic lab at MIT, engineers now are

Teaching Power Tools to Run Themselves

By Hartley E. Howe

Punches in tape code size and time of each cut.

SO JOE WORKSHOPPER figures he'd like to turn out a set of dining-room chairs—and at the same time break in his new Model 100 Super Tapemaster. Joe whips down to the hardware store and looks over photographs of different designs. He settles on a Swedish pattern popular 'way back in 1955—delicate and handsome, but full of difficult reverse curves.

That doesn't worry Joe. He plunks down $10 for a week's rental of a batch of tapes—one each for legs, arms, back and seat.

That night, he clamps a nice piece of birch into his Tapemaster, slips the tape into the control box, flips the switch, and sits back with his pipe and the new issue of *Outdoor Life*.

Forty minutes later, the rumble of the Tapemaster stops and Joe takes a look. One leg is finished. So he clamps on another piece of birch . . .

Sure it's a dream—in 1955. But the engineering basis for Joe's Tapemaster exists right now. Sitting up in the Servomechanisms Laboratory of the Massachusetts Institute of Technology in Cambridge, Mass., is a milling machine that will turn out any metal part at the command of a little roll of tape. Originally a standard, vertical 28" Cincinnati Hydro-Tel, it now has hitched to it $50,000 worth of electronics.

To conceive, design and build the MIT machine took some quarter-million

Tape is fed into computer where code is converted to electric signals.

Too big yet for home shop, this MIT milling machine is run by computer-control at left.

Signals control three-dimensional movement of cutter head, time each cut.

Figure 0.3
A page from *Popular Science* in 1955 showing the first computer numerically controlled (CNC) machine at MIT's Servomechanisms Laboratory.

1947 (which has a strikingly material and "low-tech" physical presence), we have seen rapid developments in software and hardware technologies that introduced unprecedented changes across every discipline and industry and made digital computing ubiquitous in our everyday lives.

From Ivan Sutherland's first computer-aided design (CAD) tool in 1963 and the first computer numerically controlled (CNC) machine demonstrated by MIT's Servomechanisms Laboratory in 1952 to more contemporary software and fabrication platforms, we have become able to design, analyze, and physically fabricate in ways that were previously unimaginable. These new capabilities for computational design and fabrication have sparked a renaissance in the development of materials and performance.

As we've seen, the boom in material capabilities is visible in recent developments across many disciplines. The life sciences are making rapid advances with DNA sequencing and synthesis, genetic modification tools such as CRISPR, DNA computing, microbiome research, developments in tissue engineering, and the growing field of synthetic biology. New biomaterials, synthetic biofunctionality, DNA self-assembly, drug delivery mechanisms, and bioprinting are just a few of the recent capabilities to emerge. Materials science is similarly experiencing its own bustle of activity, from the discovery of graphene in 2010 to carbon nanotubes, directed self-assembly for material formation, granular jammable matter, invisibility cloaking, and a great deal more. At the macro scale we are seeing similar shifts. Some of the recent large-scale advances include multimaterial

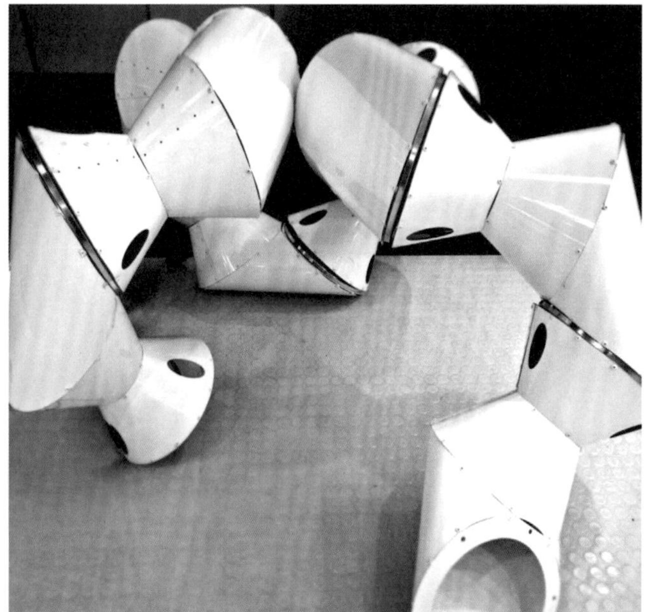

printing with metal/ceramic/glass/rubber/foams, printable electronics, 4D printing for customizable smart materials, reversible concrete-like structures with granular jamming, even building-scale automated fabrication, printable wood, programmable carbon fiber, active textiles, and many others. As new computational and digital fabrication processes are emerging, novel material capabilities have become available.

In 2007 the Defense Advanced Research Projects Agency (DARPA) initiated a program called "Programmable Matter." Programmable matter is generally understood as a material that has the ability to perform information processing much like digital electronics. The DARPA program included researchers from many universities and disciplines (many of whom are included in this book). With a few exceptions, the research fell under the category of reconfigurable microrobotics, or modular robotics: researchers developed smaller- and smaller-scale robotic modules with embedded electronics, power, actuation, sensing, and communication that would enable a variety of physical transformations and other behaviors. In somewhat traditional DARPA fashion, this program was ahead of its time. The vision was clear, but the implementation at that point was far from the dream of programming matter in an elegant and seamless way. Small robots became the stand-in for "matter," but they were not just materials; they were accumulations of software and hardware devices. The sum was perhaps not yet more than its parts. However, the vision of programmable matter laid the groundwork for today's active matter.

Since this program, a number of developments in materials science, synthetic biology, and other domains have rapidly emerged that I believe have enabled a realization of programmable matter *and more*. Now, materials can not only be programmed to compute, but can physically transform and actively self-assemble into larger aggregations. These materials aren't modules that have chips and computers or batteries in them like their predecessors; these new materials are purely material. In this sense, active matter is more than just programming matter; it is about combining programmability, transformation/adaptation, and assembly. Active matter is about matter that is literally *active*.

Figure 0.4 (top)
The Decibot, a 4-meter-long reconfigurable robot developed for DARPA's Programmable Matter program by Skylar Tibbits, Neil Gershenfeld, and the Center for Bits and Atoms.

One might ask, How does active matter relate to smart materials? As an analogy, in the history of computing we have transformed the first calculators and single-function computers into today's general-purpose, programmable machines. Similarly, the field of active matter aims to create

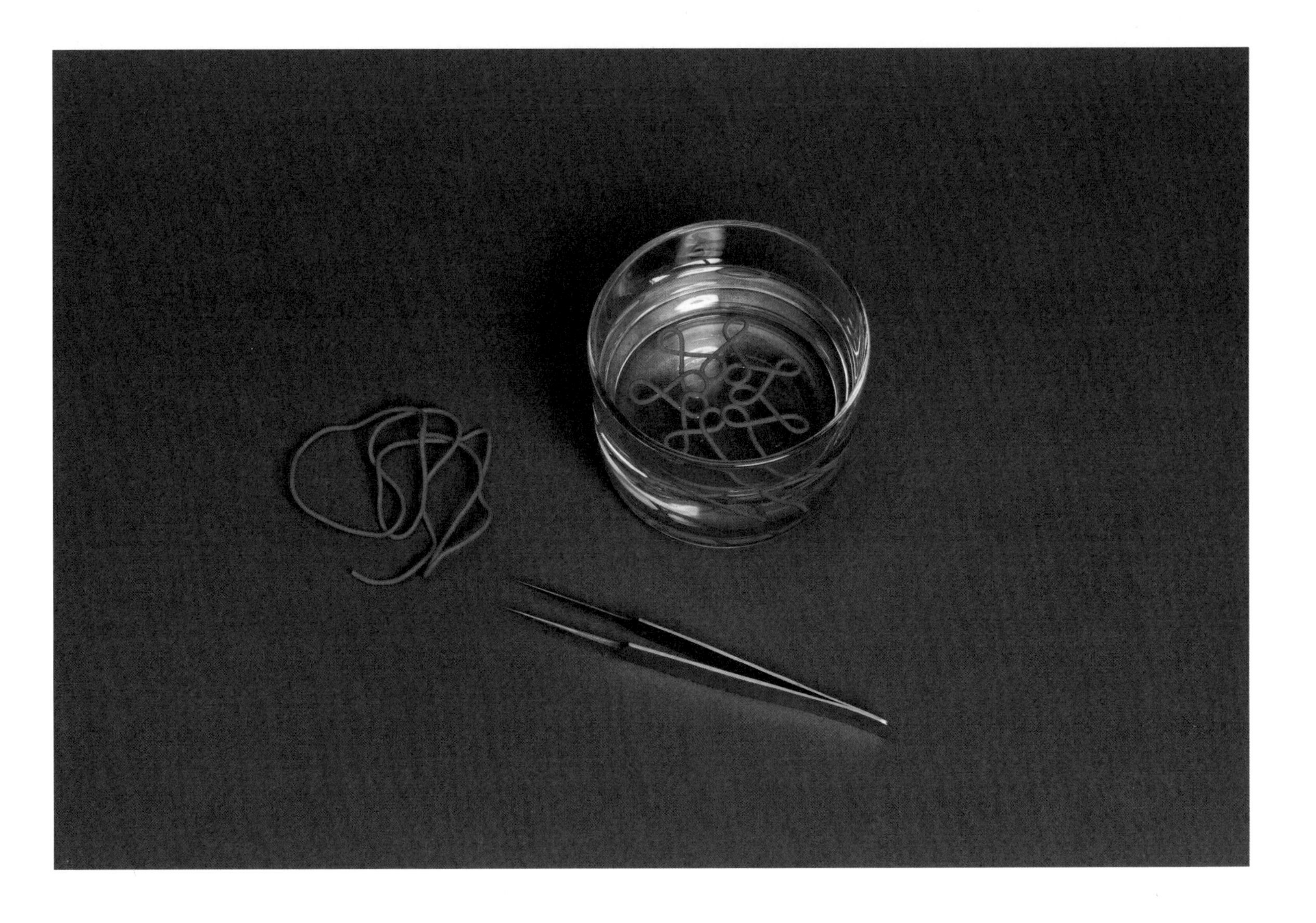

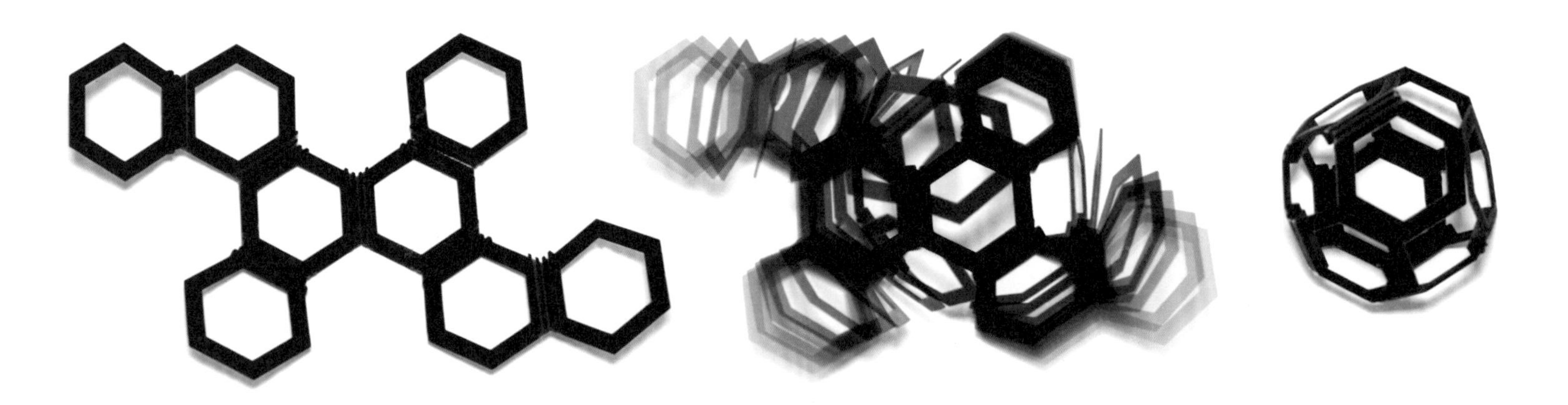

Figure 0.5 (top)
A traditional smart material, Nitinol, that can be dipped in hot water and self-transform back into a preprogrammed shape. Project by the Self-Assembly Lab, MIT, and E Roon Kang.

Figure 0.6 (bottom)
A 4D-printed flat sheet that self-folds into a truncated octahedron when activated with water. 4D printing is a process for designing and producing multimaterial prints so as to make fully customizable smart materials that transform and reconfigure over time.

general-purpose, programmable, and physically active materials. "Smart materials" or shape-memory materials also have the ability to change their property in a predetermined manner. However, active matter goes beyond today's smart materials that are only available in predetermined shapes, sizes, properties, and niche applications. Although smart materials and active matter both transform based on external input, active matter offers the freedom to design and create customized materials with unique functionality to sense, actuate, assemble, or compute. Active matter makes it possible to make any material a smart material.

If we think about it, matter has always been *active*, at least at a molecular level, yet our relationship with matter has traditionally been passive; at most, we have simply guided the growth and behavior of natural materials like bacteria, living cells, crystals, or wood. Or conversely, we have produced synthetic materials with fixed shapes and sizes to form all sorts of plasticized products—sculpting matter, rather than creating new types of matter or reprogramming its fundamental behavior. We could compare our traditional relationship with materials to that of breeding animals or plants: we didn't change the fundamental properties or capabilities of the medium, rather we recombined species in a "black box" type of way to guide the formation of useful behaviors or traits. Our new model of programming matter can be seen in CRISPR, or synthetic biology and DNA computing, where we can fundamentally change the structure, functionality, and information embedded within the medium to create new desired traits from the inside out.

A more surface-level understanding of the physical world tends to see materials as inert and slaves to our hammer and nail. Wood, a beautifully anisotropic and information-rich material, is turned into standardized lumber, as if it were a homogeneous material like plastic. However, we can redefine our relationship with matter. We can use the properties of the digital world now embedded in the physical world like logic, reprogrammability, reconfiguration, error correction, and assembly/disassembly. Or similar properties from the natural world can now be embedded in the synthetic world, like growth, repair, mutation, replication. These principles are now fundamentally available to read/write within matter itself.

How does the shift to active matter influence materials research? How will it create future products and industrial applications? What tools and design processes do we need to invent, augment, create, and discover new materials today? What are the galvanizing roles that industry, government, academia, and public institutions can play to catalyze and nurture the field of active matter? This book aims to address some of these questions by bringing together researchers, scholars, practitioners, artists, and designers, providing unique perspectives, breakthroughs in research, evocative imagery, and emerging industrial applications of active matter.

This book is organized by scale from smallest to largest, from nano to micro to planetary scale. The contributors' work varies by discipline and focus, yet they work together to stitch a comprehensive view of the emerging field of active matter, collectively telling the story of active matter, its details, visions, nuances, capabilities, pitfalls, and challenges ahead. Using scale to organize *Active Matter* allows interdisciplinary relationships and common themes/techniques/tools to emerge. This approach aims to highlight the connections and differences in research from technically or conceptually similar principles that can be applied across many fields of study. Grouping by scale makes visible the potential for insights and advances to travel laterally across the disciplines. The field is emerging literally from the bottom up, and this book attempts to provide the underlying logic, connectivity, and a perspective on the future direction of active matter.

Figure 0.7a (top)
The Active Matter Summit, held at MIT in April 2015, gathered scientists, engineers, designers, artists, government officials, and many other thought leaders in the field. Neri Oxman is shown here presenting her work at the Summit.

Figure 0.7b (below)
Student work is reviewed during the Spring 2015 design studio taught by Skylar Tibbits and Athina Papadopoulou in collaboration with Merton Flemings, in tandem to the Active Matter Summit. This fabrication-based studio was funded by the MIT Center for Art, Science & Technology (CAST) and allowed students the opportunity to experiment with a range of materials and design "active" structures.

MATTER MATTERS: A PHILOSOPHICAL PREFACE

Neil Leach

> Materialist philosophers, it is becoming increasingly clear, cannot afford to ignore the basic fact that the study of matter does matter.— Manuel DeLanda[1]

In terms of the emerging discourse of "active matter," we need to track the epistemological shift from the representational logic of postmodernity toward a more process-based way of thinking that is inflecting design today. This is a development from a discourse of symbolism and metaphors that postmodernism privileged toward a discourse of performance and material behavior that new materialism promotes and that sustains the work of the *Active Matter* contributors.

New materialism, as championed by the Mexican-American philosopher Manuel DeLanda, provides a philosophical framework for understanding a broad scientific and scholarly move away from emphasizing the subject, representation, and interpretation, which characterized twentieth-century thinking in general, toward focusing instead on the object, material processes, and expression.[2] It is through new materialism, I would claim, that we can open up an enquiry into the nonlinear logic and morphogenetic tendencies in matter and into the capacity of matter to self-organize and play an active role in its own formation. Moreover, we can learn lessons from the behavior of matter and use them to help us to understand the formation of larger-scale agglomerations, such as cities, continents, and indeed entire planets.

BEYOND POSTMODERNISM

Let us turn our attention back to 1967 when two Yale professors, Robert Venturi and Denise Scott Brown, their teaching assistant, Steve Izenour, and a group of students set off to undertake research on Las Vegas, the city of gambling in the Nevada desert that had spawned its own architectural language of billboards and neon signs. The result of their research, *Learning from Las Vegas: The Forgotten Symbolism of Architectural Form*, was published in 1972, and is commonly regarded as a radical, revolutionary text within architectural circles.[3] The message is clear enough. Architects and urban planners need to abandon their old ways and turn instead to inspiration from the street. In highlighting the representational logic of billboards and neon signs, the book became a manifesto for postmodernism.

But was this book actually as radical as is often claimed?[4] I would argue that in 1997—30 years after the famous study trip to Las Vegas—Manuel DeLanda published *A Thousand Years of Nonlinear History*, a book that is far more radical and revolutionary.[5] If *Learning from Las Vegas* became a manifesto for postmodernism, *A Thousand Years of Nonlinear History* could be read as a manifesto for new materialism. Whereas Venturi, Scott Brown, and Izenour focus on representation, DeLanda focuses on *processes*.[6] DeLanda looks at the past thousand years of history, including the history of urban development, by engaging with the logics of geological formations, biological processes, and linguistic evolution. What emerges is a picture in which the domains of the geological, biological, and linguistic are both nonlinear and isomorphic. They are nonlinear in that they all involve

Figure 0.8
Kemikism, master's design project (2016) by Albert Elias, Florida International University, tutored by Neil Leach; a fractal-based reactive landscape for a comet-like form in outer space.

feedback or interaction. And they are isomorphic in that, although they appear incommensurable, they each share similar properties that amount effectively to the same process.[7]

DeLanda further distinguishes between "hierarchies" and "meshworks": the "coagulations" or homogeneous processes of sorting and cementation, which constitute a *hierarchical* operation, as against synthesizing heterogeneous elements, which constitutes a *meshwork* operation. We can, for example, observe how sandstone can be seen to be generated hierarchically through a process of an initial "sorting" of mineral deposits/sedimentations that then become cemented over time to generate material with a relatively homogeneous consistency, sandstone. By contrast, granite can be observed to be generated through a meshwork logic, as a catalyst is deployed to bind two different substances through a form of "autocatalytic loop." Indeed, Geoffrey Winthrop-Young notes, "The interlocked heterogeneous elements, in turn, generate stable patterns of behaviour: magma cools at different speeds with one element acting as a container for those that crystallize at a later point."[8]

The key is to understand that we are dealing with far-from-equilibrium conditions, which themselves depend upon the flows of energy required to engender processes of self-organization. As DeLanda notes: "We are beginning to understand that any complex system, whether composed of interacting molecules, organic creatures or economic agents, is capable of spontaneously generating order and of actively organizing itself into new structures and forms. It is precisely this ability of matter and energy to self-organize that is of greatest significance to the philosopher."[9]

The first section of *A Thousand Years of Nonlinear History* looks at the formation and growth of cities as an instance of far-from-equilibrium conditions, where the physical urban fabric is seen as a form of "exoskeleton" to human operations, to be contrasted to the "endoskeleton" structure of human beings themselves. Moreover, the far-from-equilibrium conditions of cities contribute to processes of material self-organization. As Winthrop-Young states, "By attracting, circulating, and discharging everything ranging from money to microbes, the urban dynamics provide the energy flows necessary to induce and maintain all varieties of self-organization."[10]

One of the analogies that DeLanda deploys in order to understand the energy flows that sustain these processes of self-organization is the geological formation of "lavas and magmas." Similarly, he suggests, cities and other human settlements can be seen as sedimentations of material strata that have aggregated over time. As DeLanda himself puts it: "Human culture and society (considered as dynamic systems) are no different from the self-organized processes that inhabit the atmosphere and hydrosphere (wind circuits, hurricanes), or, for that matter, no different from lavas and magmas, which as self-assembled conveyor belts drive plate tectonics and over millennia have created all the geological features that have influenced human history. From the point of view of energetic and catalytic flows, human societies are very much like lava flows; and human-made structures (mineralized cities and institutions) are very much like mountains and rocks: accumulations of materials hardened and shaped by historical processes."[11]

The self-organizing activities of lava flows are the origins of many geological features, and this process is dependent on the speed of flow—very slow in the case of rocks and faster in the case of lava. Thus, it is not simply a question of coagulations but also of deceleration. Furthermore these same principles apply to human life itself: "Similarly, our individual bodies and minds are mere coagulations or decelerations in the flows of biomass, genes, memes, and norms."[12] And the behavior of matter is precisely the focus of *Active Matter*.

DELEUZIAN IMPULSES

In *A Thousand Plateaus*, Deleuze and Guattari offer another instance, borrowed from the field of architecture this time, of the cardinal and radical shift to process-oriented thinking.[13] They write, "Gothic architecture is indeed inseparable from a will to build churches longer and taller than the Romanesque churches. Ever further, ever higher. ... But this difference is not simply quantitative; it marks a qualitative change: the static relation, form-matter, tends to fade into the background in favor of a dynamic relation, material-forces. It is the cutting of stone that turns it into material capable of holding and coordinating forces of thrust, and of constructing ever higher and longer vaults. The vault is no longer a form but the *line of continuous variation* of the stones. It is as if Gothic conquered a smooth space, while Romanesque remained partially within a striated space (in which the vault depends on the juxtaposition of parallel pillars)."[14]

The Romanesque—or, one could argue, the logic of classicism in general—is based on a logic not of exploring experimental structures, but of following the rule book with regard to *visual* concerns, such as codified architectural details and the rules of proportions. The Romanesque obeys a rule-based, representational discourse of form, meaning, and symbolism, whereas the Gothic engages an experimental, process-based discourse of performance, expression, and material behavior.[15]

Another way to think through the distinction between the Gothic and the Romanesque is by using the distinction between *hylomorphic* and *morphogenetic* form-making. The hylomorphic approach imposes form from above regardless of the actual properties of the material. According to Deleuze and Guattari, the hylomorphic model "assumes a fixed form and a matter deemed homogeneous. It is the idea of the law that assures the model's coherence, since laws are what submit matter to this or that form, and, conversely, realize in matter a given property deduced from the form. But ... the *hylomorphic* model leaves many things, active and affective, by the wayside. On the other hand, to the formed or formable matter we must add an entire energetic materiality in movement, carrying *singularities* ... that are already like implicit forms that are topological, rather than geometrical, and that combine with processes of deformation: for example, the variable undulations and torsions of the fibers guiding the operations of splitting wood. On the other hand, to the essential properties of matter deriving from the formal essence we must add *variable intensive affects*, now resulting from the operation, now on the contrary making it possible: for example, wood that is more or less porous, more or less elastic and resistant. At any rate, it is a question of surrendering to the wood, then following where it leads by connecting operations to a materiality instead of imposing form upon a matter."[16]

By contrast, a *morphogenetic* approach teases the form out of the material in a bottom-up logic. Traditionally within architectural culture the dominant approach has been a *hylomorphic* one, with little attempt to take into account the morphogenetic capacities of the material itself. An exception to this rule can be found in the work of architects such as Antoni Gaudí and Frei Otto, whose work is now being reassessed with the shift in interest toward *morphogenetic* design as Deleuze and Guattari do in their celebration of the Gothic over the Romanesque. The idea of a bottom-up morphogenetic approach is therefore in complete contrast to the previously accepted paradigm of the top-down approach to design. As Achim Menges comments: "Architecture as a material practice is mainly based on design approaches that are characterised by a hierarchical relationship that prioritises the generation of form over its subsequent materialisation. Equipped with representational tools intended for explicit, scalar geometric descriptions, the architect creates a scheme through a range of design criteria that leave the inherent morphological and material capacities of the employed material systems largely unconsidered. Ways of materialisation, production and construction are strategised and devised as top-down engineered, material solutions only after defining the shape of the building and the location of tectonic elements. ... An alternative morphological approach to architectural design entails unfolding morphological complexity and performative capacity from material constituents without differentiating between formation and materialisation processes."[17]

This is an important difference that distinguishes contemporary, morphogenetic approaches from traditional, hylomorphic ones, just as it distinguishes new materialism from postmodernism, or the logic of process from the logic of representation. The key term here is "form" and its inclusion in terms such as "performance" and "information." The difference, then, lies in the emphasis on *form-finding* over form-making, on bottom-up over top-down processes, and on *formation* rather than form. "Formation" itself must in turn be recognized as linked to "information" and "performance." When architecture is "informed" by performative considerations, it becomes less a consideration of form in and off itself, and more a discourse of material formations.

COMPLEX MATERIALITY

The scope of DeLanda's overall enquiry into matter covers a range of scales from nanotechnology through cities and beyond.[18] In "Material Complexity," for example, he takes up the materiality of matter itself by looking at the development of metallurgy, a subject that, incidentally, Deleuze and Guattari categorize as a "minor science."[19]

The distinction between "minor" and "major" sciences, or between linear and nonlinear behavior, is not so simple. Indeed, although in his discussion of metallurgy DeLanda uses Robert Hooke as an example of an experimental bottom-up proponent of "minor science," in contrast to the more established Sir Isaac Newton as a top-down proponent of a "major science," it should be recalled that Hooke himself was responsible for Hooke's Law, surely an example of a rule-based top-down logic. Nor is it just that "major sciences" tend to privilege linear behavior in order to produce regular laws, while "minor sciences" focus more on nonlinear behavior. After all, as Hooke observed, most materials will tend to exhibit linear behavior until they reach plastic deformation. Moreover, even nonlinear systems can be linearized by studying them under conditions near or at equilibrium, and, conversely, nonlinear systems require nonequilibrium conditions to become clear. To generalize, however, one could argue that whereas linear systems tend to be characterized by a single steady state, nonlinear systems tend to display multiple steady states. Moreover, nonlinear systems have the crucial capacity to self-organize.

From a philosophical perspective, the potential for self-organization is one of the most important aspects of material behavior. This is where complexity becomes significant, in that complex materials have a tendency to self-organize.[20] But again the challenge of discerning self-organization is not so straightforward. Although every material has its own endogenous tendencies and capacities, any homogeneous material closed to energy flows will tend to "hide" its self-organization, whereas complex or heterogeneous material or material that is far from equilibrium will tend to "express" its self-organization.[21]

How, then, is matter capable of self-organization? The key is to recognize the role of emergence in the process of formation. Emergence could be defined as the principle by which properties of the whole are greater than the sum of the parts, as is seen, for example, in the case of swarming behaviors such as the flocking of birds.[22] The interesting aspect of emergence is that similar behaviors can be found in any population of multiagent systems, no matter how incommensurable their entities. Thus, as the title of Steven Johnson's book *Emergence: The Connected Lives of Ants, Brains, Cities and Software* suggests, these multiagent systems do not need to consist of living creatures, such as birds, fish, or ants. In the context of metallurgy, for example, populations of defects can cause a metal to be tough if they are allowed to be highly mobile, or rigid if they are constrained.[23]

This returns us to the Deleuzian distinction between the hylomorphic and the morphogenetic—between form imposed from above and form emerging from below—and to a recognition of the important role matter plays in its own self-organization. Matter, in other words, is not inert but *active*: "We may now be in a position to think about the origin of form and structure, not as something imposed from the outside on an inert matter, not as a hierarchical command from above as in an assembly line, but as something that may come from within the materials, a form that we tease out of those materials as we allow them to have their say in the structures we create."[24]

This leads DeLanda to conclude: "The view of the material world that emerges from these considerations is not one of matter as an inert receptacle for forms that come from the outside, a matter so limited in its causal powers that we must view the plurality of forms that it sustains as an unexplainable miracle. It is not either an obedient matter that follows general laws and that owes all its powers to those laws. It is rather an *active matter* [my italics] endowed with its own tendencies and capacities, engaged in its own divergent, open-ended evolution, animated from within by immanent patterns of being and becoming."[25]

CONCLUSION

New materialism moves us beyond a discourse of representation, symbols, and meaning that characterized postmoderism to a discourse more aware of material processes and expressions. It is not that new materialism seeks to displace or dismiss the representational logic of the late twentieth century as though it were redundant. Rather it seeks to emphasize that, after years of stressing the importance of representation, we need to redress the balance by paying greater attention to process. Indeed these two discourses—of representation and of process—are not mutually exclusive, but the one invites the other.

Importantly, new materialism is a philosophical discourse that opens up the possibility of theorizing science and appropriating it within a theoretical domain. Thanks to new materialism, science—for so long frowned upon within architectural theory circles and dismissed for being largely positivistic in its orientation—is now included in the realm of theory.[26]

Perhaps most important of all, however, is the fact that science itself is now able to offer insights to help us understand cultural life in general, and that frameworks like new materialism can help us make much-needed connections among the disciplines and in society. One such insight—that matter can be seen as active—is an insight that can help us understand not only how matter itself behaves, but also—as *A Thousand Years of Nonlinear History* reveals—how society operates. As such, the notion of *active matter* has implications beyond the strictly material world of matter itself to embrace the whole of society, from the scale of nanotechnology to that of our cities, states, continents, and entire planet. The message is clear. The behavior of matter—especially the *active* role that it plays in its own formation—can help us to understand broader questions about how the world operates. In short, matter does matter.

1 Manuel DeLanda, "Material Complexity," in *Digital Tectonics*, ed. Neil Leach, David Turnbull, and Chris Williams (London: Wiley, 2004), 14.

2 In the 1990s both DeLanda and Rosi Braidotti started using the term new materialism independently, although it is unclear who used the term first. Both DeLanda and Braidotti draw heavily on the work of Gilles Deleuze (1925—1995), who frequently collaborated with the radical psychoanalytic theorist Félix Guattari (1930—1992). However, the origins of new materialism are much deeper, and it could be argued that it operates as a retrospective manifesto for a movement whose genealogy stretches back to the work of biologist D'Arcy Wentworth Thompson, philosopher Henri Bergson, and as far as the ancient materialist philosophers such as Lucretius. For an overview of new materialism see Rick Dolphijn and Iris van der Tuin, eds., *New Materialism: Interviews and Cartographies* (Ann Arbor, MI: Open Humanities Press, 2012); Neil Leach, "New Materialism," in *De-signing Design: Cartographies of Theory and Practice*, ed. Elizabeth Greirson, Harriet Edquist, and Hélène Frichot (Lanham, MD: Lexington, 2016), 205—216.

3 Robert Venturi, Denise Scott Brown, and Steven Izenour, *Learning from Las Vegas: The Forgotten Symbolism of Architectural Form* (Cambridge, MA: MIT Press, 1972, 1977).

4 For a critique of *Learning from Las Vegas* see Neil Leach, *The Anaesthetics of Architecture* (Cambridge, MA: MIT Press, 1999).

5 Manuel DeLanda, *A Thousand Years of Nonlinear History* (New York: Zone Books, 1997).

6 Here I would argue that we need to step beyond the narrow confines of what came to be understood stylistically as "postmodernism" within architectural circles—the use of curtain walling, application of historical motifs on buildings, and the privileging of ornamental surface over structural framework—and understand postmodernism instead in terms of broader cultural issues. For, if we consult the various theorists of postmodernity in philosophy and cultural theory, we will see that the overriding concerns apply to culture at large, and refer to broader concepts such as the emphasis on the visual and the scenographic. What becames clear is that during the period known as "postmodernity" form and representation were celebrated over performance and process.

7 As Geoffrey Winthrop-Young puts it, "No matter how diverse the materials employed or the structures generated, detailed empirical investigation mixed with insights gained from the analysis of non-linear dynamics will enable us to elaborate models of structuration processes abstract enough to operate in the disparate worlds of geology, biology, and human society." Geoffrey Winthrop-Young, "Materialism at the Millennium," review of *A Thousand Years of Nonlinear History*, http://www.altx.com/ebr/reviews/rev8/r8young.htm (accessed September 27, 2016).

8 Winthrop-Young, "Materialism at the Millennium."

9 DeLanda, "Material Complexity," 17.

10 Winthrop-Young, "Materialism at the Millennium."

11 DeLanda, *A Thousand Years of Nonlinear History*, 55.

12 DeLanda, *A Thousand Years of Nonlinear History*, 258.

13 Gilles Deleuze and Félix Guattari, *A Thousand Plateaus: Capitalism and Schizophrenia*, trans. Brian Massumi (London: Athlone Press, 1988).

14 Deleuze and Guattari, *A Thousand Plateaus*, 364. Although Deleuze and Guattari offer an architectural example here, their work does not address architecture and has often been misunderstood by architects. A notable example of this is the reception of Deleuze's *The Fold: Leibniz and the Baroque* (Minneapolis: University of Minnesota Press, 1993). Not only does the term "fold" not refer to the process of *physically* folding (rather to the folding of thought), but his reference to the baroque is not an explicit reference to any architectural style. Perhaps the most extreme version of this misunderstanding comes in the issue of *Architectural Design*, "Folding in Architecture," edited by Greg Lynn (Greg Lynn, "Folding in Architecture," *AD* [1993]; repr., London: Wiley, 2004), where one of Deleuze's "diagrams" that looks initially like a drawing of a baroque house is included, with accompanying description inserted into the body of the text, as though its connection to the subject of the volume is entirely obvious. For a more informed understanding of Deleuze's thinking on the "fold" and the "baroque," see Simon Sullivan, "The Fold," in *The Deleuze Dictionary*, rev. ed., ed. Adrian Parr (Edinburgh: Edinburgh University Press, 2012), 107; Greg Lambert, *The Non-Philosophy of Gilles Deleuze* (London: Continuum, 2002), 45.

15 The real purpose of this distinction between the Gothic and the Romanesque, however, is to define not two different modes of architectural thinking, but rather two different models that serve to illustrate the difference between the "major" and "minor" sciences, the distinction, that is, between the authorized top-down "State" approach and an experimental bottom-up "nomadic" approach. DeLanda notes "two modes of conducting scientific research, a major and a minor mode: *royal science and nomad science*, the science of the Royal societies and academies at the service of the State preoccupied above all with the discovery of abstract general laws, and the humbler science of those who built the laboratory instruments and had the job of testing the validity of those laws in concrete physical situations. Indeed, the distinction between royal and nomad science is drawn more widely so that it does not coincide with the distinction between pure and applied science. In its minor mode science deals with complex material behavior, liquids not solids, heterogeneous not homogeneous matter, turbulent not steady-state (or laminar) flow." DeLanda, "Material Complexity," 15—16. As such, the Gothic and Romanesque are not simply references to approaches to architectural design, but rather offer a broader understanding of how thinking and society evolve.

16 Deleuze and Guattari, *A Thousand Plateaus*, 408 (italics in the original).

17 Achim Menges, "Polymorphism," in *Techniques and Technologies in Morphogenetic Design*, ed. Michael Hensel, Achim Menges, and Michael Weinstock, *Architectural Design* 76, no. 2 (March/April 2006): 79.

18 For DeLanda's thoughts on nanotechnology, see the series "Matter Matters" for *Domus* magazine: http://cmm.cenart.gob.mx/delanda/textos/matter.pdf (accessed September 21, 2016).

19 DeLanda, "Material Complexity"; Deleuze and Guattari, *A Thousand Plateaus*, 411.

20 "Any complex system, whether composed of interacting molecules, organic creatures or economic agents, is capable of spontaneously generating order and of actively organizing itself into new structures and forms." DeLanda, "Material Complexity," 17.

21 "Any material, no matter how simple its behavior, has endogenous tendencies and capacities, but Deleuze argues that if the material in question is homogeneous and closed to intense flows of energy, its singularities and affects will be so simple as to seem reducible to a linear law. In a sense, these materials hide from view the full repertoire of self-organizing capabilities of matter and energy. On the other hand, if the material is far from equilibrium (or what amounts to the same thing, if *differences* in intensity are not allowed to be canceled) or if it is complex and heterogeneous (that is, if the *differences* among its components are not canceled through homogenization) the full set of singularities and affects will be revealed, and complex materiality will be allowed to manifest itself." DeLanda, "Material Complexity," 19.

22 "The dynamics of populations of dislocations are very closely related to the population dynamics of very different entities, such as molecules in a rhythmic chemical reaction, termites in a nest-building colony, and perhaps even human agents in a market. In other words, despite the great difference in the nature and behavior of the components, a given population of interacting entities will tend to display similar collective behavior as long as the interactions are nonlinear and as long as the population in question operates far from thermodynamic equilibrium." DeLanda, "Material Complexity," 17.

23 James Edward Gordon, *The Science of Structures and Materials* (New York: Scientific American Library, 1988), 111.

24 DeLanda, "Material Complexity," 21.

25 Manuel DeLanda, "Emergence, Causality and Realism," in *The Speculative Turn: Continental Materialism and Realism*, ed. Levi Bryant, Nick Srnicek, and Graham Harman (Melbourne: re.press, 2011), 392.

26 See also Manuel DeLanda, *Intensive Science, Virtual Philosophy* (London: Continuum, 2002); DeLanda, *Philosophical Chemistry: Genealogy of a Scientific Field* (London: Bloomsbury, 2015).

STEELCASE: AN INDUSTRY PERSPECTIVE ON ACTIVE MATTER

Sharon Tracy and Paul Noll

Steelcase sees great potential in active matter.[1] Our working definition of active matter is the subset of materials that sense external stimuli (light, heat, electricity, etc.) and respond to them (by shape change, color change, etc.) in a predictable way. We are exploring new material compositions, constructions, behaviors, and applications that address current and potential human needs. Other practical areas of application for us include transformable packaging, adaptive manufacturing, and new methods to improve shipping and distribution. For many other industries, the future of active matter is almost limitless when all the possibilities of these materials are considered.

Steelcase was founded on materials innovation. During the early twentieth century, when all other companies were making waste receptacles out of wood, Steelcase made them out of a relatively new material in work environments—steel. Since steel is not flammable, it helped make the work environment safer during a time when smoking in the office made fire prevention a priority. The patented innovation "The Victor" (figure 0.9) is based on technological developments in bending steel; in a way, innovation in the area of materials, manufacturing, and applications is in our DNA.

Steelcase creates environments that support people and their work. Our products are based on insights derived from research about materials, technology, and people. Steelcase offers an ecosystem of spaces that allow workers to choose where and how they work, supporting both community and individual needs (figure 0.10). While people are the focus of our designs, we continue to learn and adapt over time, evaluating and exploring how new materials and technologies will impact our human experience of work. We frame the work experience through the SSI model (figure 0.11), where Social (people), Spatial (environments and materials), and Informational (computation) domains form interdependent relationships in support of how we work. Figure 0.12 shows an example of colleagues collaborating (Social) in a supportive environment (Spatial) using whiteboard and displays (Information).

Work is becoming more democratized and decentralized; it happens everywhere. This trend is driven by the forces of mobile technology supplemented by an increasing demand for authenticity and flexibility.

Figure 0.9
The Victor (1914). First product patented by Steelcase (at the time, The Metal Office Furniture Company).

Concurrently, we are becoming more connected and have more information available to us than ever before. Taken together, our lives are increasingly complex and filled with networks of relationships, information, and activities.

As humans navigate this rapidly evolving world, our expectations are that the physical environments around us will take on new characteristics to meet these new ways of working. Our spaces will have to work harder than ever to support our activities, needs, and behaviors. Creating dynamic structures and environments that sense and respond, to support our well-being (physical, cognitive, and emotional), work styles (informative, evaluative, generative), and work types (individual, collaborative; co-located, distributed), will not only be possible but also necessary and expected.

Digital technology is accelerating, becoming smaller, lighter, cheaper, and faster. Steelcase anticipates that it will become even more integrated into our environments. Advances in other scientific domains, for example, biology, synthetic biology, genetics and gene editing, nanoengineering, and molecular self-assembly, will increasingly impact our environments as well. Materials science plays a role in all of these domains by integrating advancements and creating new opportunities such as active matter.

As members of the active matter community, Steelcase plays many roles. We provide potential use cases for researchers to consider as they are investigating and developing new materials. We also apply new material solutions to products and services in support of the human experience. Additionally, we plan and conduct our own research explorations internally, where our understanding of the end user gives us a valuable perspective for matching needs with solutions.

In the progression from research to commercial products and

Figure 0.10
The office renaissance. Environments are increasingly supporting collaboration, authenticity, and well-being.

Figure O.11 (top)
The SSI model. Social (people), Spatial (environments and materials), and Informational (computation) form interdependent relationships in support of how we work.

Figure O.12 (bottom)
An example of the SSI framework, with colleagues collaborating (Social) in a supportive environment (Spatial) using whiteboard and displays (Information).

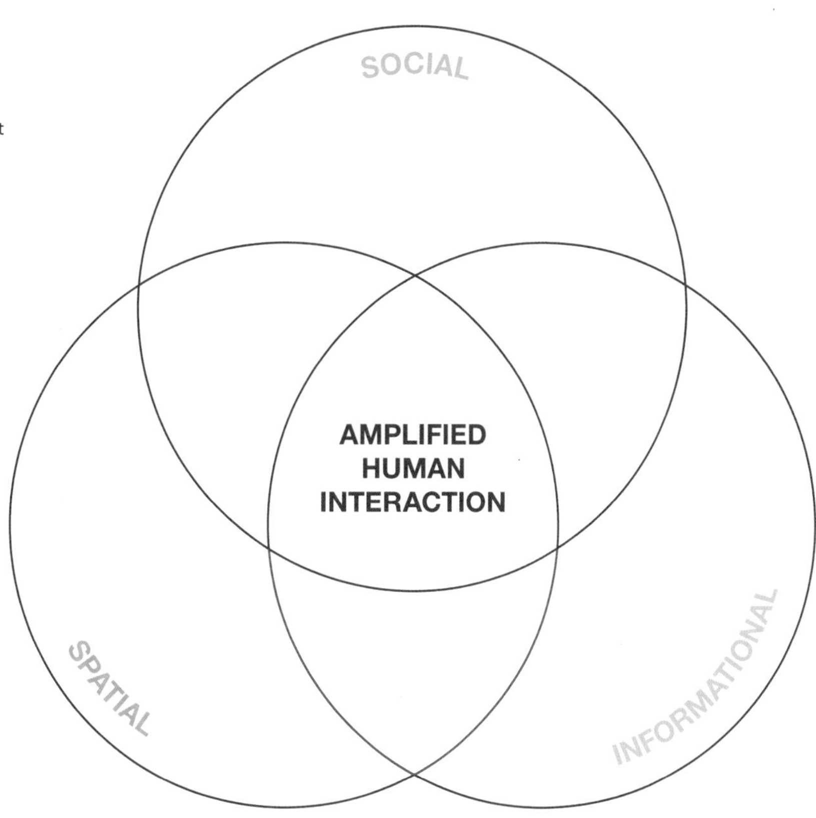

services, the bridging of boundaries (disciplinary, institutional, and industrial) increases effectiveness. Steelcase and many others have learned that successful projects depend on multidisciplinary processes. These improve communication and collaboration, change thinking, and expedite progress. Some specific questions we might ask about materials innovation in a multidisciplinary setting include: What can this new material do differently from others? What properties and functions can this new material provide? What is a tangible use (application) of this new material? We typically explore these new paradigms by creating concepts and building prototypes.

In R&D, one of the many disciplinary boundaries to be managed is associated with the scale of physical length, primarily because this relates to subject matter expertise. For example, chemists generally investigate molecules (subnanometer and smaller); materials scientists develop materials through nano- and microstructural design. Similarly, mechanical engineers and industrial designers often develop parts and products between millimeter and human scale; and architects and interior designers are looking at human scale and larger. From materials science, we know that boundaries are fruitful sources of the unexpected. We encourage researchers in their respective areas to collaborate with those working at a scale greater or smaller than theirs. Additionally, modeling, simulation, and digital visualization tools and techniques will allow deeper understanding by complementing physical/experimental/prototypical research.

These collaborative efforts yield great results, in terms of new knowledge, experience, and opportunities. The US government created a Materials Genome Initiative to increase the speed at which we discover, develop, and manufacture new materials. They are taking a "multidisciplinary approach [that] will accelerate progress as results from each aspect inform the work of the others, enhancing communication across disciplines, ... and enabling optimization."[2] It is well recognized that translating research from the bench to practice is nontrivial, and that there are benefits to shortening the time this takes. We are hopeful that the growing group of researchers in the active matter fields will take on some of these challenges to continue to make progress.

Fundamentally, Steelcase is a "user" or "applier" of materials such as active matter. We hope to continue to work with and strengthen the active matter community and encourage

— thinking about the human being as the ultimate beneficiary of new material developments,
— working across disciplinary boundaries,
— working across scale boundaries, and
— working across physical/virtual boundaries.

We are excited about the developments that are to come, and look forward to the benefits for our collective human experience.

NOTES

1 For over 100 years, Steelcase Inc. has helped create great experiences for the world's leading organizations, across industries. We demonstrate this through our family of brands—including Steelcase, Coalesse, Designtex, PolyVision, and Turnstone. Together, we offer a comprehensive portfolio of architecture, furniture, and technology products and services designed to unlock human promise and support social, economic, and environmental sustainability. We are globally accessible through a network of channels, including over 800 dealer locations. Steelcase is a global, industry-leading and publicly traded company with fiscal 2015 revenue of $3.1 billion.

2 http://www.mgi.gov

1
Interview Between Markus Buehler and Tomás Saraceno

Markus Buehler and
Tomás Saraceno

Markus Buehler and
Tomás Saraceno

Saraceno's collaboration with the MIT Department of Civil and Environmental Engineering (CEE) and Center for Art, Science & Technology (CAST) began in 2012. The collaboration has focused on optimizing the scanning unit the studio used to collect data on the materiality of spiderwebs and to apply them to art, architectural, and structural design principles. At MIT, a team of researchers and students are experimenting with Saraceno's scanning system to investigate the deformation mechanism of spiderwebs by applying different loads (e.g., point, wind, stretch, combination) to them. They calculate and analyze stresses on individual fibers in order to understand how each fiber's location in the web affects its load and the overall architecture of the spiderweb. To date, one MIT graduate student and three undergraduate students have used data collected with Saraceno's scanning system in their research. The team expects that the design rules and structural optimization derived from scanning spiderwebs can be used for innovations in engineering and material science.

Tomás Saraceno:
Taking the data of the first digitized three-dimensional black widow spiderweb in 2010[1] as a point of departure, your lab has started working on simulations such as wind load or point load (for example, the impact of prey). Interestingly enough, these simulations can be seen as attempts to scientifically realize Agamben's assertion, in his work *The Open*: "The two perceptual worlds of the fly and the spider are absolutely uncommunicating, and yet so perfectly in tune that we might say that the original score of the fly, which we can also call its original image or archetype, acts on that of the spider in such a way that the web the spider weaves can be described as 'fly-like.'"[2] Agamben was driven by his conviction that human and animal should not be conceived in completely separate terms. What led you to look at the spider's web and its construction?

Markus Buehler:
The construction of the spiderweb, whereby a liquid is turned into a material as strong as steel, is a perfect example of the merger of material and structure, and a powerful illustration of how natural systems connect molecular and nano to larger scales—the macro world we live in.

It is also a beautiful illustration of joining across the scales, which we cannot see with our eyes, whereas the effects are tangible in the performance characteristic of the web. Similar to the way the spider and the fly communicate, the web is a manifestation of the communication between the nano world and the macro world. The web is the medium by which this is made possible. For example, the strength of the silk that we can measure at the macro scale is due to intricate nanoscale assemblies of proteins into nanocrystals embedded in randomized softer structures. The play with these two elementary concepts—highly organized crystals and deliberately poorly organized protein threads—is a unifying theme across many species of spiders, and the key to their webs' remarkable properties. Nature plays with the relative composition of each to achieve tunable properties depending on the demand of the environment/the ecology.

I find it intriguing as it offers another example of how humans rediscover nature in our own engineering approaches. For instance, Feynman's challenge to create nanomanipulation of matter has already been established by natural systems such as the cells, organisms of various kinds, and the spiderweb. It should humble us and emphasize the great resource available in nature to inspire future engineering. It should also offer clues to adapt to our ever-changing world, and spring optimism that we can connect our human experience with nature of which we are so intimately part.

Tomás Saraceno:
When we first met, you told me about biomateriomics, which you defined as the relational study between the processes, properties, and function of materials

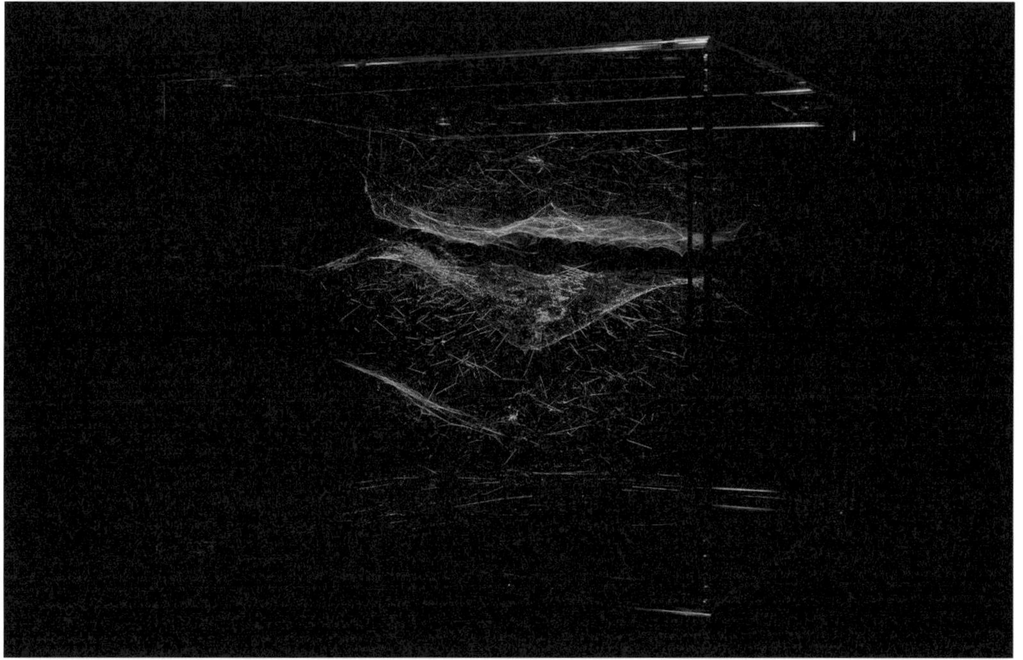

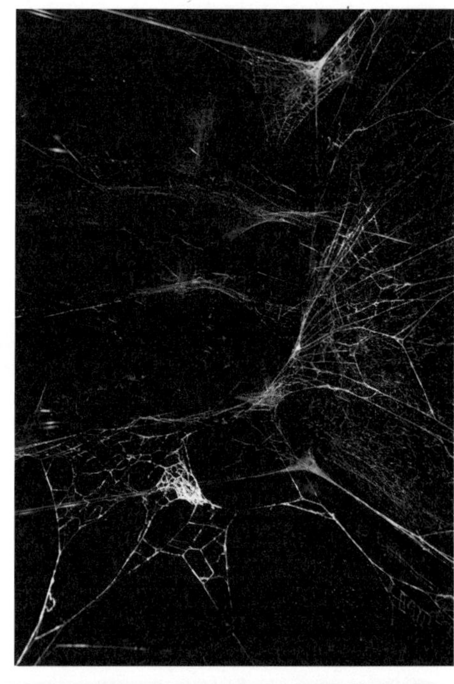

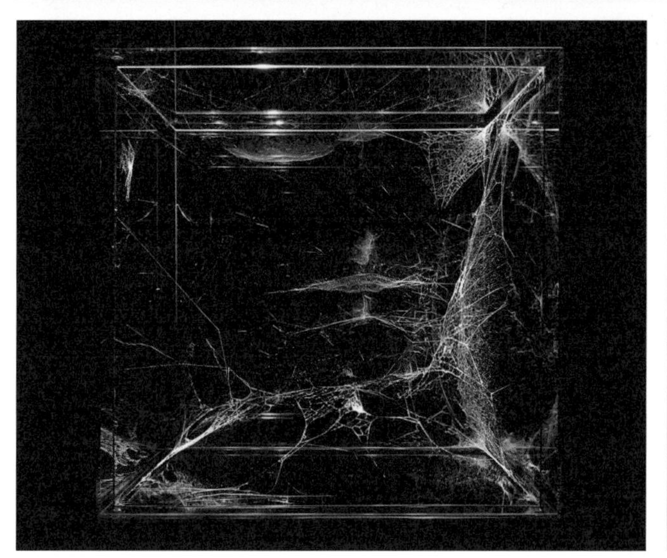

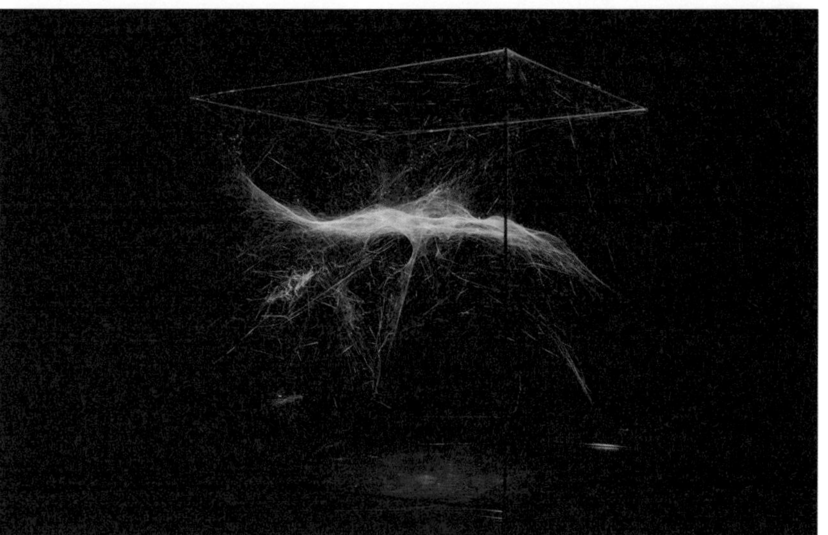

Figure 1.1 (top left)
Hybrid semisocial–semisocial musical instrument Antennae Galaxies, built by a pair of *Cyrtophora citricola* (two weeks) and a single *Cyrtophora moluccensis* (one week), 2014.

Figure 1.2 (top right)
Hybrid solitary social–semisocial musical instrument Apus, built by one *Nephila clavipes* (six days), a small community of *Stegodyphus dufouri* (four months), and six *Cyrtophora citricola* spiderlings (two weeks), 2015.

Figure 1.3 (bottom left)
Hybrid solitary social–semisocial musical instrument Apus, built by one *Nephila clavipes* (six days), a small community of *Stegodyphus dufouri* (four months), and six *Cyrtophora citricola* spiderlings (two weeks), 2015.

Figure 1.4 (bottom right)
Hybrid semisocial solitary instrument BR1202–0725 LAE, built by one *Cyrtophora citricola* (two weeks) and one *Tegenaria domestica* (six weeks), 2015.

across multiple scales, from nano to macro.

You also mentioned that silk coming from a spider's glands is liquid and turns solid through a process comparable to gas condensation in outer space, in distant galaxies. In my work, I tackle the metaphorical analogue between a spider net and the cosmic web. What other patterns, coincidences, thinking images do you see entangled in a spider's web?

Markus Buehler:
To me, a truly fascinating aspect is the concept that our human perception misguides us as we build our own models of how the world works based on what we see. To truly understand the hierarchical, multiscale aspect of the web we must let this go and consider the web in a more abstract way, to reflect the web as a mathematical "web" that reflects the different levels of the construction from the nano to macro as elements to achieve certain functions. Combining all these into a system facilitates the set of properties that it was meant to offer. This is what we call biomateriomics. The calling is for engineers to move from the geometric space to the functional space, to understand how the patterns resonate to create function.

In much of what we do, we want to tell the story about "us," and we do this in many different ways. We can discover the way the world works in many places, and the same patterns exist. Just like there are only a few amino acid building blocks that make up our body, organs,

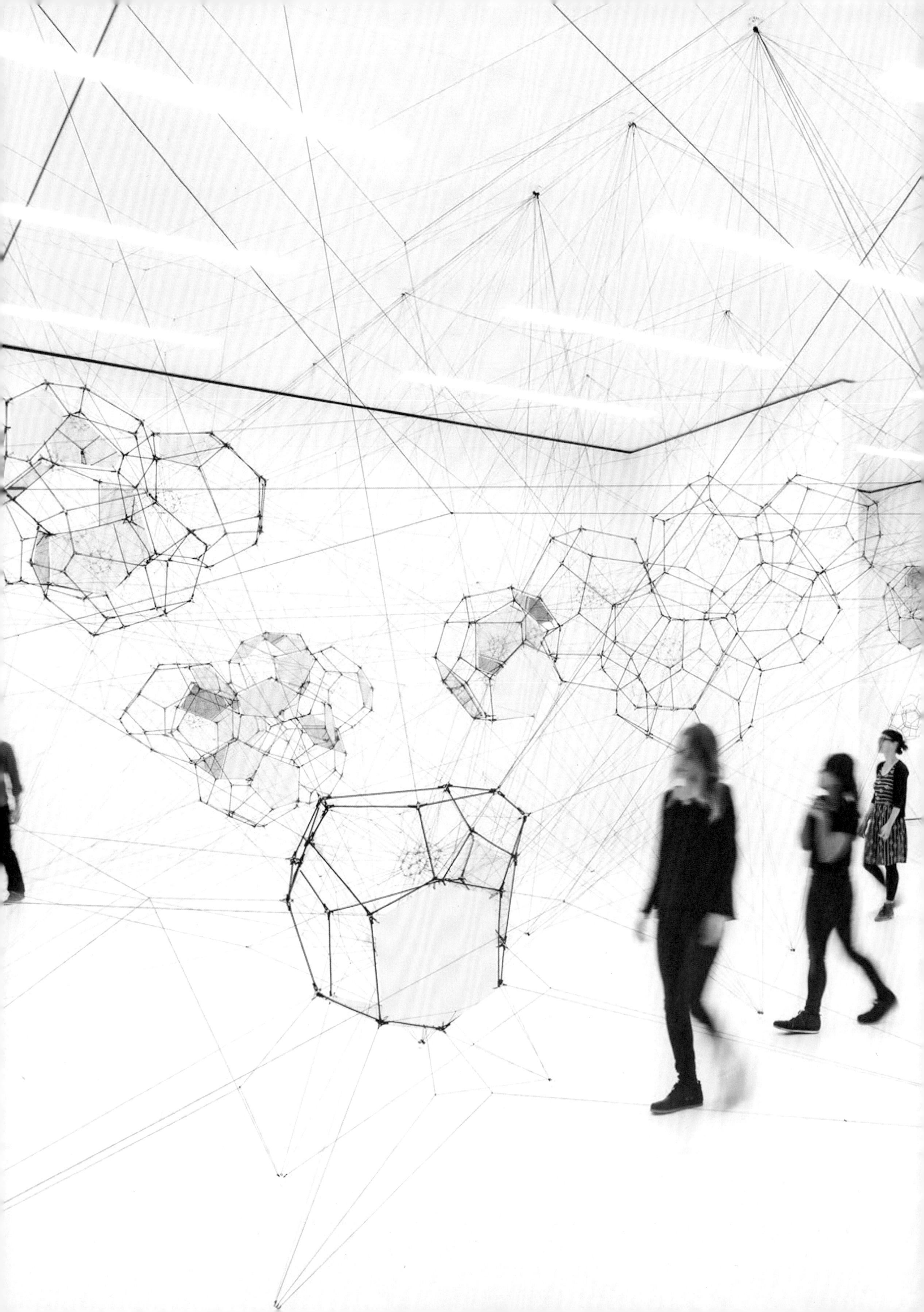

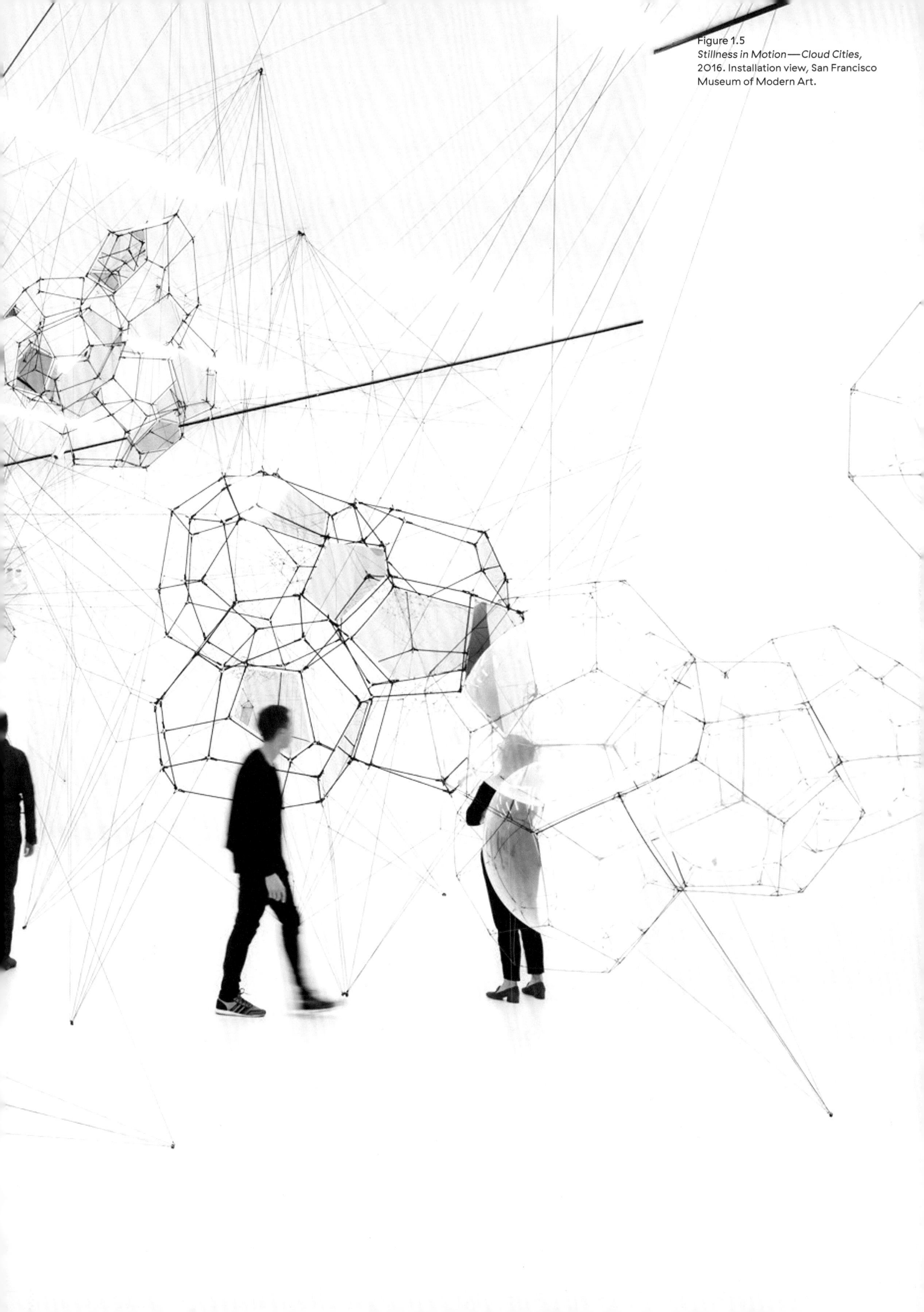

Figure 1.5
Stillness in Motion—Cloud Cities, 2016. Installation view, San Francisco Museum of Modern Art.

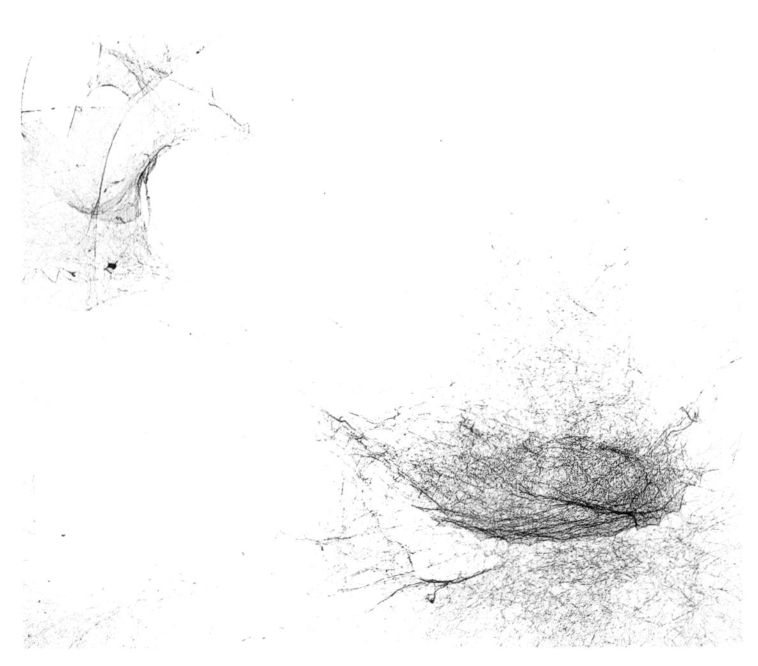

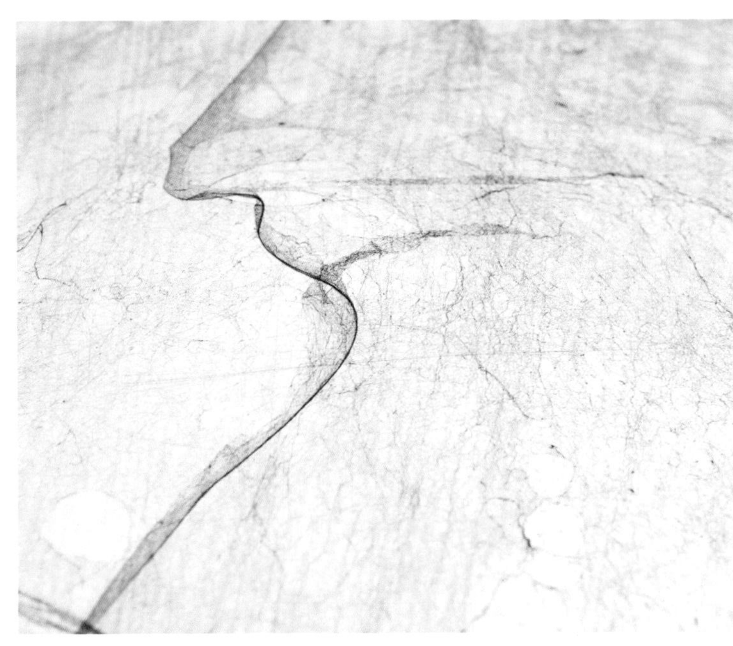

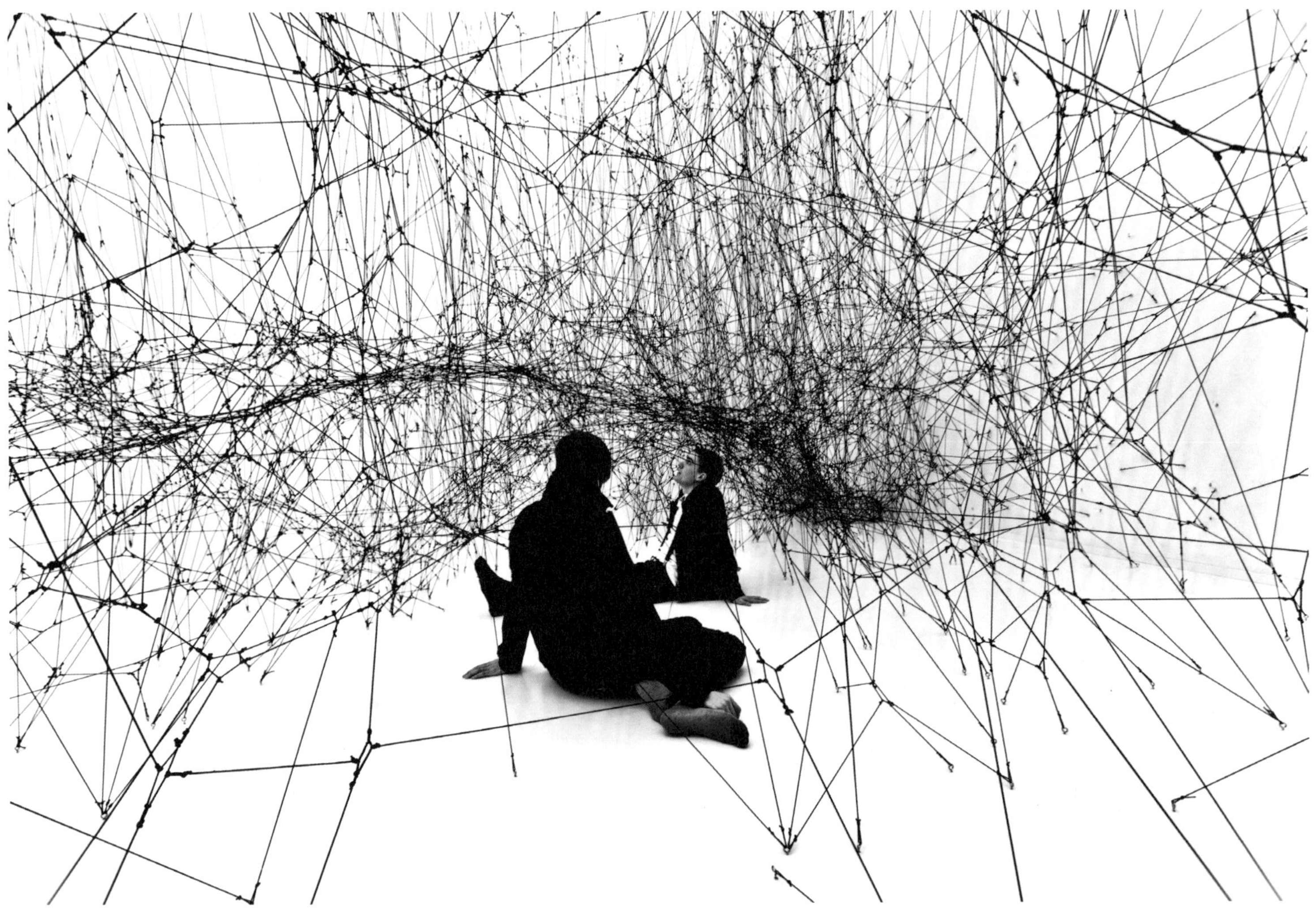

Figure 1.6 (top left)
Semisocial, solitary mapping of NGC 4676 tidal action aiming toward coalescence by two Cyrtophora citricola (one week), one Tegenaria domestica (four weeks), one Cyrtophora moluccensis (four weeks), one Agelena labirinthica (eight weeks), and one Argiope anasuja (three weeks).

Figure 1.7 (top right)
Social mapping of 1E 0657–56 merger by a colony of *Anelosimus eximius* (eight weeks).

Figure 1.8 (bottom)
14 Billions (working title), 2010. Installation view, Bonniers Konsthall, Stockholm. Commissioned by Bonniers Konsthall.

Figure 1.9
Aerocene, Flights at Salar de Uyuni, 2016. Co-commissioned by MARCO Museum of Contemporary Art, Monterrey, Mexico, on the occasion of the solo show by the artist in June 2016.

Figure 1.10
Aerocene, launches at White Sands Missile Range (New Mexico, United States), 2015. The launches in White Sands Missile Range and the symposium "Space without Rockets," initiated by Tomás Saraceno, were organized together with the curators Rob La Frenais and Kerry Doyle for the exhibition "Territory of the Imagination" at the Rubin Center for the Visual Arts.

etc., I postulate that there are a small number of hierarchical principles that are applied repeatedly in many different places. To see them, we must have the appropriate "microscope"—which requires a deep level of abstraction. Once we understand these, we will see that the world follows simple paradigms by which it is built. Perhaps we will be disappointed once we find out. Maybe not.

Humans have an extreme self-awareness, and we constantly project this to the world around us in a variety of manifestations, and it will also lead us to push the frontier of where we live beyond this planet. Perhaps we are sometimes overly self-aware and take our importance in the universe too seriously, and think that we are quite unique, when we are simply another expression of the principles of the world.

Tomás Saraceno:
Sometimes I am called a utopian artist because of my vision of Cloud Cities, which I am currently pursuing in a long-term research project called *Aerocene*, with collaborators in the MIT Department of Earth, Atmospheric and Planetary Sciences, among others in multiple institutions. These works anticipate future possibilities for more-than-human social assembly, floating up in the skies. Do you see any future utopian living spaces for spiders, spiderwebs, and spider silk?

Markus Buehler:
I believe that humans will ultimately leave our planet and try to establish civilizations on other planets, beyond Earth. The same way a spider fills space by building the web to live from/in (as visualized nicely when we watch a black widow build a web in a simple metal cage), humans can fill space by expanding what we think is the limit of the possible. Just as the spider takes a leap of faith to fill space, humans will do the same.

In the end, all human experience is natural experience, and it all falls into very similar patterns of behavior. We can discover the way the world works in many places, and the same patterns exist, and they repeat themselves over and over again. As the orbits of the planets, the orbits of electrons in the atoms, the cycle of life, or just the way we make the sounds to speak words, it's all about cyclical variation of structure which is a core structural principle (the principle of periodicity). I would go so far as to say that it's not the atom (or subatomic particles) that is the ultimate building block, but the principle of periodicity that underlies it all. Atoms and what they represent are just ways by which the principle of periodicity is expressed in one way, but there are many others. So perhaps we need to take a different look at the world to understand it.

NOTES

1 The scientific endeavor began in 2010 with a commission from Tomás Saraceno to the Photogrammetric Institute (IPK) at the Technische Universität in Darmstadt. Advised by Peter Jäger and Samuel Zschokke on arachnology-related matters, and by Rolf-Dieter Dueppe, Dieter Steineck, and Christoph Wulff on technical web capturing, Saraceno and IPK, TU-Darmstadt developed a visualization of a black widow spiderweb using laser-supported tomography combined with photogrammetric analysis, a method first devised by Tomás Saraceno. The research led to the production of a 16:1 scale black widow web exhibited at Bonniers Konsthall, Sweden. Initiated in 2012, Tomás Saraceno's ongoing collaboration with the MIT Center for Art, Science & Technology (CAST) s eeks to continue his investigations of web-building, with an emphasis on the development of measuring methods, data analysis, and modeling of the spiderwebs. In 2014, Saraceno continued working on this research in collaboration with Alessio Del Blue, Paolo Bianchini, Carlos Beltrán González, and Vittorio Murino at the Istituto Italiano di Tecnologia, from whom he commissioned a color-coded 3D visualization of a *Cyrtophora citricola* web. See Tomás Saraceno and Sara Arrhenius, *14 Billions* (Milan: Skira, 2011), 18–75.

2 Giorgio Agamben, *The Open: Man and Animal* (Stanford: Stanford University Press, 2003), 42.

REFERENCES

Agamben, Giorgio. *The Open: Man and Animal*. Stanford: Stanford University Press, 2003.

Demian, Bogdan A. "Structural and Mechanical Analysis of the Black Widow Spider Web Subjected to Stretching, Expansion and Wind." Thesis: M.Eng., Massachusetts Institute of Technology, Department of Civil and Environmental Engineering, 2014.

Ndengeyingoma, B. "Spider Web Investigation by Digitalization, Modelling and Simulation." Senior Civil and Environmental Engineering Design Capstone Project Final Report: M.Eng., Massachusetts Institute of Technology, Department of Civil and Environmental Engineering, 2015.

Nyambo, S. "Spider Web Investigation by Digitalization, Modelling and Simulating Final Report." Senior Civil and Environmental Engineering Design Capstone Project Final Report: M.Eng., Massachusetts Institute of Technology, Department of Civil and Environmental Engineering, 2015.

Wangare, Y. "Spider Web Investigation by Digitalization, Modelling and Simulation." Senior Civil and Environmental Engineering Design Capstone Project Final Report: M.Eng., Massachusetts Institute of Technology, Department of Civil and Environmental Engineering, 2015.

2
Multiscale Computational Design of Bioinspired Active Materials for Functionally Diverse Applications

Zhao Qin and
Markus J. Buehler

2
Multiscale Computational Design of Bioinspired Active Materials for Functionally Diverse Applications

Zhao Qin and
Markus J. Buehler

Natural materials including silk, bone, nacre, skin, or cellular cytoskeleton materials have advanced material functions that easily surpass those of synthetic polymers.[1, 2, 3, 4, 5, 6, 7, 8, 9] Even though advanced 3D printing is beginning to more precisely control the nanostructure of materials, these manufacturing techniques are still far from fully being able to replicate multifunctional protein materials, especially their active response to external stimuli.[3, 10, 11] Many of these active materials contain proteins as their fundamental building blocks, which are self-assembled in certain ways, resulting in structures that span many hierarchical levels.[4, 8] The structure of natural materials—defined by which building blocks they use and how these are arranged together across the relevant hierarchical levels—largely determines the material functions.[12] The knowledge of the mechanism of their exceptional material functions, including unique nonlinear mechanical behaviors in response to external mechanical loading conditions as well as distinct environmental triggers, and how atomic interactions at the chemical scale lead to large-scale material functions, provides fruitful resources for the design of innovative functional materials that can generate impacts broadly in engineering fields. Such applications range from structural properties to water filtration or the design of advanced sensors, and they offer new routes for directed material discovery.

The primary goal in frontier research activities in this field is to obtain materials with innovative functions by design.[1, 3, 10, 11, 13] We can learn from biological materials that feature advanced functions and generate synthetic composite materials with comparable functions, as shown in figure 2.1.[1, 2, 5, 14, 15] To achieve this goal, multiscale computational modeling methods with the capability of studying the structure-features at all hierarchical levels are being applied, enabling us to

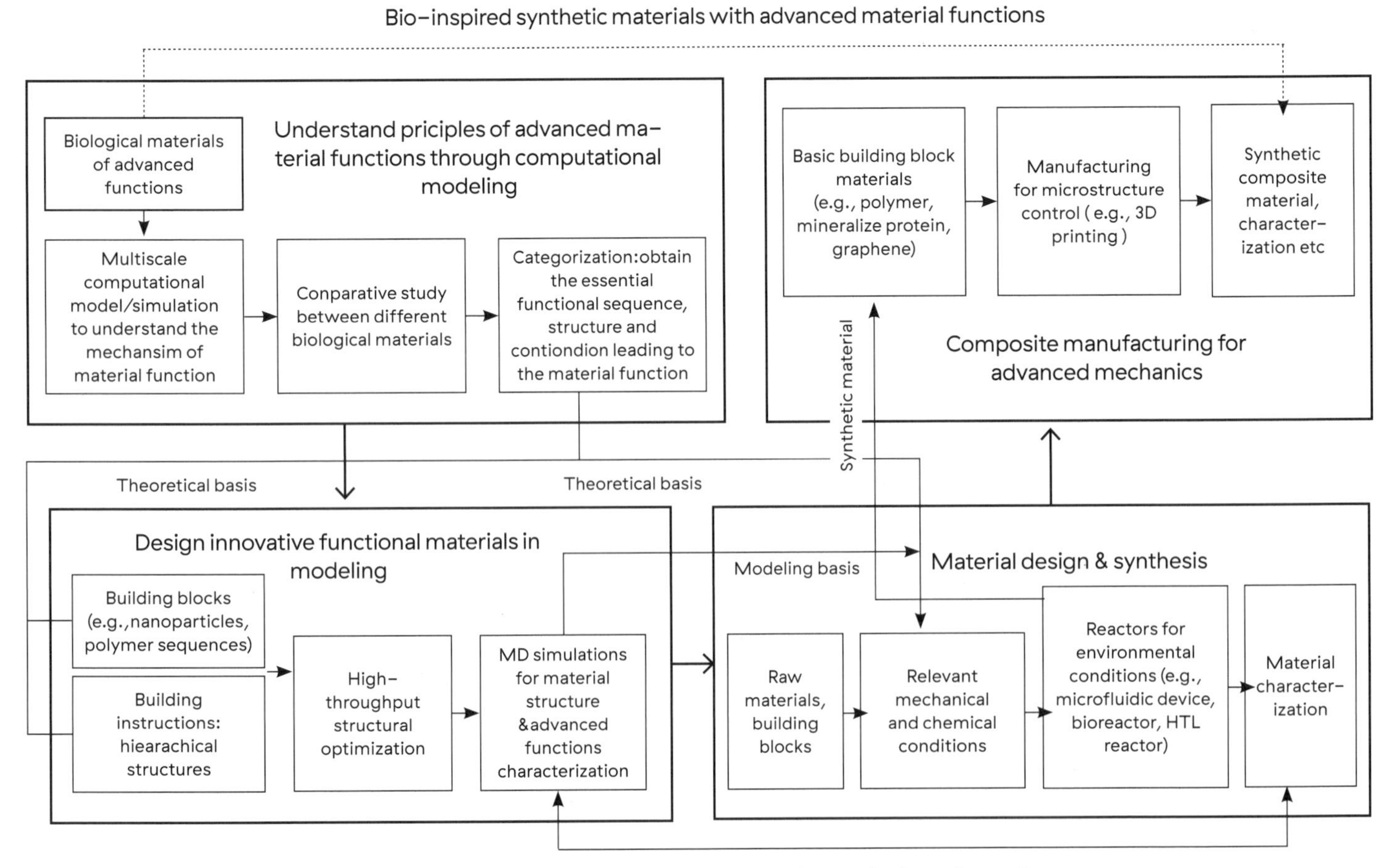

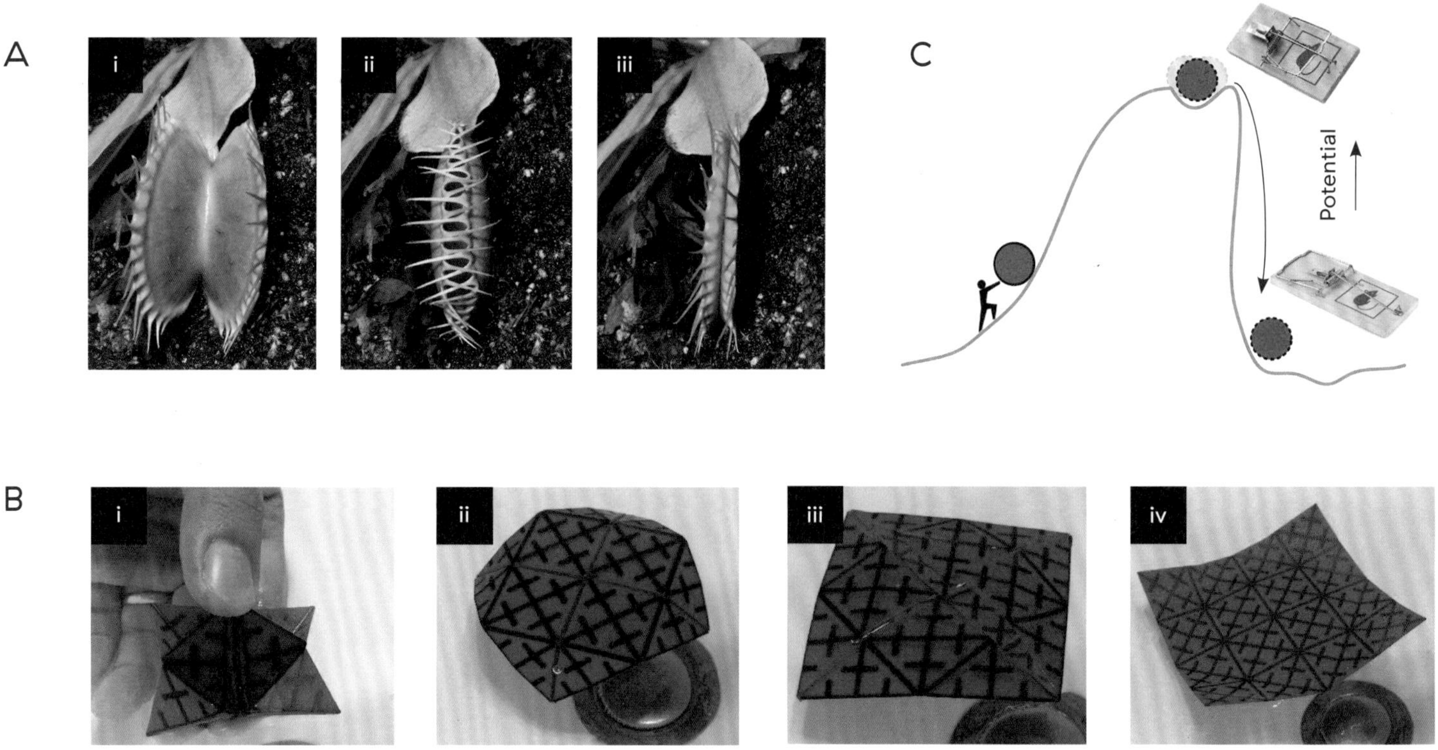

Figure 2.1 (left page)
A blueprint of rational material design that starts from biological materials and leads to synthetic-protein composite materials with advanced mechanical and multifunctionalities. Major efforts focus on the design of innovative functional materials in both modeling and experimental work. Our effort includes protein design, synthesis, and bulk manufacturing with controllable microstructures. Such a design strategy, by integrating computational, theoretical, and experimental tools, is the key to exploring the space of functional material design by looking into all the scale effects in building hierarchical materials inspired by natural counterparts.

Figure 2.2 (top)
Bioinspired design and manufacturing of active materials. (A) Snapshots of the closing process of a Venus flytrap (images reproduced from Libiakova et al. (note 17) under the Creative Commons Attribution (CC BY) license). (B) A 3D-printed foldable structure of heterogeneous material distribution; the black material is rubber-like extensible material and the magenta material is plastic-like rigid material. (C) A schematic figure for the design of the energy aspect of active materials and structures: it is important that the design has structure that can store sufficient energy to drive the active motion as well as a mechanism to trap the structure at the quasi-static state before releasing the energy.

more fully simulate complex material responses to environmental stimuli.[15] With high-resolution multimaterial 3D printers, we can further define the material property and geometry of a structure according to computational modeling and optimization results.[3] Such an integrated methodology allows us not only to duplicate the structure in natural materials, but also to scan for suitable raw materials in order to make a product that has a similar material function to that of a natural material of interest, created in a fully controllable process according to computational modeling.

Learning from biological materials and using such recipes to design new materials with complex hierarchical structures provides great opportunities for material innovations. This learning from and reverse-engineering of nature is nontrivial and requires multidisciplinary efforts from biology, structural and mechanical engineering, mathematics, computational science, and chemical science. For example, the actively controlled open-to-close process of the Venus flytrap (figure 2.2A) makes it a very unique plant that catches insects.[16, 17] It is very interesting that the trap closes instantly only when the insect contacts two *different* hairs, in *sequence*. Although it is still not clear how the plant receives the signal and triggers this motion, it is clear that the closing process involves a large deformation in the trap material that associates with the structural transition between two equilibrated states (figure 2.2B). As the state of higher potential energy becomes the state that is farther from equilibrium, the energy release causes high-speed motion similar to what is observed in other mechanical traps. To make synthetic structures that fully mimic the behavior of the Venus flytrap, we need foldable structures that can store significant amounts of potential energy. This goal, for instance, could be achieved by using our 3D printer (figure 2.2C), but also by using active materials that generate instant reactions to environmental stimuli.

The multiscale investigation of protein materials has now allowed us to more thoroughly understand the structure–function relations of many protein materials that feature hierarchical structures. More and more evidence has emerged to suggest that many of them have switchable material functions, with the transition actively triggered by interacting with environmental factors. These factors include chemical conditions such as pH, ion concentration, enzymatic conditions, and mechanical conditions like temperature, osmotic pressure, and mechanical stress states. For example, as shown in figure 2.3, the mechanics of collagen protein material is defined by the hierarchical structures and their interaction with all the environmental factors, which can drastically control the mechanics of collagen materials at different length scales.[7, 8, 18, 19, 20, 21, 22]

Just as muscle cells can generate contraction force that accounts for body motion, several protein materials have been identified to have similar active mechanical functions. Collagen, for example, the main component of skin, tendon, bone, and cartilage, contracts during the drying process, producing a force surpassing those generated by contractile muscles by over two orders of magnitude.[8] In contrast, spider silk contracts during the wetting process, leading to supercontraction, which explains the tightening-up of the spider web within the morning dew environment (figure 2.4).[4, 23] These new mechanisms provide new functions to the protein materials that used to be considered as only passively responding to mechanical loading. They allow us to design and build biological materials with active functions. By considering the new mechanisms during the design of structural materials, it may now be possible to make shape-tunable materials for applications ranging from soft robotics to humidity-sensing.

To amplify the active function of protein materials, such as by turning a small deformation into a large deformation, we will need to design skeleton structures that provide proper confinements to allow synergistic

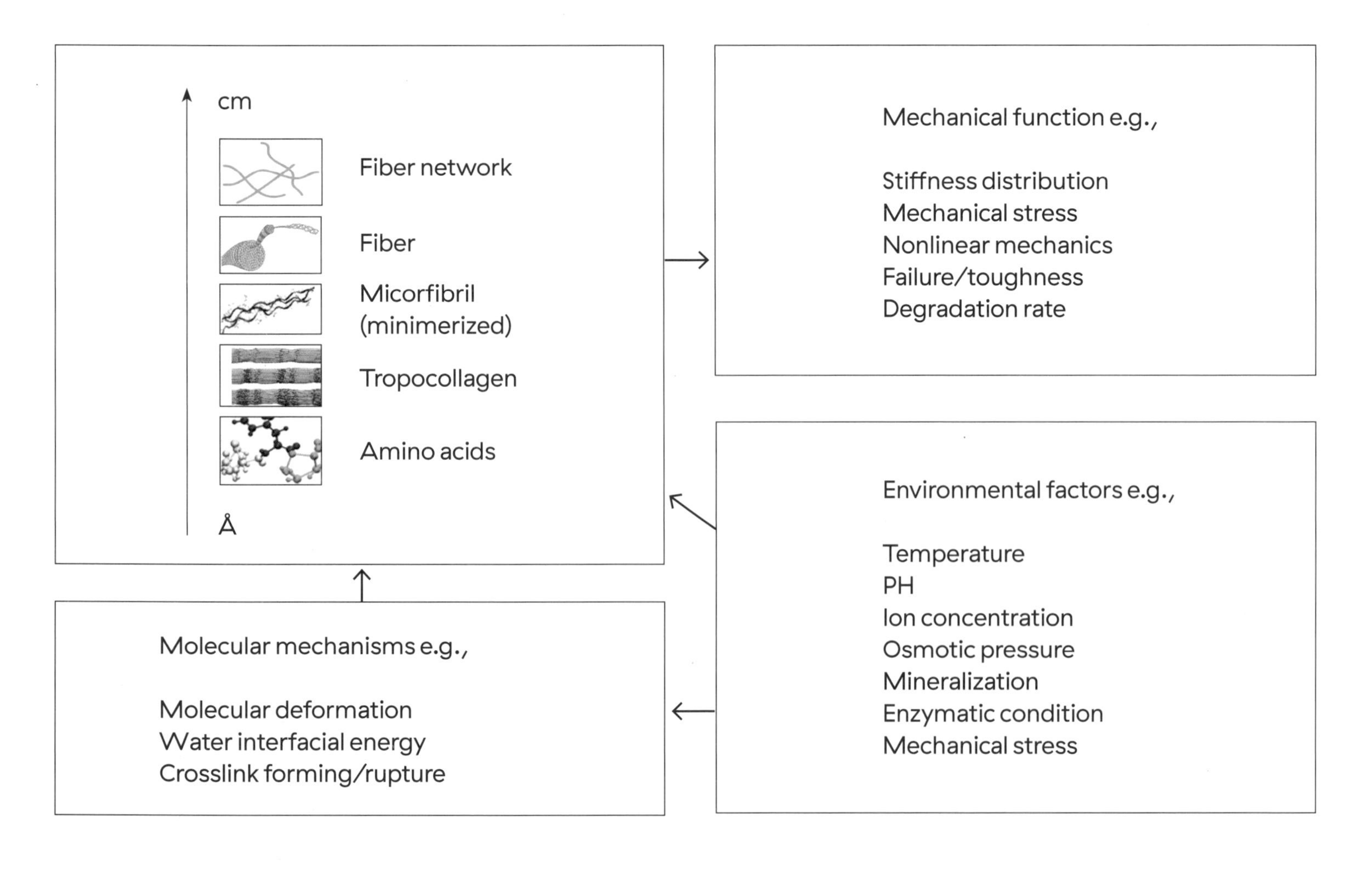

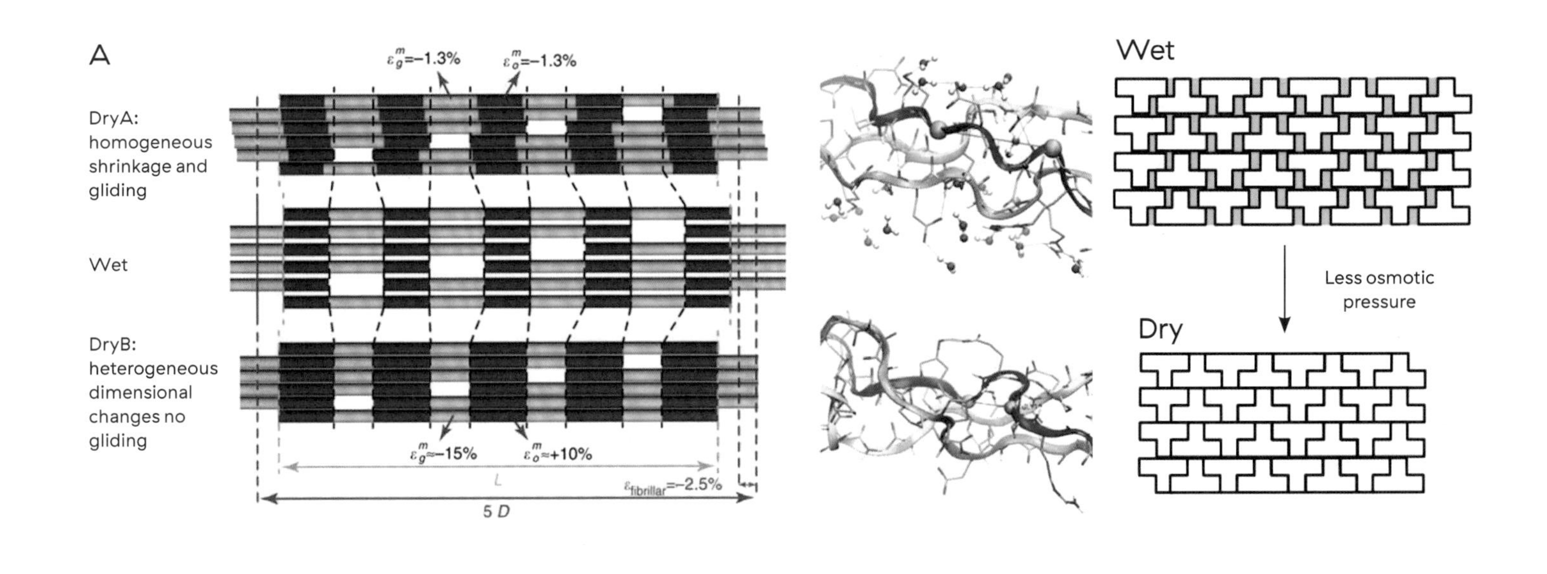

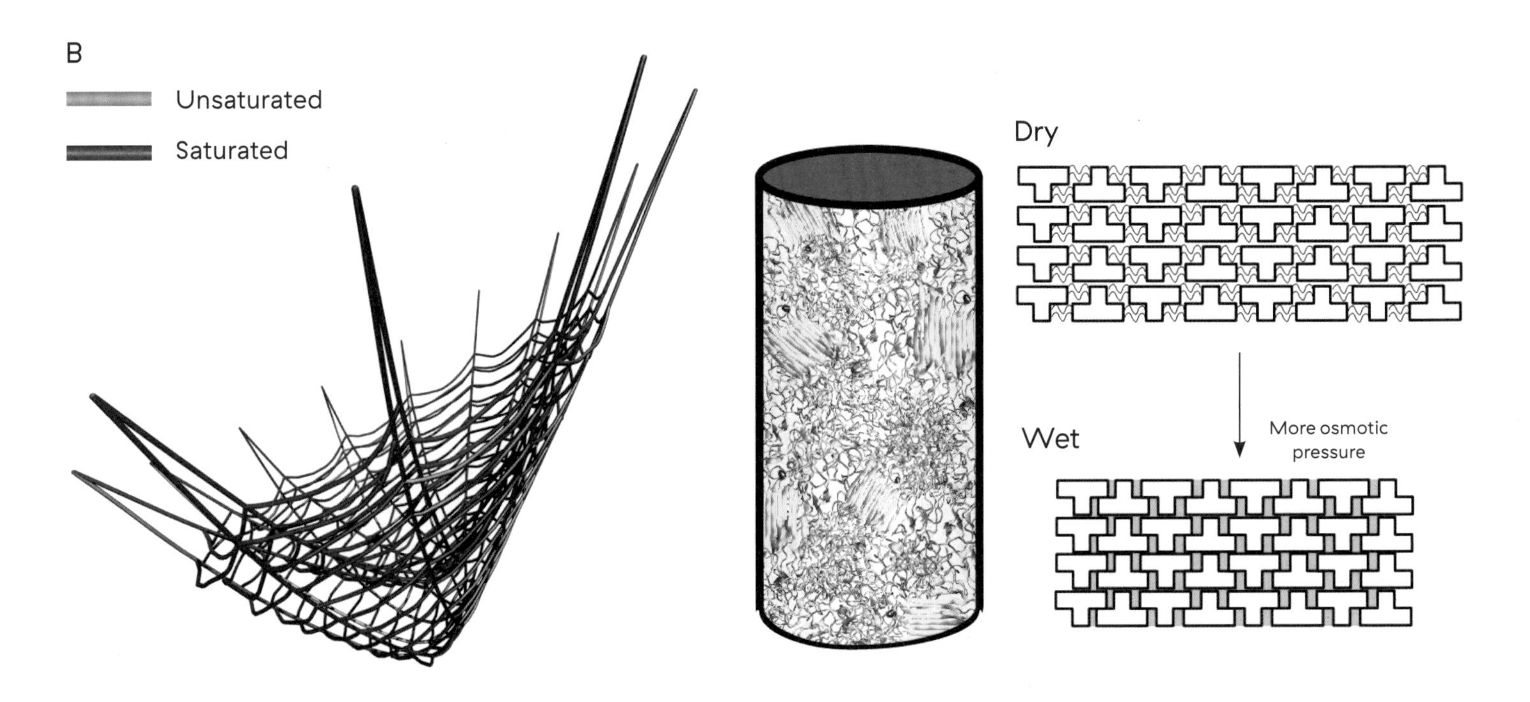

Figure 2.3 (left page)
Multiscale computational modeling, here exemplified for active collagen material design and functionalization. Collagen is a key structural protein material that offers not only strength, toughness, and resilience to tissues such as bone, tendon, skin, or cartilage, but also features an amazing range of active properties.

Figure 2.4 (top)
Active behavior of protein materials (collagen and silk) in response to osmotic pressure. (A) A collagen microfibril (main component of bone, tendon, and other extracellular matrix materials) under low osmotic pressure will tend to lose its interaction with water molecules at the molecular level and forms a more compact stacking configuration to generate contractile force. (B) In contrast, water molecules can enter the amorphous region of spider silk and cause supercontraction of the material, that is, a significant shrinking under exposure to wet conditions.

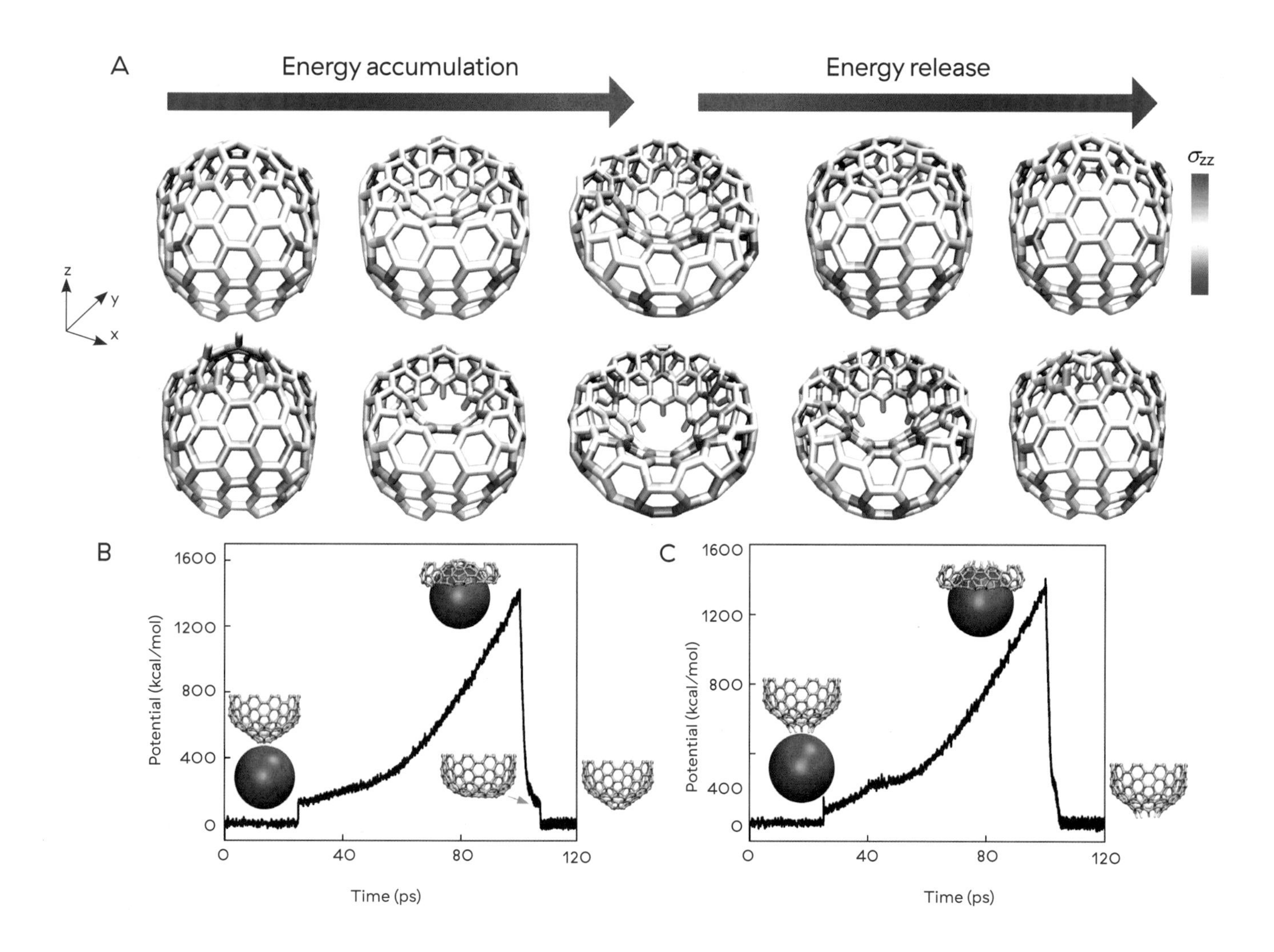

Figure 2.5
Energy storage and release in nanostructures of graphene in molecular dynamics simulations based on a first-principle-based reactive force field. (A) The potential energy accumulation and release in the nanostructure controlled by a nanoindenter. Atoms are colored according to atomic stress in the indenting direction, with red for tensile stress and blue for compressive stress. (B) The energy history of the indenter during the simulation, as the indenting process causes gradual increasing of the potential energy; retracting of the indenter leads to rapid energy release by deforming the nanostructure back to its original shape. (C) The same process as panel B, except for the existence of a vacancy at the top of the nanostructure with free edges terminated by hydrogen atoms.

deformation of the materials. The development of additive manufacturing will allow for the mixing of protein and other synthetic materials during a single print. Two-dimensional nanomaterials including graphene and boron nitride can be ideal materials to construct structures that host the protein materials to maximize their active material functions. For example, using molecular dynamics simulation based on a first-principle-based reactive force field,[24] we can simulate the deformation process of a graphene hemisphere structure under indentation and record the potential energy change, as shown in figure 2.5. Our calculation shows that such a nanostructure is greatly capable of absorbing and storing impact energy, and such energy can be released instantly by removing the constraint. The structure is also robust for mechanical energy storage, supported by the evidence that, even with the existence of vacancy, the structure is still capable of storing a similar amount of energy and releasing the energy by recovering the original form of its geometry. Indeed, the specific energy storage of the structure is computed to be ~3 MJ/kg, which is higher than that of many traditional electrochemical and electrical energy sources, including batteries and capacitors, and it is of the same order as food and explosives (figure 2.6).

Combining nanostructures for energy storage and protein materials that play an active role in responding to environmental change and effectively “charge” their underlying nanostructures, we are working toward designing stand-alone functional active materials that are sensitive to the changes of environmental factors without using

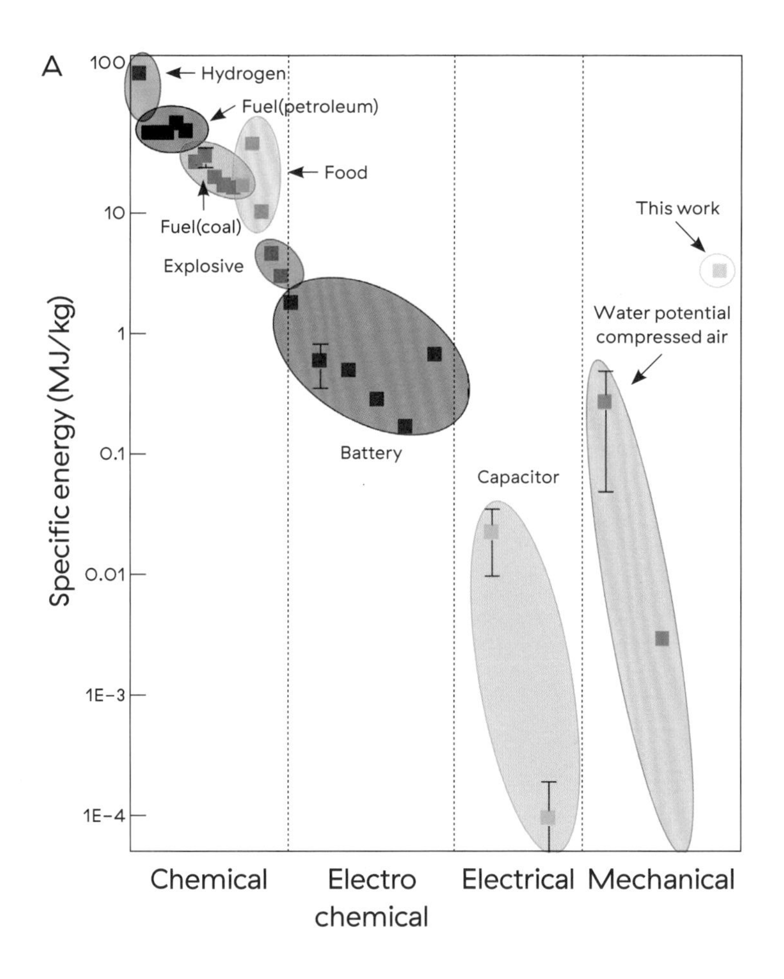

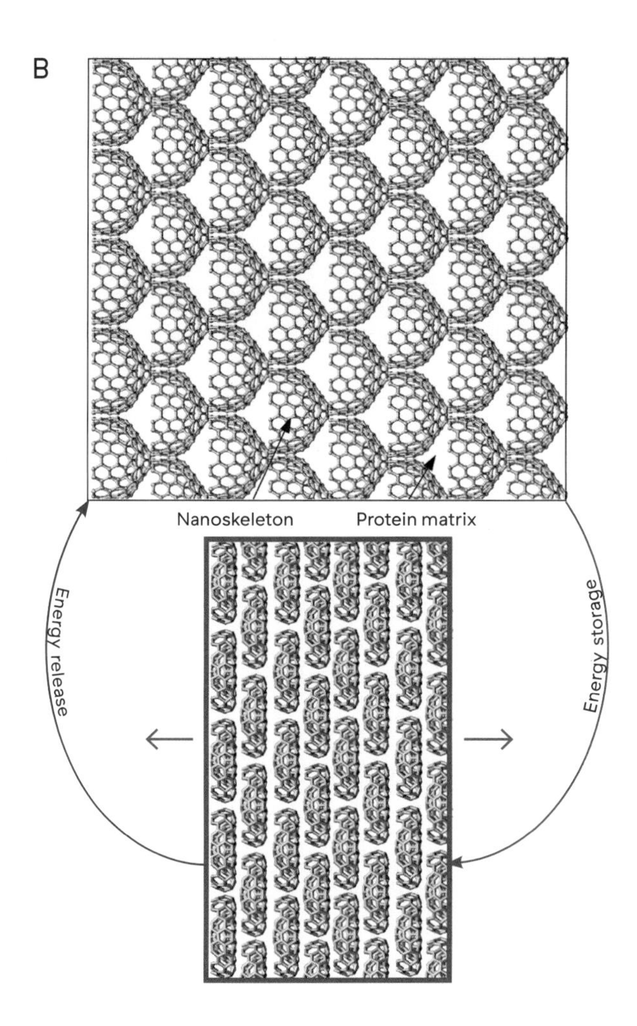

Figure 2.6
Specific energy storage and release during the deformation and recovery of the graphene nanostructure considered here. (A) Comparison of the energy storage capability of different materials, including the current nanostructure and other conventional energy sources including chemical, electrochemical, electrical, and mechanical. It is noted that the high energy density of the nanostructure in deformation is of similar order to that of food and explosives. (B) The schematic figure of the design of a composite material composed of graphene nanostructures and protein matrix material. While the protein material plays the active role in harvesting the mechanical energy by reacting to changes of environmental factors, the nanostructures play the role in rapidly releasing the energy and returning the composite back to the equilibrated state

complex electronic devices.[25] We expect such materials, as in the example schematically shown in figure 2.6, will be of high sensitivity, repeatability, and capacity for energy storage, and thus will be able to actively respond to environmental changes in a very short time. Multiscale computational models will enable us to simulate the material behavior from quantum to macro scale *in silico*, which provides an efficient way to design and optimize the material structure and composition. Some of the mechanical response and biological function of the optimized designs can be tested by performing experimental tests on 3D printed samples. In total, such an integrated design scheme sets up a paradigm that enables us to design active materials rationally from the most fundamental scale and up.

NOTES

1 S. Keten, Z. P. Xu, B. Ihle, and M. J. Buehler, "Nanoconfinement Controls Stiffness, Strength and Mechanical Toughness of Beta-Sheet Crystals in Silk," *Nature Materials* 9, no. 4 (2010): 359–367.

2 M. J. Buehler, "Tu(r)ning Weakness to Strength," *Nano Today* 5, no. 5 (2010): 379–383.

3 Z. Qin, B. G. Compton, J. A. Lewis, and M. J. Buehler, "Structural Optimization of 3d-Printed Synthetic Spider Webs for High Strength," *Nature Communications* 6 (2015).

4 Z. Qin and M. J. Buehler, "Spider Silk: Webs Measure Up," *Nature Materials* 12, no. 3 (2013): 185–187.

5 Z. Qin, and M. J. Buehler, "Impact Tolerance in Mussel Thread Networks by Heterogeneous Material Distribution," *Nature Communications* 4 (2013): 2187.

6 Z. Qin and M. J. Buehler, "Molecular Dynamics Simulation of the Alpha-Helix to Beta-Sheet Transition in Coiled Protein Filaments: Evidence for a Critical Filament Length Scale," *Physical Review Letters* 104, no. 19 (2010): 198304.

7 A. K. Nair, A. Gautieri, S. W. Chang, and M. J. Buehler, "Molecular Mechanics of Mineralized Collagen Fibrils in Bone," *Nature Communication* 4 (2013): 1724.

8 A. Masic, L. Bertinetti, R. Schuetz, S. W. Chang, T. H. Metzger, M. J. Buehler, and P. Fratzl, "Osmotic Pressure Induced Tensile Forces in Tendon Collagen," *Nature Communication* 6 (2015).

9 T. Giesa, M. Arslan, N. M. Pugno, and M. J. Buehler, "Nanoconfinement of Spider Silk Fibrils Begets Superior Strength, Extensibility, and Toughness," *Nano Letters* 11, no.11 (2011): 5038–5046.

10 S. J. Ling, Q. Zhang, D. L. Kaplan, F. Omenetto, M. J. Buehler, and Z. Qin, "Printing of Stretchable Silk Membranes for Strain Measurements," *Lab on a Chip* 16, no. 13 (2016): 2459–2466.

11 L. S. Dimas, G. H. Bratzel, I. Eylon, and M. J. Buehler, "Tough Composites Inspired by Mineralized Natural Materials: Computation, 3d Printing, and Testing," *Advanced Functional Materials* 23, no. 36 (2013): 4629–4638.

12 M. J. Buehler and Y. C. Yung, "Deformation and Failure of Protein Materials in Physiologically Extreme Conditions and Disease," *Nature Materials* 8, no. 3 (2009): 175–188.

13 T. Ackbarow, X. Chen, S. Keten, and M. J. Buehler, "Hierarchies, Multiple Energy Barriers, and Robustness Govern the Fracture Mechanics of Alpha-Helical and Beta-Sheet Protein Domains," *Proceedings of the National Academy of Sciences* 104, no. 42 (2007): 16410–16415.

14 M. J. Buehler, "Nature Designs Tough Collagen: Explaining the Nanostructure of Collagen Fibrils," *Proceedings of the National Academy of Sciences* 103, no. 33 (2006): 12285–12290.

15 Z. Qin, L. Kreplak, and M. J. Buehler, "Hierarchical Structure Controls Nanomechanical Properties of Vimentin Intermediate Filaments," *Plos One* 4, no. 10 (2009): e7294.

16 Y. Forterre, J. M. Skotheim, J. Dumais, and L. Mahadevan, "How the Venus Flytrap Snaps," *Nature* 433, no. 7024 (2005): 421–425.

17 M. Libiakova, K. Flokova, O. Novak, L. Slovakova, and A. Pavlovic, "Abundance of Cysteine Endopeptidase Dionain in Digestive Fluid of Venus Flytrap (Dionaea Muscipula Ellis) Is Regulated by Different Stimuli from Prey through Jasmonates," *Plos One* 9, no. 8 (2014).

18 B. Depalle, Z. Qin, S. J. Shefelbine, and M. J. Buehler, "Large Deformation Mechanisms, Plasticity, and Failure of an Individual Collagen Fibril with Different Mineral Content," *Journal of Bone and Mineral Research* 31, no. 2 (2016): 380–390.

19 S. W. Chang, S. J. Shefelbine, and M. J. Buehler, "Structural and Mechanical Differences between Collagen Homo- and Heterotrimers: Relevance for the Molecular Origin of Brittle Bone Disease," *Biophysical Journal* 102, no. 3 (2012): 640–648.

20 Z. Qin, A. Gautieri, A. K. Nair, H. Inbar, and M. J. Buehler, "Thickness of Hydroxyapatite Nanocrystal Controls Mechanical Properties of the Collagen-Hydroxyapatite Interface," *Langmuir* 28, no. 4 (2012): 1982–1992.

21 B. Depalle, Z. Qin, S. J. Shefelbine, and M. J. Buehler, "Influence of Cross-Link Structure, Density and Mechanical Properties in the Mesoscale Deformation Mechanisms of Collagen Fibrils," *Journal of the Mechanical Behavior of Biomedical Materials* 52 (2015): 1–13.

22 S. G. M. Uzel and M. J. Buehler, "Molecular Structure, Mechanical Behavior and Failure Mechanism of the C-Terminal Cross-Link Domain in Type I Collagen," *Journal of the Mechanical Behavior of Biomedical Materials* 4, no. 2 (2011): 153–161.

23 K. J. Koski, P. Akhenblit, K. McKiernan, and J. L. Yarger, "Non-Invasive Determination of the Complete Elastic Moduli of Spider Silks," *Nature Materials* 12, no. 3 (2013): 262–267.

24 Z. Qin, M. Taylor, M. Hwang, K. Bertoldi, and M. J. Buehler, "Effect of Wrinkles on the Surface Area of Graphene: Toward the Design of Nanoelectronics," *Nano Letters* 14, no. 11 (2014): 6520–6525.

25 Z. Qin, G. S. Jung, M. J. Kang, and M. J. Buehler, "The Mechanics and Design of Light-Weight Three-Dimensional Graphene Assembly" (2016), in revision.

3
Silk Materials at the Intersection of Technology and Biology

Fiorenzo G. Omenetto

3
Silk Materials at the Intersection of Technology and Biology

Fiorenzo G. Omenetto

The natural world is a colorful canvas of materials whose performance rivals or exceeds many of their inorganic counterparts that are used in our daily lives. Biomimicry is rooted in the observation and inspiration derived from these many forms and their associated functions, in order to move toward the next generation of high-performance materials. Examples of functions that are found in nature and are not available in the common materials used today point to a unique and distinguishing feature, namely the material interaction with the environment that surrounds it. Among these are materials that are antibacterial and antifouling, that manipulate light for heat management and dissipation, that promote water and nutrient sequestration, that exhibit superhydrophobicity, self-cleaning or self-healing properties, physical adhesiveness, and enhanced mechanical properties.

A particular challenge (and opportunity) of bioinspired strategies is to develop universal fabrication strategies to generate new structural materials that can be used in a variety of fields, ranging from the biomedical to the technological and architectural.

Naturally occurring materials are generated through a bottom-up generative process that involves a nontrivial interplay of mechanisms acting across scales from the atomic to the macroscopic. The self-assembly of structural biopolymers, the fundamental building blocks of natural materials, leads to hierarchically organized architectures that impart unique functionality to the end material formats.

Among the several structural proteins that have been studied, silk fibroin was recently shown to be suited for the generation of a number of biopolymer-based advanced material formats leveraging control of form (through self-assembly) and function (through material modification).

The ability to generate functional materials based on water-based silk self-assembly rests on the ability to control and direct the sol-gel-solid transition of silk fibroin materials in ambient conditions. Silk is extracted from natural sources (i.e., natural silk fibers from *Bombyx mori* cocoons) with a previously developed protocol that yields a water suspension of the fibroin protein. Controlling the dynamics of solvent removal makes it possible to direct the bottom-up process of self-assembly and generate a large collection of end material formats. These material outcomes are wide and disparate, ranging from plastic-like transparent sheets to optical fibers, ceramic-like monoliths, inkjet-printed materials, 3D-printed geometries, sponges, nanofibrillar lattices, nanostructured lattices, and molded objects, to name a few.

What particularly distinguish silk from other biopolymers are its robust mechanical properties, its

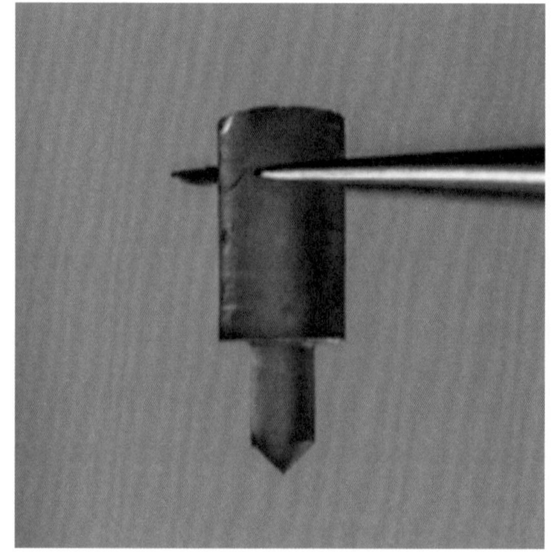

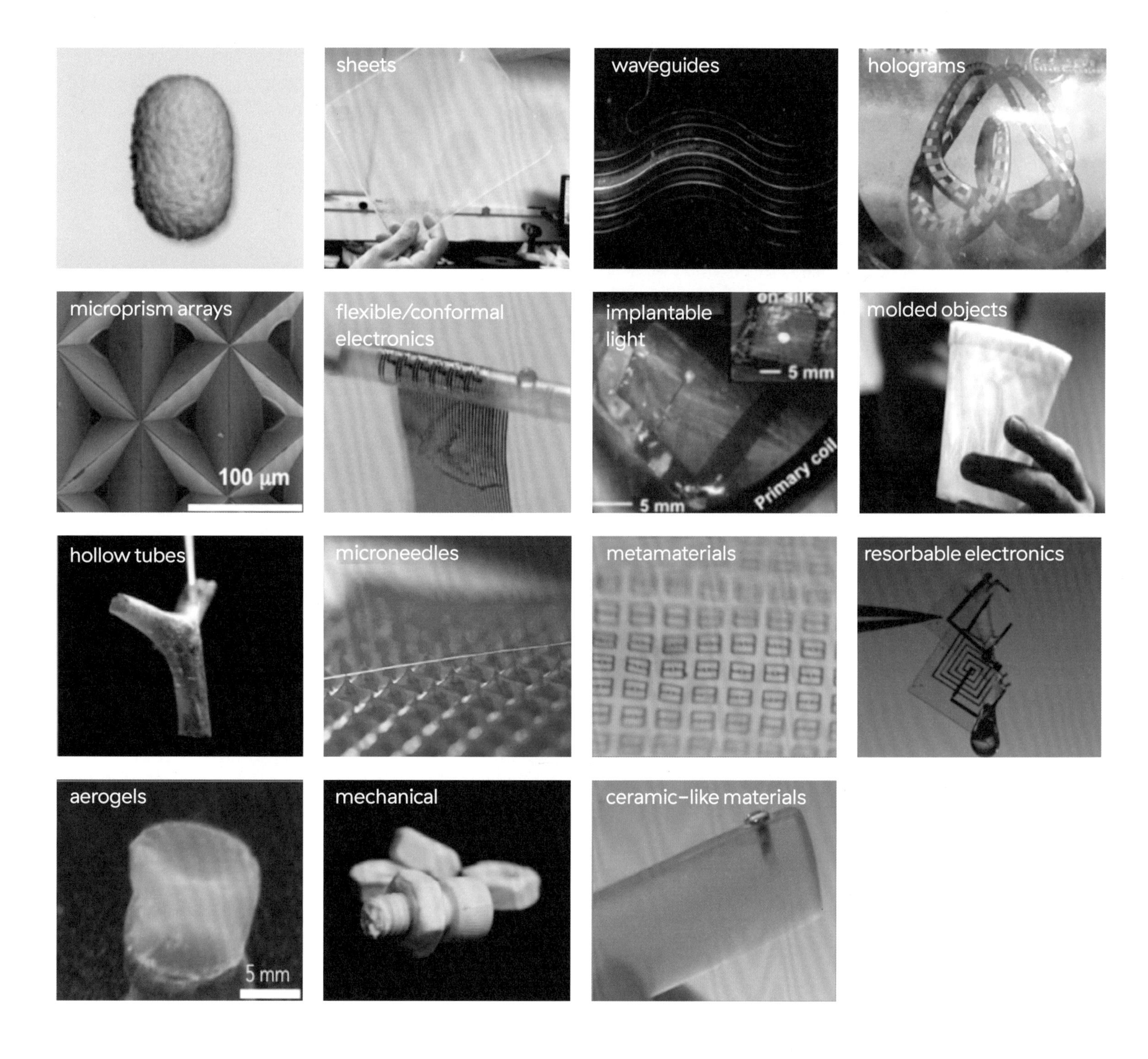

Figure 3.1 (left page)
Functional silk mechanical components. Silk can be reformulated into hard material formats that respond to external forces with a colorimetric reaction thanks to the ease of incorporating chemistries in the water-based silk solution used to generate these material outcomes. These hard silk monoliths can be machined into desired mechanical forms such as screws, bolts, or pins. Polydiacetylene (PDA) vesicles are added to silk fibroin suspensions at the point of material self-assembly to generate this hybrid material that "transduces" internal strain by changing color. For example, as shown here, silk fibroin–PDA pins undergo a blue-to-red chromatic transition when the force applied reaches the yield point (image on the right) of the material, making these hard formats "interactive."

Figure 3.2 (top)
The multiple forms of silk. Controlling the assembly of fibroin molecules in a water solution derived from natural fibers from the *Bombyx mori* silkworm cocoon (pictured in the upper left corner) and the dynamics of water removal makes it possible to obtain multiple material outcomes. The materials' utility is amplified by their ability to dissolve (or not) controllably and embed living components within them (e.g., microneedles that directly embed therapeutic compounds). The favorable properties of silk and its biocompatibility allow it to be interfaced with photonics and electronics, expanding the horizon of technological materials and narrowing the gap between high technology and the natural sciences.

facile control of material properties through the control of water content during processing, its programmably controllable (from instantaneous to years) degradation lifetime, and its unique applicability of top-down transformation techniques to modify materials from the nano to the macro scale. The ambient environment during silk-processing allows for the incorporation of labile biological components without loss of function and with retention of bioactivity over extended time frames. Furthermore, the material is comestible and implantable, opening a true avenue for a library of "living materials" with unusual forms and outcomes. Silk (and, more broadly, structural proteins) allows the redefinition of structure-function relations through the combination of top-down and bottom-up assembly strategies, continuously fueled by the inspiration that nature provides.

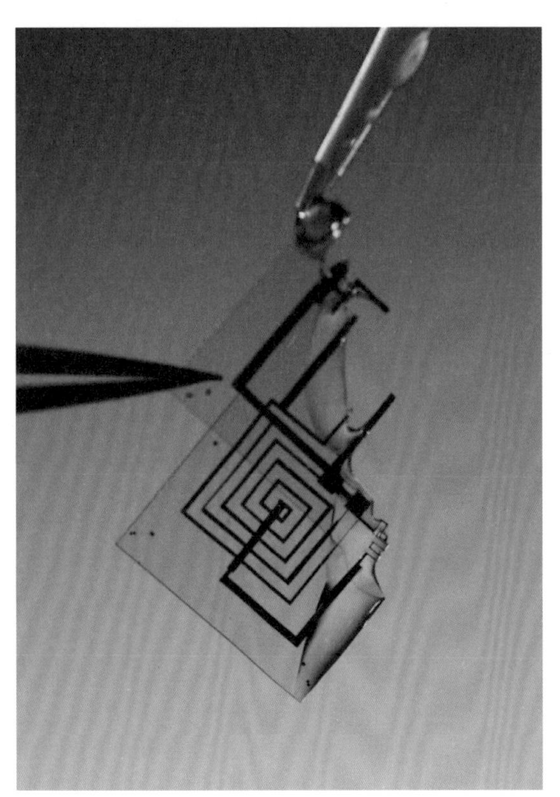

Figure 3.3 (left)
Biodegradable high technology. Biopolymers such as silk possess favorable properties for the integration/interface of photonic components and electronic circuits. These systems can be designed to fold, bend, or disappear after a predefined amount of time. The image shows an ultrathin microelectronic circuit on a silk film dissolving in a drop of wat

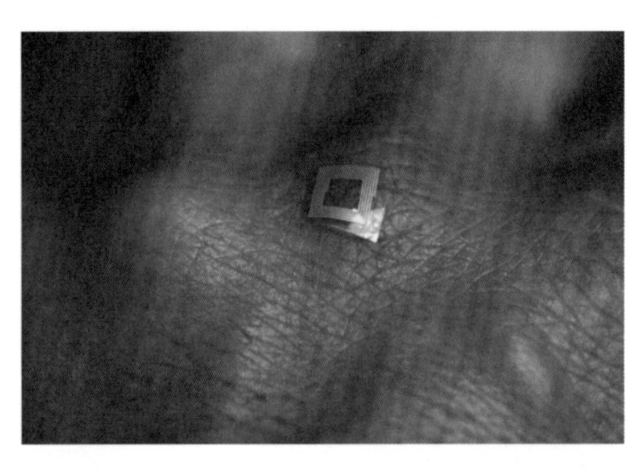

Figure 3.4 (right, top)
The internet of living things. Magnesium and silk make up the small pictured device (measuring approximately 1 cm × 1 cm) that is a wireless heater. When implanted in living tissue, it can be remotely activated and used to kill local infections. After a prescribed amount of time, the device harmlessly disappears in the tissue, eliminating the need to retrieve the device. Similar formats of these devices can be used to wirelessly control the release rate of drugs contained in the silk substrate that supports the electronic circuit.

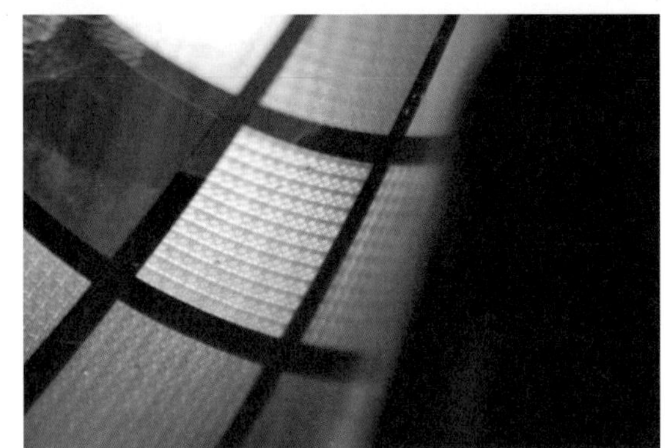

Figure 3.5 (right, bottom)
Reshaping a protein's surface to reshape light propagation. Films of natural biopolymers can be reshaped to have fine features on the nano and micro scales. The image shows an optically patterned silk film with diffractive optics—the business-card-sized transparent silk has features as small as a few hundred nanometers that allow the all-protein film to manipulate, control, reshape, and transform light propagating through it.

REFERENCES

Hwang, S. W., H. Tao, D. H. Kim, H. Y. Cheng, J. K. Song, E. Rill, M. A. Brenckle, B. Panilaitis, S. M. Won, Y. S. Kim, Y. M. Song, K. J. Yu, A. Ameen, R. Li, Y. W. Su, M. M. Yang, D. L. Kaplan, M. R. Zakin, M. J. Slepian, Y. G. Huang, F. G. Omenetto, and J. A. Rogers. "A Physically Transient Form of Silicon Electronics." *Science* 337, no. 6102 (2012): 1640–1644.

Kim, S., B. Marelli, M. A. Brenckle, A. N. Mitropoulos, E. S. Gil, K. Tsioris, H. Tao, D. L. Kaplan, and F. G. Omenetto. "All-Water-Based Electron-Beam Lithography Using Silk as a Resist." *Nature Nanotechnology* 9, no. 4 (2014): 306–310.

Kim, S., A. N. Mitropoulos, J. D. Spitzberg, H. Tao, D. L. Kaplan, and F. G. Omenetto. "Silk Inverse Opals." *Nature Photonics* 6 (2012): 817–822.

Marelli, B., N. Patel, T. Duggan, G. Perotto, E. Shirman, C. Li, D. L. Kaplan, and F. G. Omenetto. "Programming Function into Mechanical Forms by Directed Assembly of Silk Bulk Materials." *Proceedings of the National Academy of Sciences* 113 (2016).

Omenetto, F. G., and D. L. Kaplan. "New Opportunities for an Ancient Material." *Science* 329, no. 5991 (2010): 528–531, 101.

Tao, H., M. A. Brenckle, M. M. Yang, J. D. Zhang, M. K. Liu, S. M. Siebert, R. D. Averitt, M. S. Mannoor, M. C. McAlpine, J. A. Rogers, D. L. Kaplan, and F. G. Omenetto. "Silk-Based Conformal, Adhesive, Edible Food Sensors." *Advanced Materials* 24, no. 8 (2012): 1067–1072.

Tao, H., S. W. Hwang, B. Marelli, B. An, J. E. Moreau, M. M. Yang, M. A.

Brenckle, S. Kim, D. L. Kaplan, J. A. Rogers, and F. G. Omenetto. "Silk-Based Resorbable Electronic Devices for Remotely Controlled Therapy and In Vivo Infection Abatement." *Proceedings of the National Academy of Sciences* 111, no. 49 (2014): 17385–17389.

Tao, H., J. M. Kainerstorfer, S. M. Siebert, E. M. Pritchard, A. Sassaroli, B. J. B. Panilaitis, M. A. Brenckle, J. J. Amsden, J. Levitt, S. Fantini, D. L. Kaplan, and F. G. Omenetto. "Implantable, Multifunctional, Bioresorbable Optics." *Proceedings of the National Academy of Sciences* 109, no. 48 (2012): 19584–19589.

Tao, H., D. L. Kaplan, and F. G. Omenetto. "Silk Materials—A Road to Sustainable High Technology." *Advanced Materials* 24, no. 21 (2012): 2824–2837.

Tao, H., B. Marelli, M. M. Yang, B. An, M. S. Onses, J. A. Rogers, D. L. Kaplan, and F. G. Omenetto. "Inkjet Printing of Regenerated Silk Fibroin: From Printable Forms to Printable Functions." *Advanced Materials* 27, no. 29 (2015): 4273–4279.

4
Single-Stranded DNA Brick and Tile Assemblies

Luvena Ong and Peng Yin

4
Single-Stranded DNA Brick and Tile Assemblies

Luvena Ong and
Peng Yin

Deoxyribonucleic acid (DNA) is the building block of life. The sequence of nucleic acids codes for how biology functions—from the tiniest bacteria to complex multicellular organisms. When programmed appropriately, DNA can also be used as a powerful building block to physically construct materials.

DNA is a unique nanomaterial capable of self-assembling into a well-defined double-helix structure comprised of two complementary strands. Through assigning specific sequences, we can program different strands into precise architectures. Thus, DNA self-assembly is a powerful technique for creating a diversity of complex nanoscale structures.

Complex discrete 2D and 3D DNA brick structures can be assembled modularly by respectively using DNA tiles or bricks. A DNA tile or brick consists of a single-stranded oligomer with four binding domains. When incorporated in a structure, each tile or brick adopts a U-shaped configuration where two binding domains lie along two parallel helices connected by a single crossover. These DNA tile and DNA brick structures allow for the design of structures with prescribed shape and size. Completely synthetic structures with highly complex features can be assembled in one-pot annealing reactions. The modularity of the DNA tile and DNA brick structure architecture allows for facile design of complex structures through selection of desired tiles or bricks.

2D STRUCTURES

To assemble 2D structures, strands of length 42 bases are used with domain lengths of 10 or 11 nucleotides (nt). Because each helical turn corresponds to 10.5 base pairs (B), a domain length of alternating 10 or 11 nt would result in tiles that would bind and lie adjacent to one another, occupying a 3 nanometer (nm) by 7 nm area. By assigning unique sequences to each domain, structures of prescribed size and dimensions can be created.

Rectangular structures can be used as molecular canvases for patterning complex 2D shapes. By treating each strand as a pixel, one can easily design shapes by selecting the necessary strands to form the desired structure. Using a 310-pixel canvas from a 24-helix (H) by 28-turn rectangular structure, we demonstrated the construction of 107 shapes (100 of which are shown in figure 4.2) measuring approximately 100 nm by 100 nm by selecting subsets from a library of strands.

3D STRUCTURES

Three-dimensional structures can be assembled using DNA bricks containing domain lengths of 8 nt, which correspond to three-fourths of a helical turn. As a result, a 90° dihedral angle

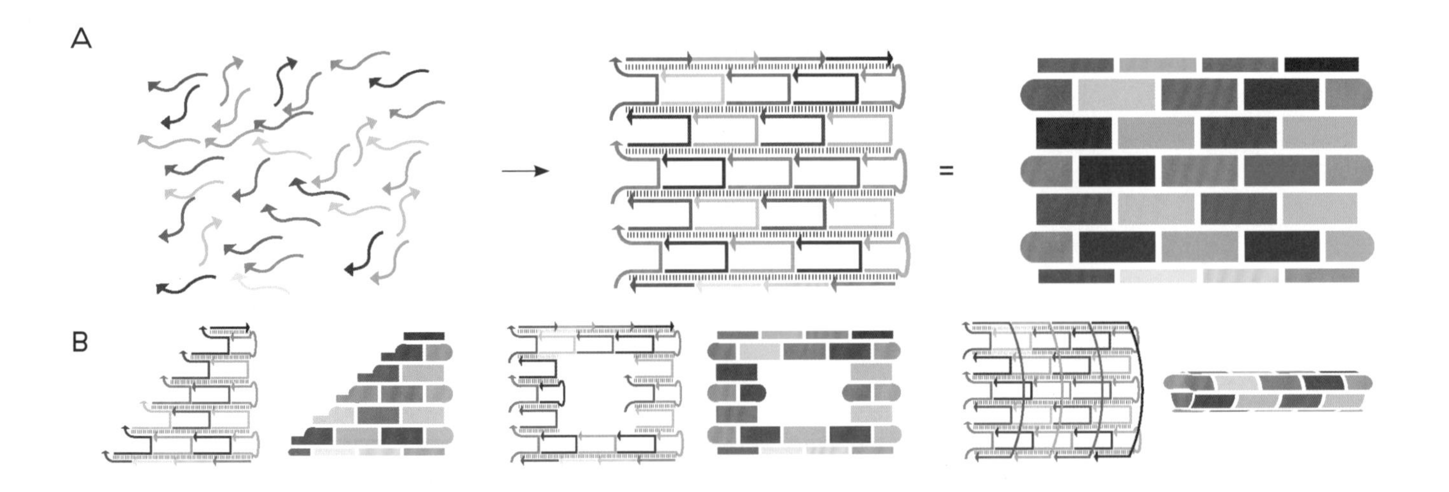

Figure 4.1 (left page)
Assembly of discrete 2D single-stranded DNA tile structures into a molecular canvas (A). Selection of tiles from the canvas allows for discrete structures of different shapes to be formed, including a tube (B).

Figure 4.3
DNA brick structures self-assemble via 8 base pair interactions (A) to form discrete cuboidal structures (B). Voxels can be selected from molecular canvases to form structures of different designs (C).

exists between two hybridized bricks. Using DNA bricks method, cuboids containing hundreds of strands have been assembled in single-pot reactions.

A DNA cuboid of size 8 helices (H) by 8 H by 80 B was used as a molecular canvas by treating each binding domain as a voxel of size 2.5 by 2.5 by 2.7 nanometers. Different shapes can be designed by selecting voxels from the 1000-voxel canvas. We selected strands from a master brick collection to demonstrate the assembly of 102 distinct 3D structures (100 of which are shown in figure 4.4) measuring approximately 20 nm in each dimension.

We can extend the capability of DNA beyond its native genetic use by creatively engineering its properties. When treated as modular building blocks, DNA can be programmed to self-assemble into precise, tunable nanostructures, including letters, numbers, and symbols in both 2D and 3D. These DNA tile and DNA brick approaches offer a simple platform for nanoscale prototyping. With further development, DNA-based systems will likely catalyze many new applications by providing a convenient means for building new nanotechnologies from the bottom up.

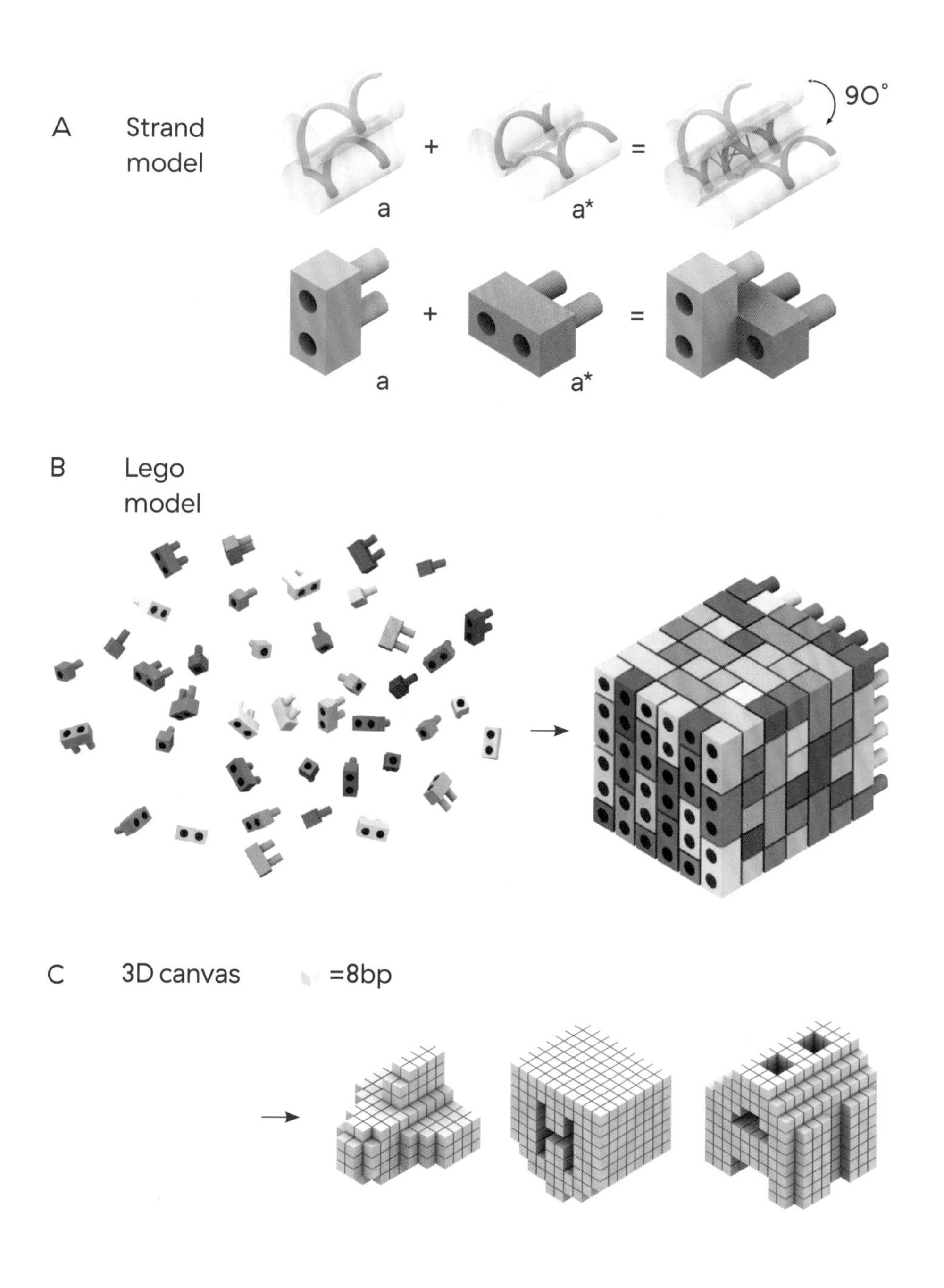

Figure 4.2 (right)
Atomic-force microscopy images of complex shapes designed using 2D DNA single-stranded tiles.

Figure 4.4
Complex shapes designed using 3D DNA bricks. The top shows a 3D model of the design and expected projection. The bottom shows transmission electron microscopy images the shapes.

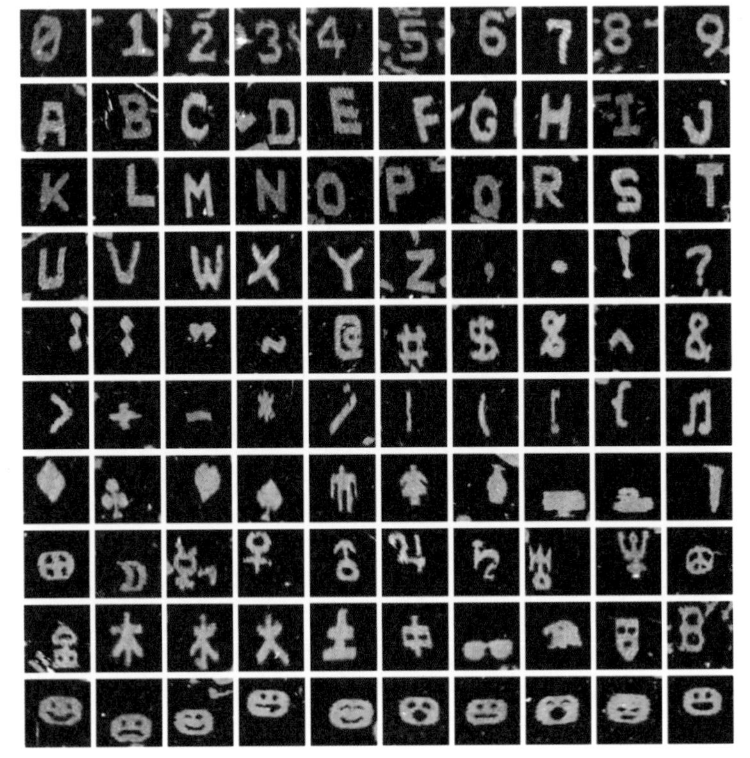

AFM image width:150mm

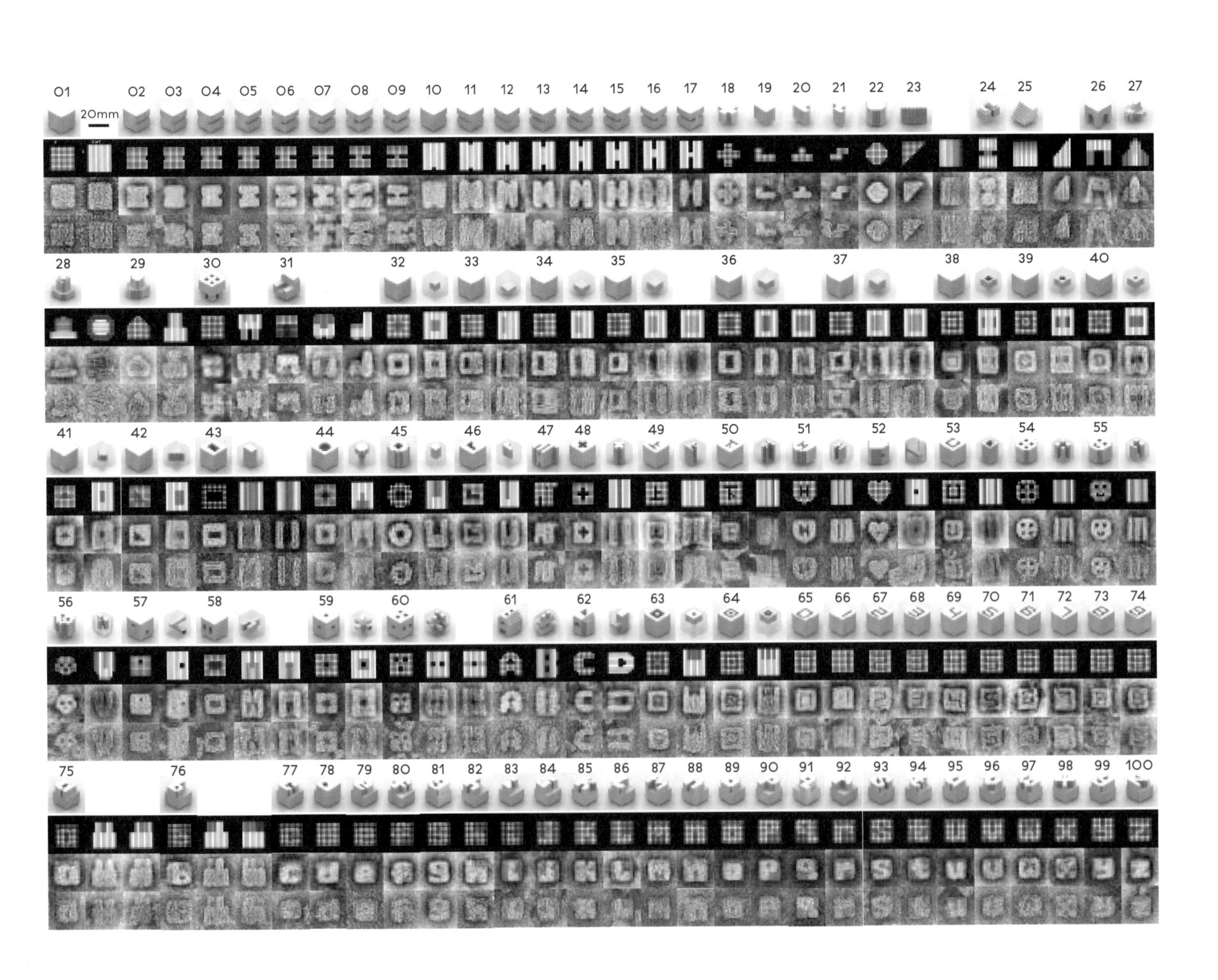

5
Kirigami with Atom-Thick Paper

Paul McEuen

2 μm

5 Kirigami with Atom-Thick Paper

Paul McEuen

Can we push the paper arts of origami and kirigami down to the atomic scale? This question has been the obsession of a team of a dozen or so nanoscientists and engineers at Cornell University over the last few years. The answer, as we will see, is a resounding *yes*.

The starting point is graphene, a one-atom-thick sheet of paper composed of carbon atoms that has been the subject of intense study for the last decade.[1] Figure 5.1 shows the atomic structure of graphene—a honeycomb lattice of carbon atoms that form a structure analogous to tiling the 2D plane with benzene-like tiles.[2] These sheets are grown by heating a copper sheet to about 1000° Celsius while flowing methane (which is composed of carbon and hydrogen) over the sheet. If the conditions are right, a one-atom-thick layer of graphene grows on top of the copper. The copper can then be dissolved away, leaving only the graphene.

A single sheet typically consists of micrometer-sized crystallites of oriented tiles joined by complex boundaries connected together with five- and seven-sided tiles, as shown schematically in figure 5.1.[2] Figure 5.2 shows a larger-scale image, in which dozens of microcrystallites of different orientation join together to make what resembles a patchwork quilt. Remarkably, these sheets are very robust and strong despite their patchwork structure.

These sheets can be patterned using a variety of lithographic techniques and incorporated into other lithographically patterned devices. Figure 5.3 shows a graphene sheet (colorized purple) with electrical contacts (colorized gold) stretched over a pair of adjacent trenches. The sheet can vibrate like the head of a drum, and our team pioneered the measurement of these devices for both fundamental physics and electromechanical applications.[3, 4] These suspended membranes can be further patterned to create kirigami devices. Examples are shown in figure 5.4. Here, a focused ion beam was used to cut the membrane after it was stretched over the holes. The tension on the membrane causes the structure to bend out of the plane, as seen in the images.

Even more control can be realized in the devices shown in figure 5.5. These graphene sheets were patterned into the same shapes as the paper models shown above them and then later placed in water and chemically released from the substrate.[5] They can then be manipulated using the external probes. Figure 5.6 is a 3D reconstruction of one of these kirigami devices. The atom-thin graphene device behaves nearly identically to the paper model. This is not an accident: the bending of a sheet of paper is to a first approximation scale invariant, meaning the same design works no matter how big (or small) the paper is.

While the structures built so far are very simple, we envision an entirely new class of micro- to nanoscale functional machines built on the designs of paper arts. In the same way as the miniaturization of electronics revolutionized the last 50 years, we expect the miniaturization of machines to revolutionize the next 50. And paper arts with atomic paper may be one of the key ways we achieve such miniature machines.

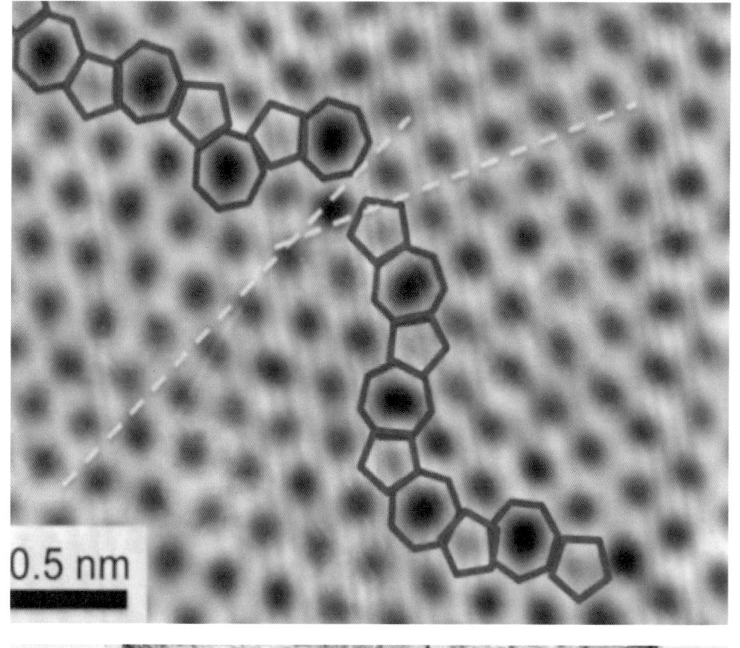

Figure 5.1 (left, top)
Atomic-resolution image of a graphene sheet from a scanning transmission electron microscope. The vertices of the honeycomb pattern are the individual carbon atoms. Two domains of graphene with different orientation (indicated by the dotted yellow lines) are joined by a boundary of five- (blue) and seven- (red) sided polygons. Image adapted from Huang et al. (see note 2).

Figure 5.2 (right)
Graphene patchwork quilt. Transmission electron micrograph of the grain structure of a single graphene sheet. Each region of a given orientation of the crystal lattice is colorized differently. Image adapted from Huang et al. (see note 2).

Figure 5.3 (left, bottom)
Stretched graphene membrane. False-color scanning electron microscope image of a graphene membrane suspended across two trenches.

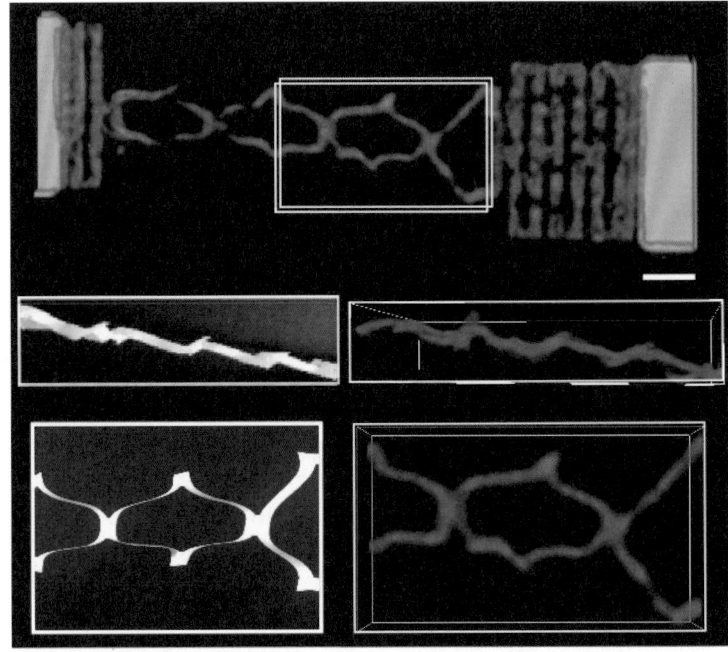

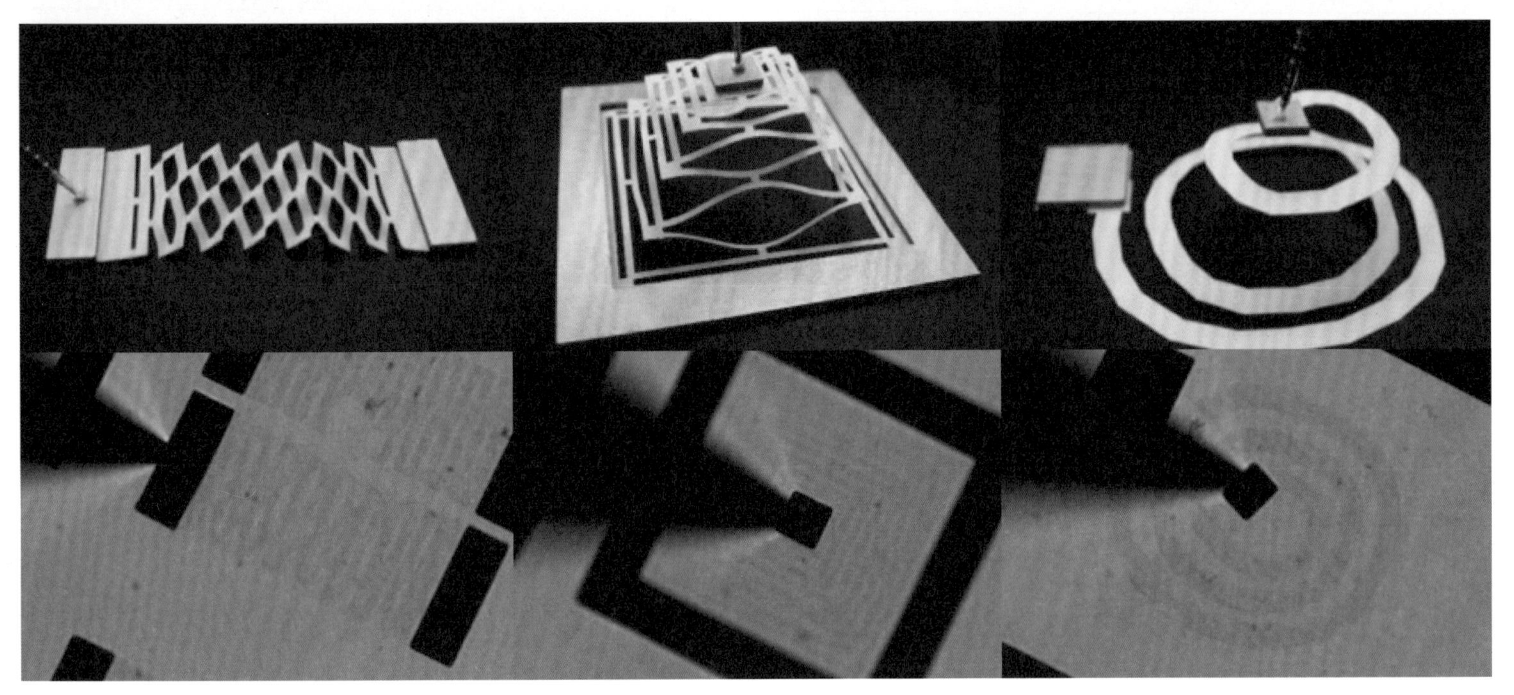

Figure 5.4 (top left)
Simple kirigami structures in graphene made using a focused ion beam to pattern the suspended membranes.

Figure 5.5 (bottom)
Graphene kirigami. Top: Paper models of the graphene kirigami devices. Bottom: Optical images of graphene patterned into kirigami structures. Image adapted from Blees et al. (see note 5).

Figure 5.6 (top right)
Stretched graphene kirigami spring. The blues images are 3D optical reconstructions of the device from a series of images. The white images are the paper model. Image adapted from Blees et al. (see note 5).

NOTES

1 For a review, see: K. S. Novoselov, V. I. Fal'ko, L. Colombo, P. R. Gellert, M. G. Schwab, and K. Kim, "A Roadmap for Graphene," *Nature* 490 (2012): 192.

2 P. Y. Huang, C. S. Ruiz-Vargas, A. M. van der Zande, W. S. Whitney, S. Garg, J. S. Alden, C. J. Hustedt, Y. Zhu, J. Park, P. L. McEuen, and D. A. Muller, "Imaging Grains and Grain Boundaries in Single-Layer Graphene: An Atomic Patchwork Quilt," *Nature* 469 (2011): 389–392.

3 J. S. Bunch, A. M. van der Zande, S. S. Verbridge, I. W. Frank, D. M. Tanenbaum, J. M. Parpia, H. G. Craighead, and P. L. McEuen, "Electromechanical Resonators from Graphene Sheets," *Science* 315 (2007): 490–493.

4 R. A. Barton, B. Ilic, A. M. van der Zande, W. S. Whitney, P. L. McEuen, J. M. Parpia, and H. G. Craighead, "High, Size-Dependent Quality Factor in an Array of Graphene Mechanical Resonators," *Nano Letters* 11 (2011): 1232–1236.

5 M. K. Blees, A. W. Barnard, P. A. Rose, S. P. Roberts, K. L. McGill, P. Y. Huang, A. R. Ruyack, J. W. Kevek, B. Kobrin, D. A. Muller, and P. L. McEuen, "Graphene Kirigami," *Nature* 524 (2015): 204–207.

6
Fouling Resistance: Controlling Nanoscale Adhesion with Optical Properties

Max Carlson,
Alex Slocum, and
Michael Short

6 Fouling Resistance: Controlling Nanoscale Adhesion with Optical Properties

Max Carlson, Alex Slocum, and Michael Short

INTRODUCTION

Fouling, or the undesired deposition of materials onto any surface, is a problem that affects fields ranging from energy to transportation to medicine. Energy production and distribution systems are particularly susceptible to fouling, as they rely on the continued cleanliness of their functional surfaces to transfer heat, catalyze chemical reactions, and resist corrosion. Particulate fouling, or the adhesion of particulate scales to surfaces, is present to some extent in almost all geothermal plants, oil refineries, nuclear plants, chemical processing facilities, and marine systems.[1, 2, 3] Fouling of internal pipe surfaces increases pressure drops across components, reduces heat transfer efficiency, and may block coolant channels entirely, necessitating the replacement of components. The costs of increased energy consumption, reduced throughput, and maintenance associated with fouling gives an economic impact of billions of dollars.[4]

Most engineers turn to solutions like Teflon™ or other slippery polymers to combat fouling, as they resist the buildup of just about everything. However, many energy systems operate in conditions far too harsh for Teflon, or any organic material, to remain stable. A more general solution is required that gets at the heart of what makes things stick to each other. Casting aside any unusual forces, such as charge buildup (permanent dipole interactions) or magnetic forces, the well-established Lifshitz theory of van der Waals (vdW) adhesive forces[5] describes how some sticky surfaces really work. It is then up to the scientist or engineer to take on a complex materials selection problem, using this theory to design slick surfaces and to ensure that these surfaces will not melt, corrode, dissolve, wear away, or otherwise cease to exist.

In this study, we explain particulate fouling using the Lifshitz theory of vdW adhesion, and propose methods of preventing fouling altogether, thus controlling the nanoscale interactions of matter to achieve an engineering-scale objective. We analyze Lifshitz theory as a way of predicting adhesion, suggest materials that may be fouling-resistant, and test them in the lab with atomic force microscopy (AFM) force spectroscopy (FS) measurements of adhesive force on seven candidate materials using a SiO_2 microsphere-functionalized AFM probe.

BACKGROUND

In particulate fouling, as shown schematically in figure 6.1, the adhesion of the first fouling particles to the clean pipe surface must be initiated by an attractive force between the particle and the pipe surface. This adhesive force is assumed to be predominantly due to van der Waals interactions between the materials. The vdW force arises due to coupling of electron motion in materials and is thus applicable to all physical systems.[6, 7] It is the underlying cause of adhesion at the atomic level (short of chemical bonding). The vdW force between two materials *a* and *b* in a fluid *f* with separation *r* is of the form $1/r^3$: (1)

$$F_{afb}^{vdW} = \frac{A_{afb}^{Ham}}{6\pi r^3}$$

The force is directly proportional to A_{abf}^{Ham}, known as the Hamaker constant. This constant defines the magnitude of the force and whether the force is attractive or repulsive (the latter being very uncommon). Since the vdW force arises from coupled electron motion creating induced dipoles, the Hamaker constant is calculated by taking into account a material's response to oscillating electric fields—in other words, optical properties like index of refraction or reflectivity. It is most direct to calculate the Hamaker constant from the material's imaginary-frequency dielectric response $\varepsilon(\zeta)$, but this difficult-to-interpret quantity is directly related to more conventional measurements such as reflectivity or ellipsometry. It may also be obtained to high accuracy by more advanced measurements such as valence electron energy loss spectroscopy (VEELS).[8] The formula for a full-spectrum relativistic

Hamaker constant (the formula with the fewest assumptions and limitations, but also the most complex) is: (2)

$$A_{afb}^{Ham} = \frac{3}{2} k_B T \sum_{n=0}^{\infty} R_n(r) \Delta_{af}(\zeta_n) \Delta_{bf}(\zeta_n)$$

where k_B is Boltzmann's constant in $\frac{eV}{K}$, T is the temperature in Kelvin, $R_n(r)$ is an optical retardation factor (which accounts for differing path lengths for differently polarized photon propagation), and Δ_{jk} is the difference in dielectric response to a *virtual photon* at an imaginary (complex) frequency $\zeta_n = i\nu_n$, where n is a discrete energy level from 0 to ∞. The Δ_{jk} variables can be thought of as contributions to adhesion energy based on differing polarizability at different frequencies, or differences in electron vibrations at different frequencies. Each of these can be expressed as follows: (3)

$$\Delta_{jk}(\zeta) = \frac{\varepsilon_j(\zeta) - \varepsilon_k(\zeta)}{\varepsilon_j(\zeta) + \varepsilon_k(\zeta)}$$

where $\varepsilon_j(\zeta)$ is the dielectric response function of material j at imaginary frequency ζ.

Equations (2)—(3) are too complex to apply directly to the initial candidate material search. Thus we sacrifice theoretical accuracy and use the Tabor–Winterton approximation (TWA), which is valid for materials with similar absorption frequencies ω and low refractive indices n in the visible spectrum:[9] (4)*

In this way, we can use the visible spectrum indices of refraction n_i along with zero-frequency dielectric constant (polarizability) ε_i to take first guesses at which materials should be slick. The goal of finding a material that reduces adhesion is thus methodically

* $A_{afb}^{Ham} \approx A_{afb}^{TWA} = \frac{3}{4} kT \left(\frac{\varepsilon_a - \varepsilon_f}{\varepsilon_a + \varepsilon_f}\right)\left(\frac{\varepsilon_b - \varepsilon_f}{\varepsilon_b + \varepsilon_f}\right) + \frac{3\pi\hbar\nu_e}{4\sqrt{2}} \frac{\left(n_a^2 - n_f^2\right)\left(n_b^2 - n_f^2\right)}{\sqrt{\left(n_a^2 + n_f^2\right)\left(n_b^2 + n_f^2\right)}\left(\sqrt{\left(n_a^2 + n_f^2\right)} + \sqrt{\left(n_b^2 + n_f^2\right)}\right)}$

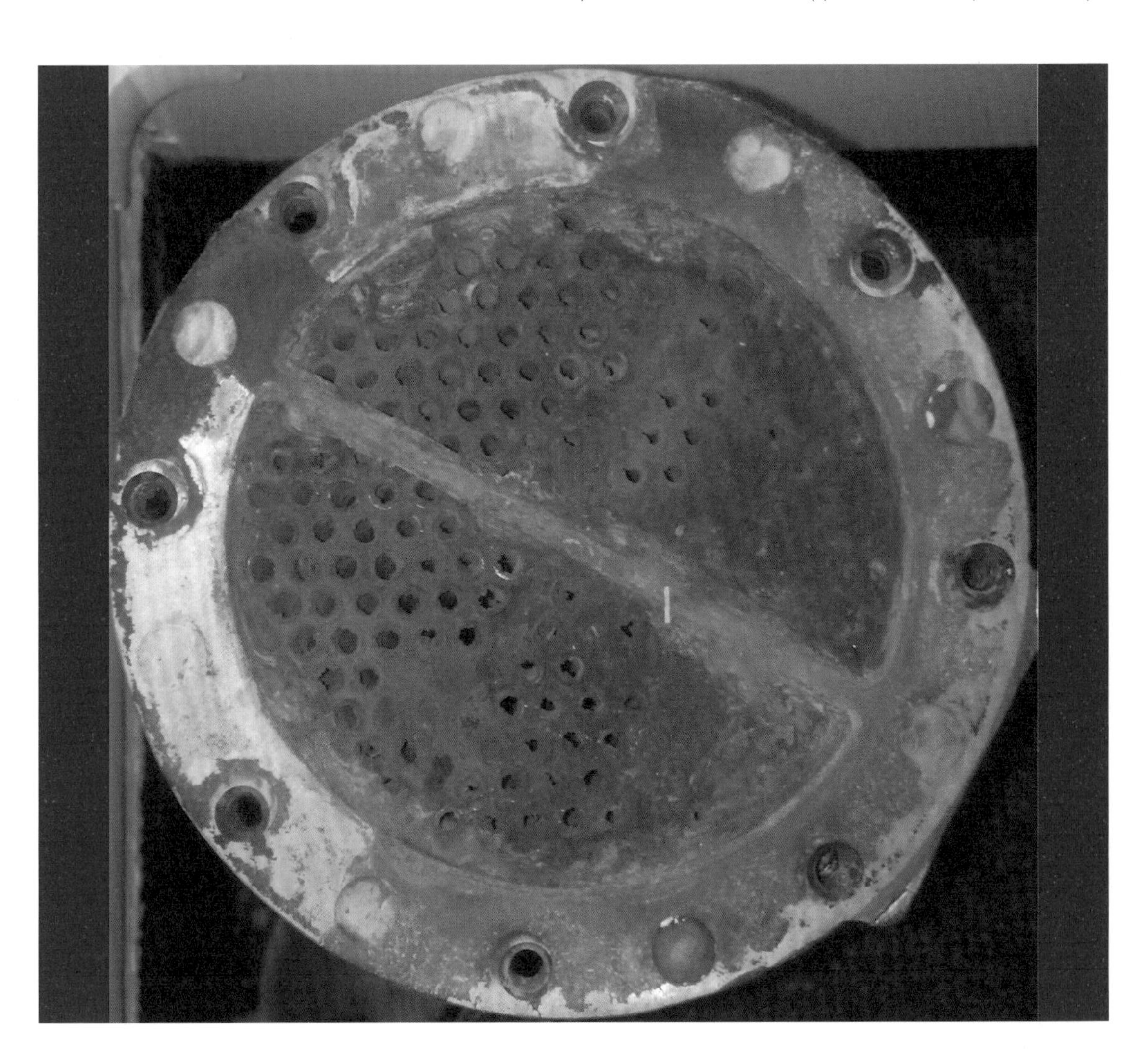

Figure 6.1a
An example of particulate fouling on a small heat exchanger, with some tubes entirely clogged.

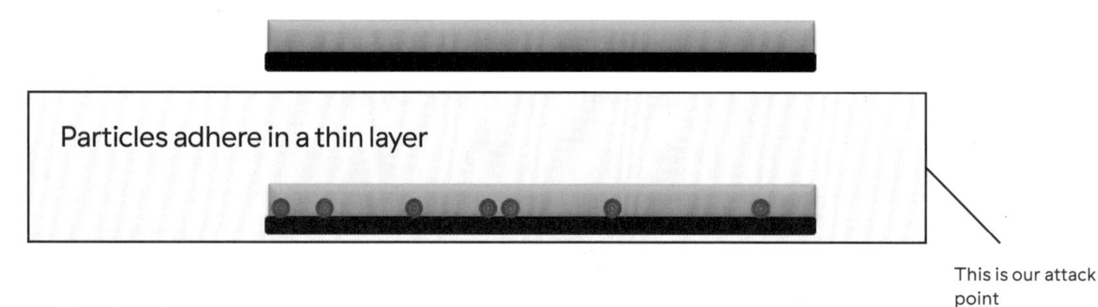

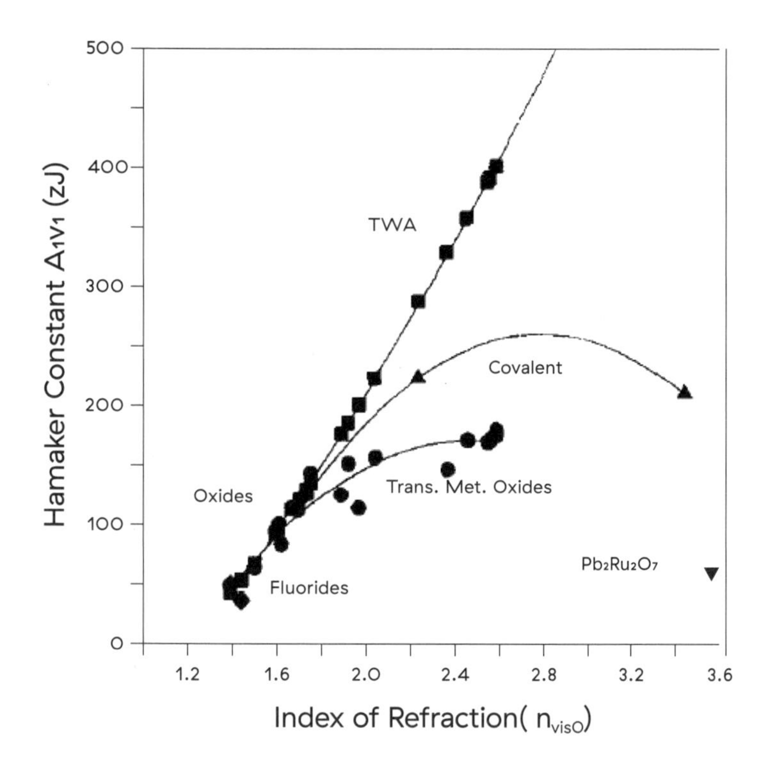

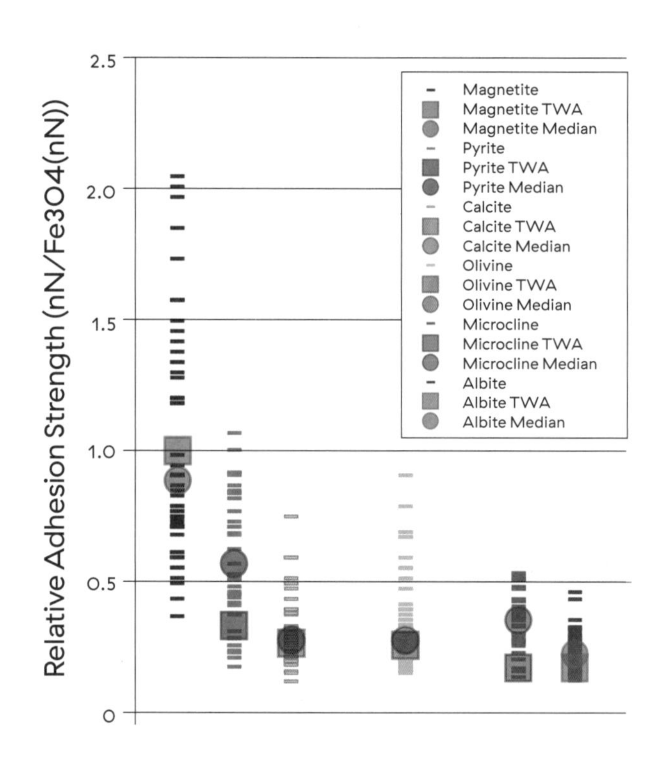

Figure 6.1b (top)
Schematic showing the fouling process, beginning with a clean surface, then a monolayer of attached particles, followed by continued scale growth. We focus on the monolayer formation and seek to eliminate this process.

Figure 6.2 (bottom left)
Hamaker constants of various materials, using full-spectral and Tabor-Winterton calculations.[12]

Figure 6.3 (right page)
SEM image of a 4μm microsphere of SiO2 affixed to a SiN AFM-FS cantilever.

Figure 6.4 (bottom right)
Most recent AFM-FS measurements compared to the TWA model results show overall agreement, despite some variability. Outlier data have not yet been removed.

approached by finding a material with specific optical properties suited for the system, since the properties of the fluid and of the fouling particles are already given. The relatively tedious experimental measurements may be significantly expanded by computational analysis of the extensive literature containing measured spectra of materials (applying equations (2)—(3)), as well as by modeling of new materials where spectra are calculated based on energy bands from DFT simulations[10]. Due to the vast number of permutations of sample materials to simulate and a complex dependence of adhesive force on material properties, a genetic algorithm would be appropriate to explore this computational design space.

It is important to note that many materials with ultralow indices of refraction, many of which are metamaterials, do *not* fit the criteria for fouling-resistant materials. Even though their *overall* indices of refraction may be very low or even negative, this is a geometrical and not an *intrinsic* material property. Therefore we should not be tempted to gravitate toward this solution, as vdW forces operate on the nanoscale, below the length scales of these metamaterials.

EXPERIMENTAL

From the TWA we see that one way to minimize vdW interaction is to minimize the difference terms in the numerator of equation (3). Substituting realistic values for the operating conditions, we see that the second term of equation (4) dominates the first term. Therefore we use the criterion that the initial candidate fouling-resistant materials should have an index of refraction n_a close to that of the fluid, and we feel justified in neglecting the first term of equation (4) for materials whose dielectric properties are not yet known.

We begin by focusing on systems of water-insoluble fluorides,

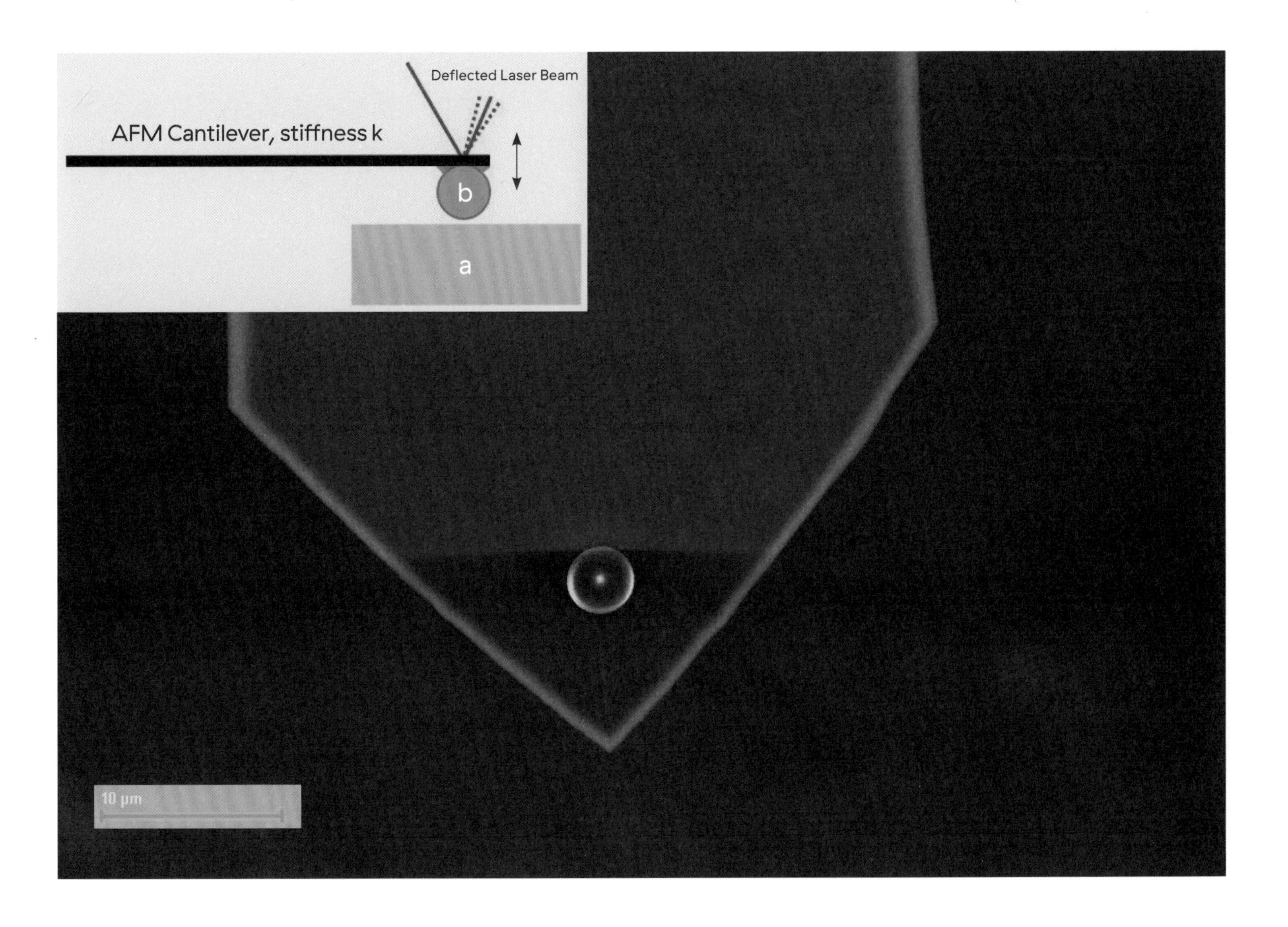

Material Type	Name	Formula	Mohs Hardeness	Dielectric Constant ε	Refraction Index n	A_{abf}^{TWA} (zJ)	Refer-ence
Foulant (*b*)	Silica	SiO_2	6.5 – 7.0	3.8	1.448	—	
Fluid (*f*)	Water (25C)	H_2O	—	80	1.333	—	[9]
Natural oxide surfaces (*a*)	Magnetite	Fe_3O_4	5.0 – 6.5	20	2.42	1695	[9]
	Chromia	Cr_2O_3	8.0 – 8.0	13.3	2.551	1980	[14]
Proposed anti-fouling surface coatings (*a*)	Calcite	$CaCO_3$	4	8.67	1.66	448	[14]
	Microcline/Albite	$KAlSi_3O_8$	6.0 – 6.5	???	1.52	(288)	[14]
	Olivine	$(Mg,Fe)_2SiO_4$	6.5 – 7.0	???	1.65	(435)	N/A
	Pyrite	FeS_2	6.0 – 7.0	???	1.75	(560)	N/A

Name	A_{abf}^{TWA} (zJ) Calculated	$A_{abf}^{TWA}/A_{Fe_3O_4}$ Calculated	$A_{afb}/A_{Fe_3O_4}$ Measured
Magnetite	1695	1.000	1.000
Olivine	435	0.265	0.270
Calcite	448	0.264	0.276
Microcline	288	0.170	0.355
Pyrite	560	0.330	0.568
Albite	441	0.170	0.225

Table 6.1 (top)
Initial candidates for slick surface coatings, and indices of refraction compared to those of water and typical passive oxide scales[23]

NOTE

Parentheses indicate that dielectric constant data were not available for the approximation.

Table 6.2 (left)
Initial AFM–FS results of six materials selected from table 6.1

NOTE

All measurements were carried out in room-temperature, deionized water.

Figure 6.5
Photograph (left) and diagram (right) of our vacuum AFM–FS facility that enables sample cleaning using an argon ion gun within a clean atmosphere with no exposure to air.

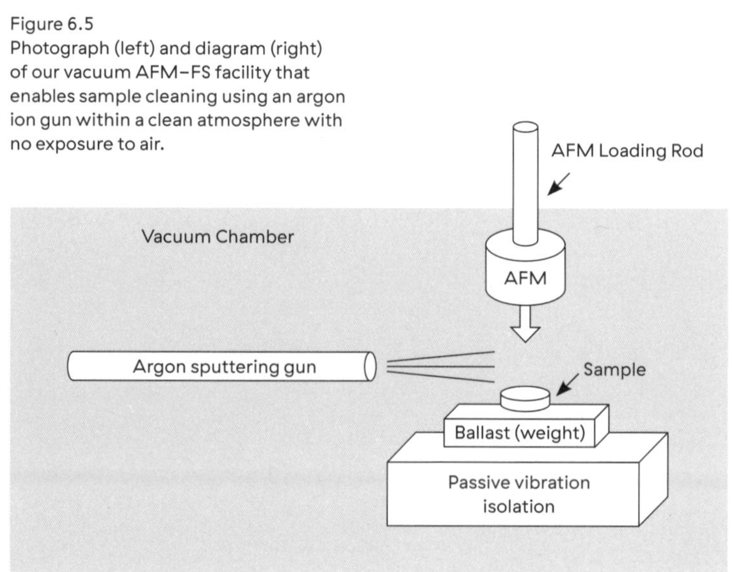

as they tend to be quite hard, earth-abundant, and insoluble in water. Figure 6.2 shows that many of the fluorides possess both low Hamaker constants and low indices of refraction,[11] while table 6.2[12] tabulates many of these Hamaker constants with their TWA values. We specifically focus on the case of geothermal power plants, which use high-temperature water as the fluid and suffer from SiO_2 particulate fouling.[13] Using tabulated data to find minerals that are water-insoluble, relatively hard (to avoid erosion by the SiO_2), and have low refractive indices, the TWA has been applied to calculate expected vdW adhesive forces on a few materials in water at room temperature.

Table 6.1 summarizes the materials chosen for this first investigation. Note how each of them have indices of refraction very close to that of water, compared to the passive oxide layers which normally form on carbon steel (Fe_3O_4) and stainless steels or nickel-based superalloys (Cr_2O_3). As can be seen in table 6.1, quite a number of materials sufficiently close to water's index of refraction are predicted to pose a significant improvement over the passive oxides that naturally grow on structural materials in geofluid.

Confirmation of our adhesion model comes from directly measuring the force of adhesion using AFM-FS. In this technique, a particle of the foulant/scale material, in this case SiO_2, is affixed to the end of a thin SiN cantilever. An example of one of the cantilevers being used in this study is shown in figure 6.3. This cantilever is brought into contact with the surface to be measured, and a laser is bounced off the back of the cantilever, as shown in the figure inset. The deflection of this laser is proportional to the bending of this cantilever, yielding the force required to bend the cantilever (following calibration of its spring constant k). Then the cantilever is pulled up from the surface, and the force required to dislodge the particle is measured using the same technique. [Table 6.1]

RESULTS AND DISCUSSION

Initial tests using our AFM have shown that contamination from moisture in the air, ions in the water, and ambient hydrocarbons must be taken into account and subtracted in order to acquire accurate AFM-FS data. This was realized once tests in laboratory air were found to be nonrepeatable on some of these surfaces, because adherent hydrocarbon contamination from the air instantly attaches to free surfaces, as confirmed by X-ray photoelectron spectroscopy (XPS) measurements. As a starting point, we thus carried out AFM-FS measurements in a droplet of water, which improves the reproducibility but still results in a high spread in data due to surface roughness and cleanliness. The data, along with comparisons between predicted and measured adhesion forces, are tabulated in table 6.2 and graphically shown in figure 6.4.

As mentioned earlier, an absolute force measurement requires calibration of the AFM probe spring constant, but this was not carried out due to the unusual cantilever geometry. Therefore our initial results are relative, scaled such that the dimensionless force measured for magnetite is defined to be unity. Despite the variability, we see a clear pattern that roughly follows the expected TWA results, except for the slight discrepancy with microcline, which is likely due to impurities in the naturally occurring sample (evident by discoloration of the sample). Agreement between theory and experiment is expected to be enhanced once a suitable method is developed to remove outlying data points without physical significance.

To further reduce the variability and obtain a reliable baseline for confirming our adhesion model, a custom vacuum-inert AFM chamber is under construction. Figure 6.5 shows a diagram of its major functional parts. Here, the vacuum chamber allows for pumping the entire system down to 10^{-3} Torr using a turbomolecular pump, while a differentially pumped argon ion sputtering gun will be used to sputter-clean the surface of the material to remove oxides and organic contaminants. Then, without exposure to air, AFM-FS will be performed on the fresh surface in either vacuum or dry inert gas atmosphere. Along with improved surface preparation to eliminate the effect of roughness, we expect this method to yield highly reproducible measurements to further establish the validity of our model.

VISION FOR THE FUTURE

At present, fouling resistance is a property empirically found by testing candidate materials (as has been done in this study). The reason for this is twofold: fouling resistance is difficult to put in mathematical terms useful to theory, and computational design of material structures from first principles is a multidimensional problem with many variables and minimal analytic tractability.[14] With the initial confirmation of the TWA adhesion model by our AFM measurements, we propose that fouling in general can be controlled through the optical properties of a coating material. Based on these findings, the first issue is resolved by using the Hamaker constant as a numerical measure of fouling propensity. While the second issue is still pertinent, recent (and ongoing) advances in computational materials design[15] make simulation and analysis of novel materials practical with sufficient computational resources. Specifically, we propose to use an evolutionary algorithm approach such as USPEX,[16] which was optimized to explore the multidimensional space of possible crystal structures and find the most energetically favorable ones. Numerous matches of USPEX crystal structure predictions with experiment,[15] along with its use to optimize indirect parameters such as density and hardness[17, 18] (and in our case, adhesion), support our choice of this code.

The final link to enable computational discovery of fouling-resistant materials is the ability to calculate the Hamaker constant of crystal structures proposed by simulations like those in USPEX. We use density functional theory (DFT) packages, such as VASP (Vienna Abinitio Simulation Package), to find frequency-dependent dielectric response as supported by earlier works.[19, 20, 21, 22] Then a script solves the full-spectrum sum (Eq. (2)), resulting in a calculated Hamaker constant. The DFT calculation also provides a measure of the energy of the crystal structure, which is an indication of whether the structure is realistic and stable at the operating conditions.

In the USPEX algorithm, crystal structures are optimized in an evolutionary fashion. The algorithm begins by forming random structures that satisfy the initial "hard" constraints, such as the number and type of atoms. These structures are then evaluated to obtain a fitness value for each, which determines whether the structure survives in subsequent generations. In standard usage of USPEX, this fitness value is the overall energy of the structure, which results in the most chemically stable (lowest-energy) crystal. For our purposes, the fitness value is instead the Hamaker constant, favoring the structures with lowest adhesion forces. In this case the local energy minimization, which is still required to find realistic materials, is implemented as a relaxation of the structure in the DFT simulation.[18] The surviving structures are then modified by three operators: heredity, mutation, and permutation.[14] Some of the resulting structures will have more favorable fitness values, and the process is repeated again until a satisfactory number of high-performing structures is found. This is shown schematically in figure 6.6. We propose to begin the search with fluorine-containing arrangements, since fluorine seems to effectively reduce adhesion (Teflon, fluorite, cryolite) due to its valence. We then plan to computationally design materials that have an index of refraction very close to the process fluid. If these steps are successful, we will have a path to scientifically combat fouling in any system, especially for those with a liquid coolant whose optical properties are known.

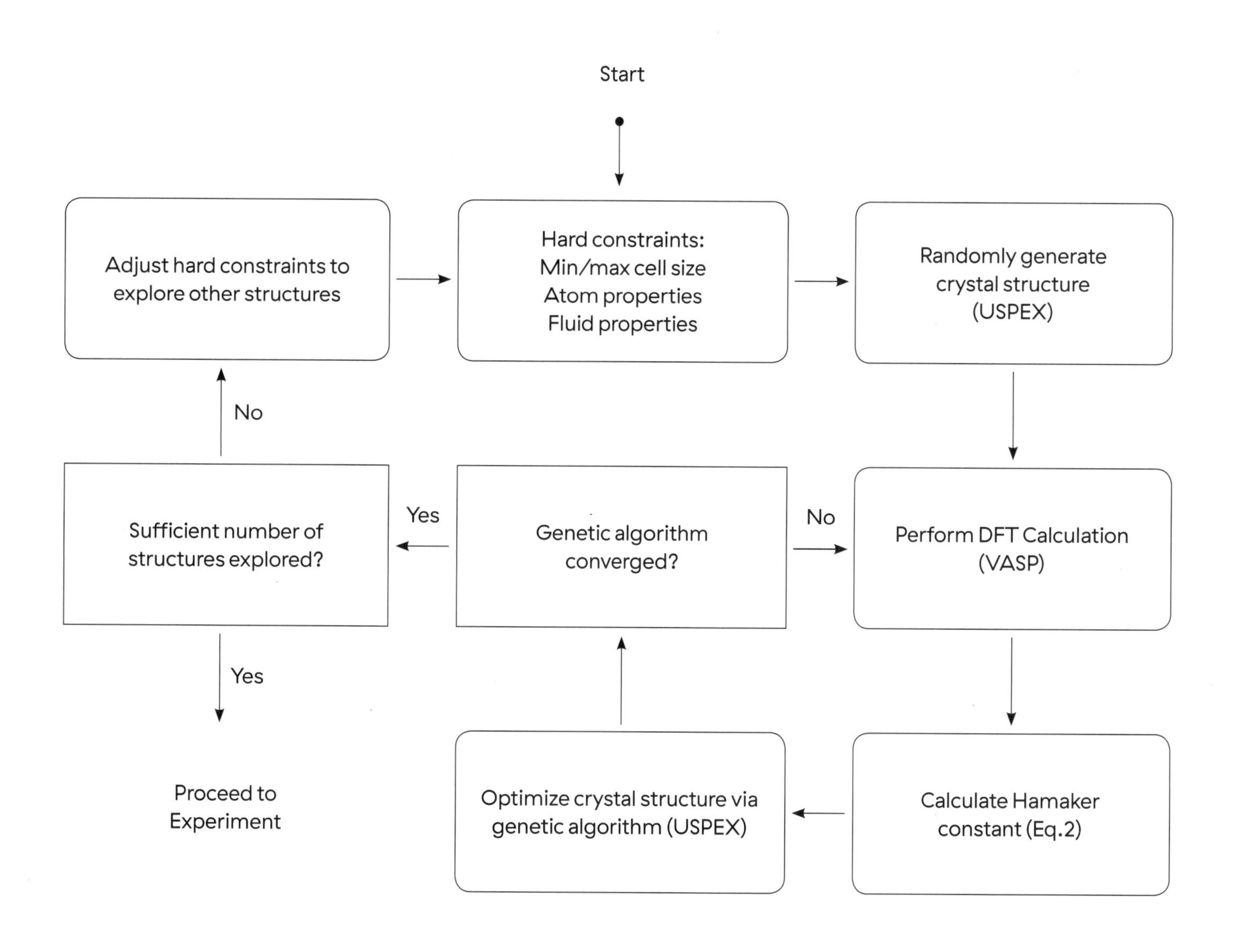

Figure 6.6
Pathway to computational discovery of fouling-resistant materials

ACKNOWLEDGMENTS

This material is based upon work supported by Statoil, the Norwegian state oil company, through the MIT Energy Initiative (MITEI) under Grant No. 022784-0003. This work was also made possible in part by the National Science Foundation Graduate Research Fellowship under Grant No. 1122374

NOTES

1 L. F. Melo, T. R. Bott, and C. A. Bernardo, eds., *Fouling Science and Technology*, NATO ASI Series E: Applied Sciences (Dordrecht: Kluwer Academic Publishers, 1987).

2 J.-W. Yeon, I.-K. Choi, K.-K. Park, H.-M. Kwon, and K. Song, "Chemical Analysis of Fuel Crud Obtained from Korean Nuclear Power Plants," *J. Nucl. Mater.* 404, no. 2 (2010): 160–164.

3 K. Ingason et al., "Iceland Deep Drilling Project (IDDP): Fluid Handling and Evaluation," Proceedings, World Geothermal Congress 2010, Bali, Indonesia, April 25–29, 2010.

4 L. Lin, "Controlling CRUD Vapor Chimney Formation in LWRs through Surface Modification," master's thesis, Massachusetts Institute of Technology, 2014.

5 L. Bergstrom, "Hamaker Constants of Inorganic Materials," *Adv. Colloid Interface Sci.* 70 (1997): 125–169.

6 V. A. Parsegian, *Van der Waals Forces: A Handbook for Biologists, Chemists, Engineers, and Physicists* (Cambridge: Cambridge University Press, 2005).

7 R. H. French, "Origins and Applications of London Dispersion Forces and Hamaker Constants in Ceramics," *J. Am. Ceram. Soc.* 83, no. 9 (2000): 2117–2146.

8 B. Da, Y. Sun, S. F. Mao, Z. M. Zhang, H. Jin, H. Yoshikawa, S. Tanuma, and Z. J. Ding, "A Reverse Monte Carlo Method for Deriving Optical Constants of Solids from Reflection Electron Energy-Loss Spectroscopy Spectra," *J. Appl. Phys.* 113, no. 21 (2013).

9 J. N. Israelachvili, *Intermolecular and Surface Forces*, 3rd ed. (Burlington, MA: Academic Press, 2011).

10 F. Kootstra, P. L. de Boeij, and J. G. Snijders, "Application of Time-Dependent Density-Functional Theory to the Dielectric Function of Various Nonmetallic Crystals," *Phys. Rev. B* 62 (2000):7071–7083.

11 H. D. Ackler, R. H. French, and Y.-M. Chiang, "Comparisons of Hamaker Constants for Ceramic Systems with Intervening Vacuum or Water: From Force Laws and Physical Properties," *J. Colloid Interface Sci.* 179, no. 2 (1996): 460–469.

12 L. K. DeNoyer, Y. M. Chiang, R. H. French, and R. M. Cannon, "Full Spectral Calculation of Non-Retarded Haymaker Constants for Ceramic Systems from Interband Transition Strengths," *Solid State Ionics* 75 (1995): 13–33.

13 T. Sugama and K. Gawlik, "Anti-Silica Fouling Coatings in Geothermal Environments," *Mater. Lett.* 57, no. 3 (2002): 666–673.

14 A. R. Oganov and C. W. Glass, "Crystal Structure Prediction Using Evolutionary Algorithms: Principles and Applications," *J. Chem. Phys.* 124 (2006): 244704.

15 S. Curtarolo, G. L. W. Hart, M. B. Nardelli, N. Mingo, S. Sanvito, and O. Levy, "The High-Throughput Highway to Computational Materials Design," *Nat. Mater.* 12, no. 3 (2013): 191–201.

16 C. W. Glass, A. R. Oganov, and N. Hansen, "Uspex: Evolutionary Crystal Structure Prediction," *Comp. Phys. Comm.* 175, no. 11–12 (2006): 713–720.

17 Q. Zhu, A. R. Oganov, M. A. Salvado, P. Pertierra, and A. O. Lyakhov, "Denser than Diamond: Abinitio Search for Superdense Carbon Allotropes," *Phys. Rev. B* 83 (2011): 193410.

18 A. O. Lyakhov and A. R. Oganov, "Evolutionary Search for Superhard Materials: Methodology and Applications to Forms of Carbon and tio2," *Phys. Rev. B* 84 (2011): 092103.

19 J. Muscat, A. Wander, and N. M. Harrison, "On the Prediction of Band Gaps from Hybrid Functional Theory," *Chem. Phys. Lett.* 342, no. 3–4 (2001): 397–401.

20 J. Paier, M. Marsman, and G. Kresse, "Dielectric Properties and Excitons for Extended Systems from Hybrid Functionals," *Phys. Rev. B* 78 (2008): 121201(R).

21 F. Sottile, V. Olevano, and L. Reining. Parameter-free calculation of response functions in time dependent density-functional theory. *Phys. Rev. Lett.* 91 (2003): 056402.

22 M. Gajdoš, K. Hummer, G. Kresse, J. Furthmüller, and F. Bechstedt, "Linear Optical Properties in the Projector-Augmented Wave Methodology," *Phys. Rev. B* 73 (2006): 045112.

23 K. F. Young and H. P. R. Frederikse, "Compilation of the Static Dielectric Constant of Inorganic Solids," *J. Phys. Chem.* Ref. Data, 2:2 (1973).

7

Programmable Bacteria: An Emerging Therapeutic

Candice Gurbatri, Tetsuhiro Harimoto, and Tal Danino

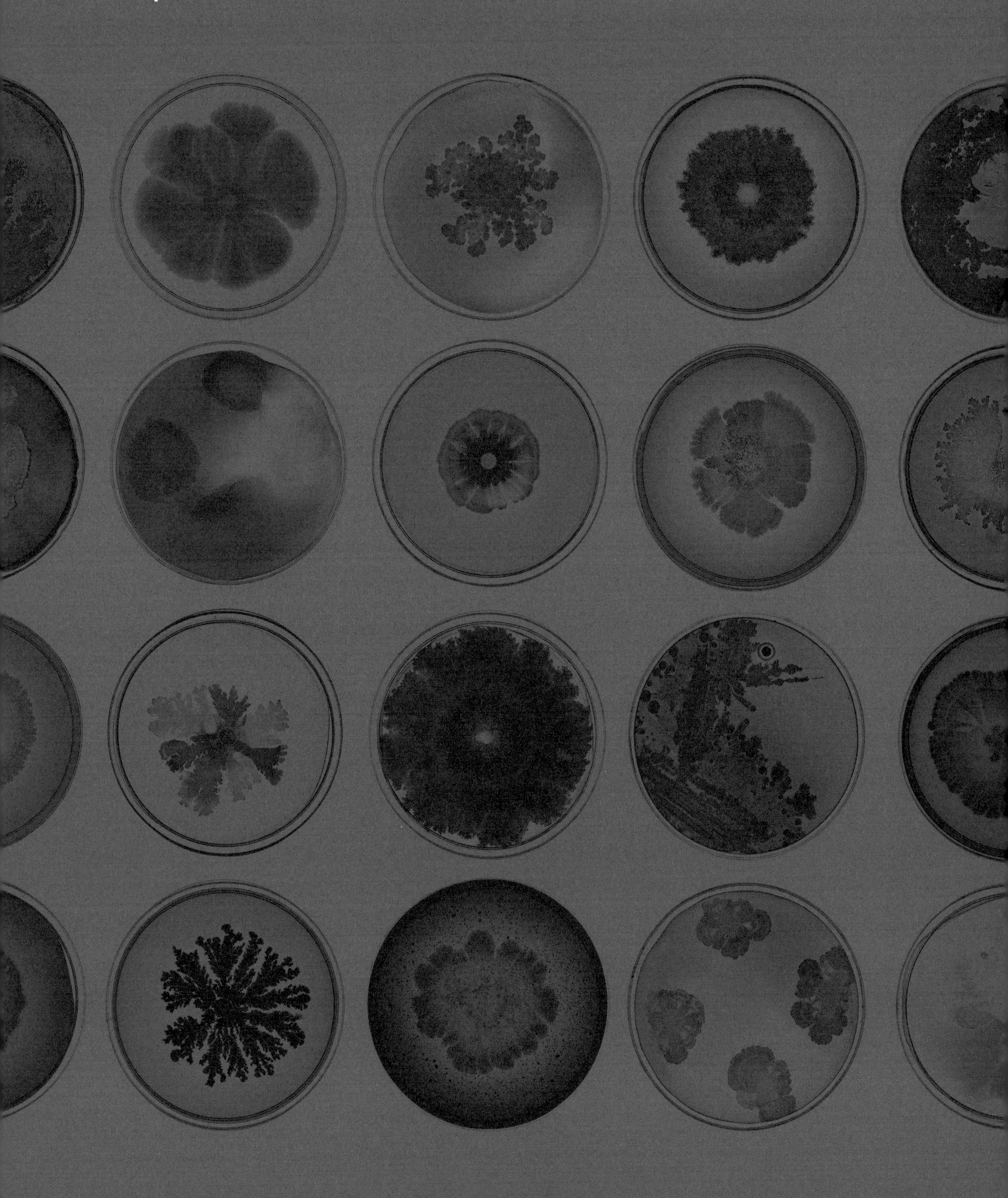

7 Programmable Bacteria: An Emerging Therapeutic

Candice Gurbatri, Tetsuhiro Harimoto, and Tal Danino

Are all bacteria harmful to us? Bacteria are often associated with illness and it is common to find people constantly using hand sanitizer or spraying disinfectants to kill surrounding germs. For centuries after Antoine van Leeuwenhoek first discovered bacteria under the microscope in 1673, the relationship between bacteria and humans was studied extensively, and experimentation focused on the pathogenic properties of bacteria and their role as the source of human disease. However, in recent years the long-held view of bacteria as pathogens has been transformed by microbiome data revealing the prevalence of trillions of functional microbes in the human body. More recently, engineers have adopted their language to the field of biology, applying the same logic to living matter and allowing for the unique manipulation of bacterial behavior for a desired functionality. With the rapid advancement of microbiome research and development of biotechnologies, there is an emerging paradigm to program living matter and activate microorganisms to fight disease and benefit human health.

I. INTRODUCTION

Gregor Mendel's 1865 experiments on pea plants set the beginnings of basic genetic principles.[1] In the next one hundred years, scientists would coin the word "gene," define chromosomes, begin to understand recessive and dominant traits in accordance with Mendelian rules, and eventually discover the structure of the "molecule of life," or deoxyribonucleic acid (DNA). In just over a century, the scientific world was transformed not only by rapid advancements in our understanding of human life at its core, but also by the advancing biotechnologies that allow for such fast-paced innovation.

Equipped with knowledge about DNA and its replication, structure, and function, researchers began to experiment with cutting, pasting, and rearranging its parts. By modulating its sequence, one can alter the resulting protein and biological function.
With this emerging biotechnology, a subdiscipline of genetics was formed: genetic engineering. Defined as the manipulation of an organism's DNA, genetic engineering serves the purpose of editing genes to produce a desired characteristic. Historically, humans have indirectly modified genomes of animals and plants through domestication and the selection of seeds from high-yield produce. Through selection and preference, selective pressures were created such that certain genes remained in the gene pool, thus indirectly modulating future genetic

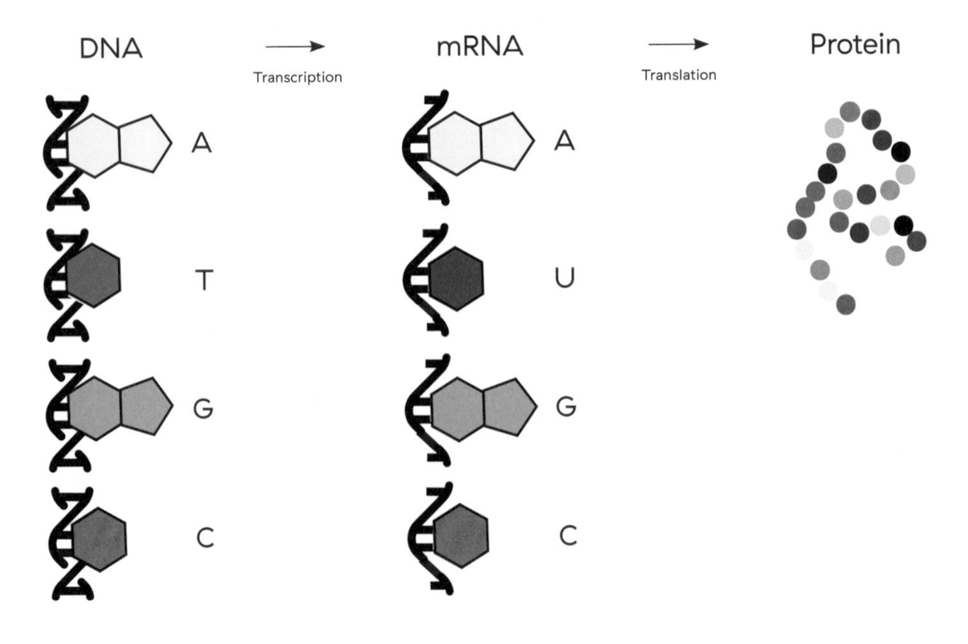

Figure 7.1 (left page)
Central dogma of molecular biology. Double helix DNA with nucleotides A, T, C, and G is transcribed into a single-stranded mRNA with nucleotides A, U, C, and G. The sequence is then translated into a string of amino acids that folds into a protein.

Figure 7.5
Microuniverse slides. Stained bacteria placed into photo slides.

code. More recently, however, genetic engineering has become a fast-paced process resulting in a deliberately altered organism through the use of recombinant DNA, or DNA formed by combining constituents from multiple organisms.

In 1973, Herbert Boyer and Stanley Cohen performed the first recombinant DNA cloning experiment in which a gene conferring antibiotic resistance was cut and pasted into the plasmid, or circular DNA, of bacteria. This plasmid was then transferred back into the bacteria and cloned as the bacteria replicated. Bacteria served as an ideal vehicle for gene cloning due to its short replication rate of 20 minutes, and allowed Boyer and Cohen to confirm the retaining of gene function over several generations of bacterial populations.[2, 3] Furthermore, their extensively studied genetic system allowed for increased feasibility in programming bacterial behavior. As the idea of modulating bacterial functions for human health emerged, there was a greater need for evolving biotechnologies and understanding of bacterial regulation.

Early studies by Monod and Jacob investigating lactose regulation in bacteria laid the foundation for the development of a quantitative framework of regulatory networks that govern bacterial responses to the environment.[4] With the rapid expansion of automated high-throughput sequencing technologies and improved computation in the 1990s, researchers were able to characterize a handful of networks in bacteria that could be deconstructed into simple components.[5] In the year 2000, two seminal papers published in *Nature* forward-engineered two modules from a few well-characterized components—a toggle switch and an oscillator—marking the beginning of the field of synthetic biology.[6, 7] Soon thereafter, multiple modules, termed genetic circuits, had been designed that allowed bacteria to sense light, invade cancer cells, and alter metabolic processes.[8, 9, 10, 11] Applying these same principles to other microorganisms, synthetic biology has contributed to the development of anti-tuberculosis compounds, conversion of biomass into biofuels, environmental and biosensors, and the larger-scale production of therapeutics.[12, 13, 14, 15] Currently, scientists are designing and synthesizing entire genomes of bacteria by building minimal cells.[16] In just over a decade, humans have learned how to

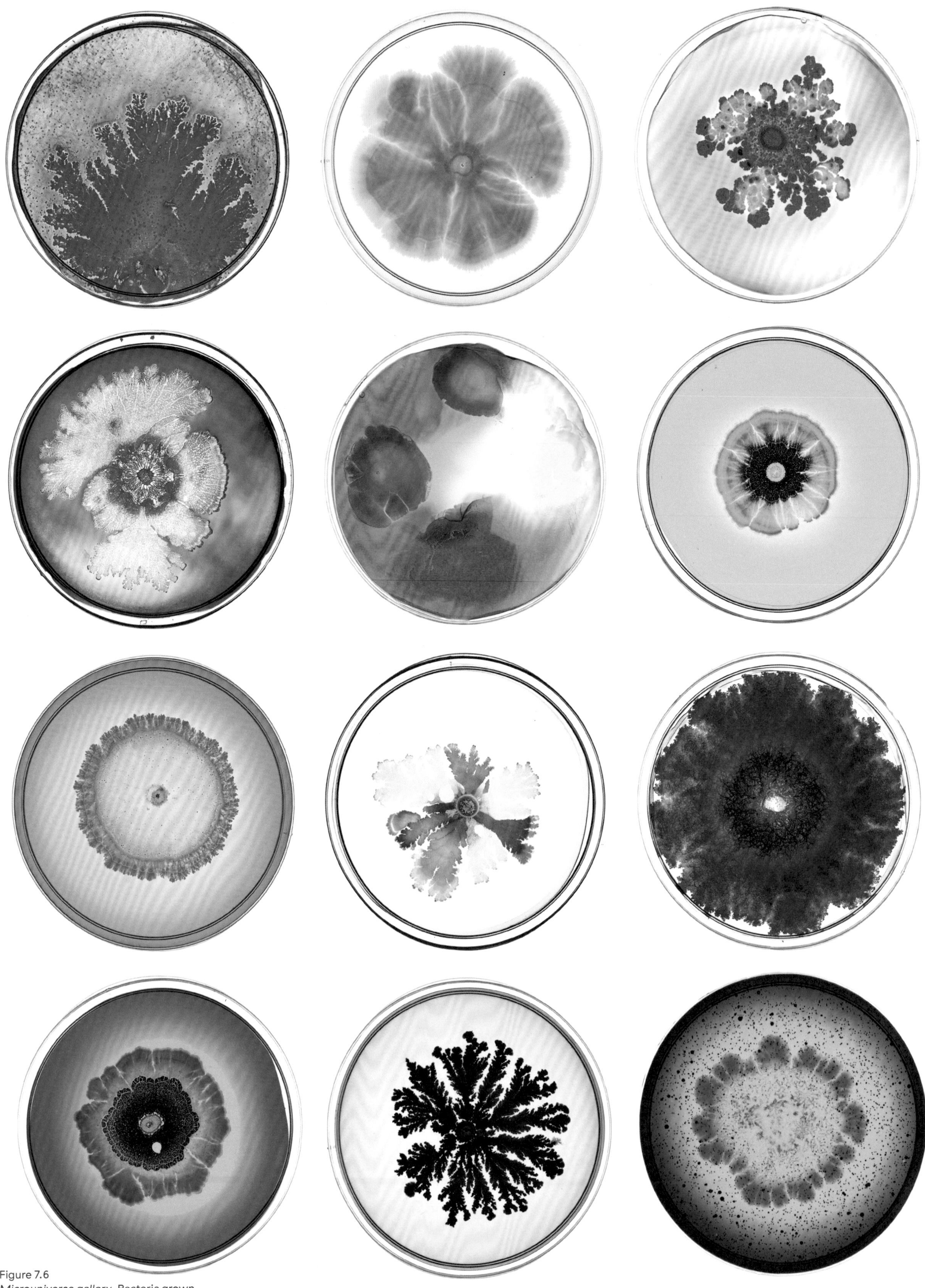

Figure 7.6
Microuniverse gallery. Bacteria grown and stained in petri dishes.

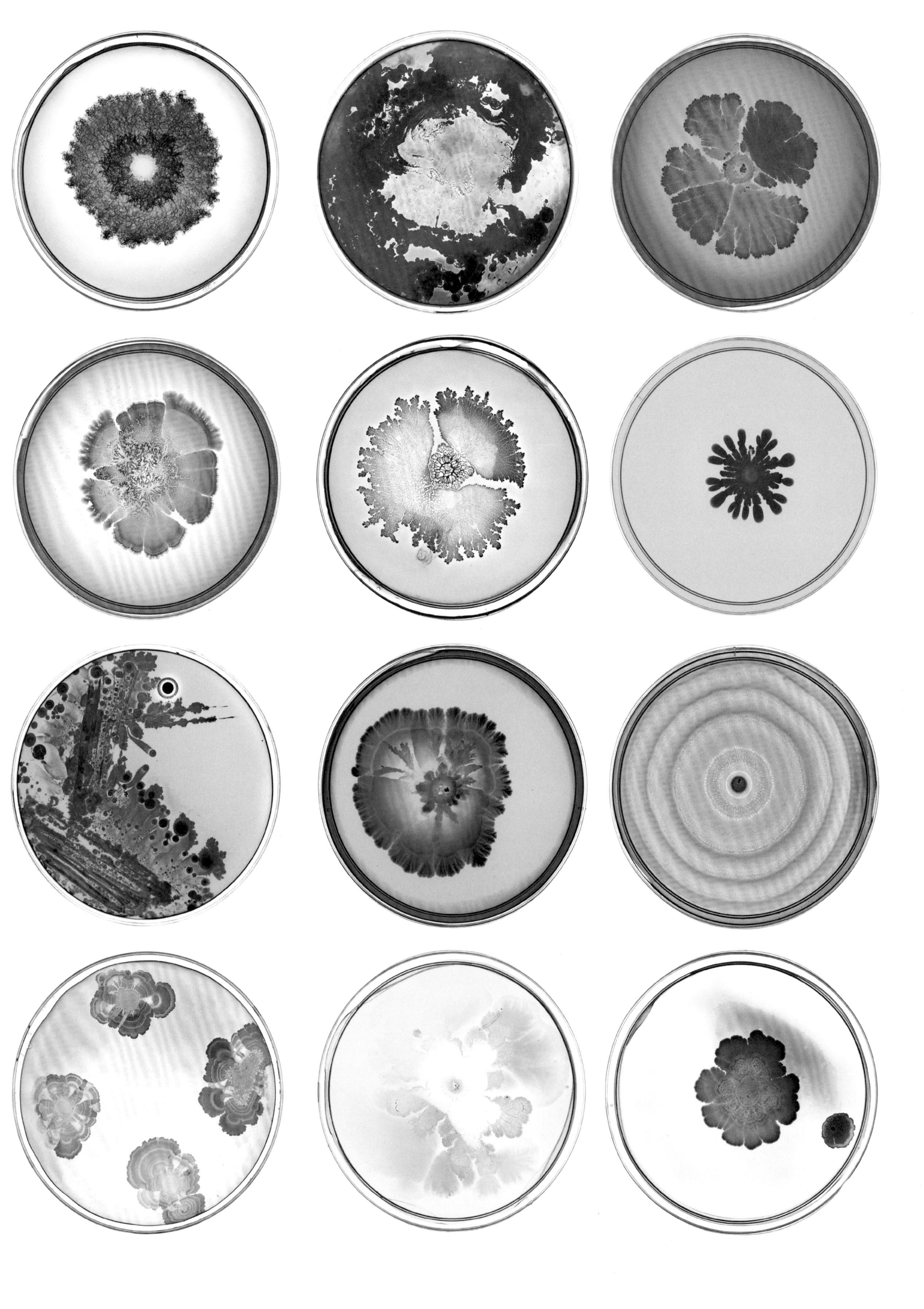

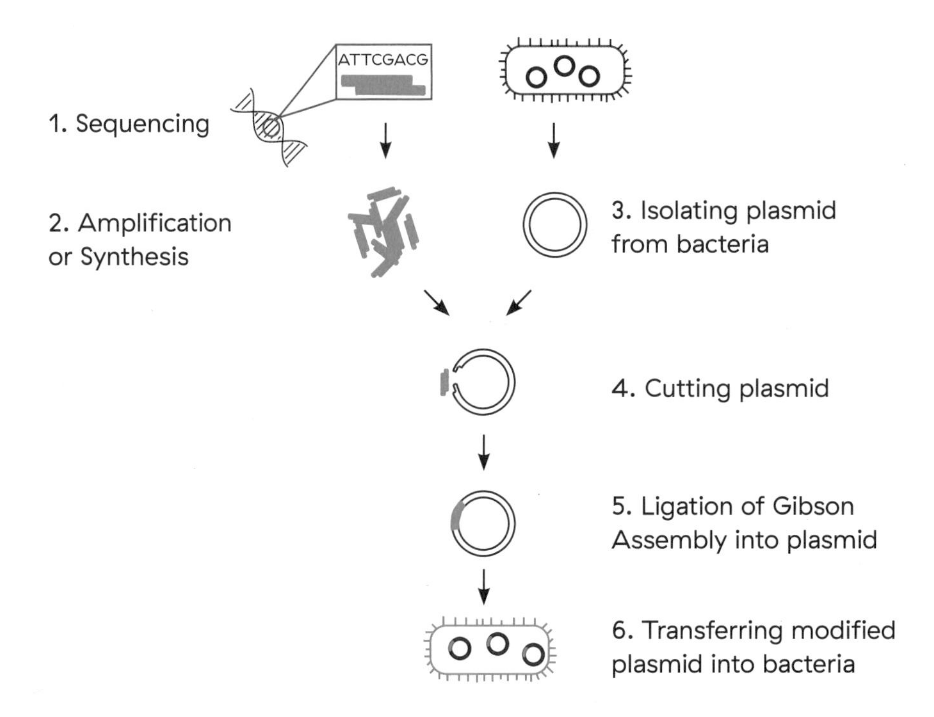

rewrite, rewire, and reprogram genomic code to control living organisms for a myriad of applications. Below we will discuss the core principles of molecular biology and technologies of genetic engineering, how to engineer bacteria, and emerging therapeutic applications for diseases, including cancer.

II. MOLECULAR BIOLOGY OF THE CELL

Cells contain numerous molecules that cooperate with one another to sustain life. On the scale of nanometers, molecular machinery called proteins carry out specific tasks within a cell. All of the tasks are processed with surprising precision and efficiency, ranging from simply capturing and transporting molecules to synthesizing complex structures and generating energy. A single micron-sized bacterium such as *Escherichia coli (E. coli)* contains over 4,000 types of discrete proteins that constantly work to keep the cell functioning.[17] Not only do proteins process their own tasks, but their dynamic interactions are what enables a cell—and hence a whole living organism—to adapt to a multitude of situations. In fact, proteins are constantly made and degraded to respond to a changing environment in a matter of seconds. For example, when a cell senses sugar in the environment (e.g., lactose), it can produce proteins to transport additional sugar into the cell and degrade proteins not necessary at the moment.

Information encoded in the cell flows from gene to protein, a process referred to as the central dogma of molecular biology (figure 7.1). A cell regulates intracellular protein concentrations by reading the code of DNA and transcribing it to ribonucleic acid (RNA), which is eventually translated to protein. All of the information necessary to perform molecular tasks is stored in DNA and copied to the temporary information storage of RNA. Proteins are made based on the RNA sequence; their parts fit together, move, and interact to perform essential cellular functions.

DNA is encoded by a sequence of four different nucleotides: adenine (A), thymine (T), cytosine (C), and guanine (G). Two strands of DNA are paired (A to T, C to G), forming a double helix. Within the entire sequence of nucleotides, or genome, there exist functional units called genes that provide instructions for each life-sustaining task. When a cell needs to make a protein, a gene that codes for a specific protein is

Figure 7.2 (left page)
Overview of molecular cloning. (1) Gene of interest is sequenced or identified, (2) and amplified or synthesized using PCR. (3) Plasmid isolated from bacteria, (4) and cut using restriction enzymes. (5) Gene of interest is ligated or assembled into the plasmid, and (6) plasmid with genetic insert is transformed back into the bacteria. As the bacteria replicates, the gene of interest is cloned.

first "transcribed" into a single-stranded messenger RNA (mRNA). The level of gene activation, termed "expression," is initiated by regulatory regions of DNA preceding the gene called promoters. When a cell senses changes in its environment, a complex called RNA polymerase recognizes a specific promoter and transcribes the gene by producing complementary base pairs to the provided DNA code. Chemically, mRNA slightly differs from DNA in that it contains a uracil (U) nucleotide instead of a thymine (T). Functionally, however, mRNA serves only as a temporary information cassette that includes additional levels of regulation as it is transported, edited, and eventually discarded. Once a gene is copied from DNA to mRNA, the nucleotide sequence is "translated" via the ribosome into a protein in which every three nucleotides form a codon that corresponds to a distinct amino acid. In human cells, the protein size can range from a mere 29 amino acids like that of hormones to 34,000 amino acids like that of muscle.[18,19] The diverse characteristics of amino acids can produce a wide variety of protein functions.

III. SEQUENCING, AMPLIFICATION, AND ASSEMBLY OF DNA

To engineer complex genetic circuits in bacteria, one must (1) know the DNA sequences, (2) produce enough DNA to manipulate the DNA pieces, and (3) assemble DNA together into the desired location (figure 7.2). The first step in this process is DNA sequencing. In 1977, a team led by Drs. Frederick Sanger and Walter Gilbert developed Sanger sequencing, a then novel DNA sequencing technique that became the most widely used sequencing technology for the next 40 years.[20] Sanger sequencing uses a system in

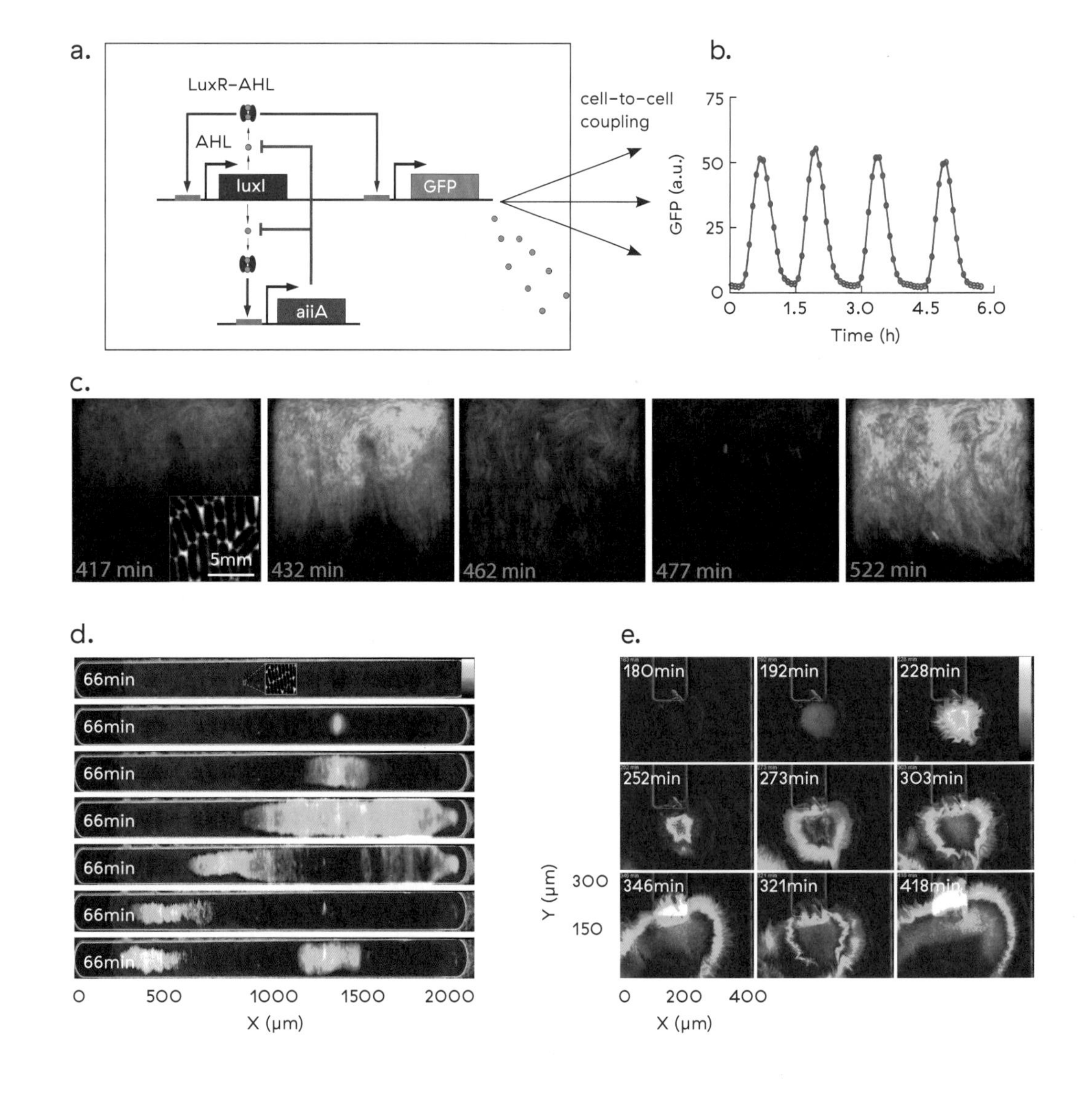

Figure 7.3 (bottom)
Synchronized genetic clocks. (A) Network diagram. The luxI promoter drives production of the three molecules (luxI, aiiA, and yemGFP) in three identical transcriptional modules. LuxI produces a small molecule AHL, which can diffuse outside of the cell membrane and into neighboring cells, activating the luxI promoter. AiiA negatively regulates the circuit by acting as effective degradation machinery called protease for AHL. (B) Bacteria population (measured by fluorescence) as a function of time. (C) Fluorescence slices of a typical experimental run demonstrate synchronization of oscillations in a population of *E. coli*. Inset in the first snapshot is a 3100x magnification of cells. (D) Snapshots of the GFP fluorescence superimposed over brightfield images of a densely packed monolayer of *E. coli* cells, shown at different times after loading. Traveling waves emerge spontaneously in the middle of the colony and propagate outward. (E) Snapshots of the GFP fluorescence superimposed over the brightfield images of a three-dimensional growing colony of *E. coli* cells at different times after loading. Bursts of fluorescence begin when the growing colony reaches a critical size of about 100 mm. These bursts are primarily localized at the periphery of the growing colony.

which DNA is replicated with a portion of fluorescently labeled nucleotides that terminate replication and allow for specific size fragments to be read one by one. While Sanger sequencing became the standard method in laboratories around the world, it remained slow and costly to read an organism's entire genome. Recently, high-throughput sequencing methods termed next-generation sequencing (NGS) have been more commonly employed for large-scale synthesis.[21, 22] In particular, Illumina Inc. developed HiSeqX™Ten, which parallelizes the sequencing process and assembles multiple sequencing results computationally. One NGS machine can read over 45 human genomes in a day, at a cost of approximately $1,000 for each sequenced genome, making it both a rapid and a cost-efficient technology.[21]

Amplifying DNA is a necessary step to create a sufficient amount of DNA for assembly and reinsertion into the host organism. In 1983, Dr. Kary Mullis utilized thermostable DNA replication machinery called DNA polymerase to amplify desired DNA segments.[23] The method, called polymerase chain reaction (PCR), separates double-stranded DNA and initiates replication using short primers of DNA that uniquely bind to a given position. With multiple reaction cycles, larger amounts of a specific DNA can be produced. A disadvantage of PCR is that this method requires template DNA and causes errors in each cycle, hence making construction of a large, reliable, novel DNA sequence challenging. As technology evolved, methods to chemically synthesize DNA sequences from constituent molecules improved rapidly, allowing for the creation of new sequences for which no template exists. By 2010, the entire genome of a bacterium *Mycoplasma mycoides* was computationally designed and chemically synthesized by a team led by Dr. J. Craig Venter. By replacing the genome of another bacterium, *Mycoplasma capricolum*, with synthesized *M. mycoides* genome, they were able to demonstrate that DNA can be designed and printed to construct life from scratch.[24] With increasing capacity and lowering costs of DNA synthesis, it is now possible to order DNA sequences from commercial companies around the world.

With either amplified or synthesized DNA in hand, the next step is the assembly of DNA into functional pieces. There are many approaches to assembling DNA. Traditionally, restriction enzymes that recognize specific sequences and cut DNA have been used;[25, 26] subsequently, a different DNA fragment cut with the same restriction enzyme can be joined together with a DNA ligase.[2] To date, over 3,000 restriction enzymes have been discovered and many of them are commercially available, allowing researchers to cut and paste DNA into a variety of sites. While restriction enzymes have served as a standard technique for gene editing, new tools have been recently invented for faster and more accurate genetic engineering. For example, Gibson assembly, named after Dr. Daniel Gibson at the J. Craig Venter Institute, was introduced in 2009.[27] This method allows for an assembly of multiple DNA fragments with a single experimental step, making insertion of multiple genes more efficient. More recently, the clustered regularly interspaced short palindromic repeats (CRISPR) system has emerged as a precise, effective, and flexible method. CRISPR consists of two components: a "guide" that recognizes a specific sequence of DNA and an enzyme (Cas9) that cuts at the recognized site.[28, 29, 30] CRISPR is not limited by the sequence, so cut sites can be tailored to a specific gene with precision. As the tools for engineering genetic materials have emerged, these successes have stimulated interest in reprogramming organisms.

IV. ENGINEERING GENE CIRCUITS

Analogous to programming software or engineering electrical circuits, DNA sequences can be designed to introduce genes and novel regulatory networks into a cell. At the foundation of synthetic biology design is the building of genetic circuits that perform logic functions, created with two essential components: a promoter that senses an input, and a gene that produces an output. For multiple inputs, these are typically described as Boolean logic functions like AND (all inputs satisfied) and OR gates (some inputs satisfied). One of the earliest works in genetic circuit design incorporating logic functions was published in 2000 when Dr. James Collins and colleagues produced a genetic toggle-switch circuit. In their work, *E. coli* was programmed to toggle between two stable states based on external cues. To do so, two genes (*lacI* and *cI*) were designed to repress each other's expression; either gene could be disengaged by an external stimulus, letting the unrepressed gene dominate the state of the cell.[6] This allowed the cells to maintain a stable ON or OFF state after a transient induction.

Since then, more complex circuitry has been created. Dr. Chris Voigt has used multiple logic gates to create gene circuits that detect light-dark edges.[11] In this edge detection circuit, unilluminated bacteria are programmed to sense the absence of light and secrete a signaling molecule, N-acyl homoserine lactone (AHL), whereas illuminated bacteria receive AHL as an "input" and produce a black pigment. Because the AHL receiver cells cannot produce AHL by themselves and can only receive it from neighboring cells, a black outline is formed at the light-dark interface. Another recent example of a logic gate is a safeguard gene circuit, known as the "kill switch."[31] In this system, the bacterial gene is designed to produce toxins to self-destruct in the absence of the molecule anhydrotetracycline (ATc). Since ATc is rarely found in nature, this switch ensures that the engineered bacteria can only survive if provided with the supplement.

Oscillatory circuits are another type of genetic circuit that has attracted synthetic biologists' attention. In 2000, Drs. Michael Elowitz and Stanislas Leibler engineered the first oscillatory circuit, termed the "repressilator."[7] (Their genetic circuits are composed of three promoters driving three repressors that cyclically repress one another. Due to the interaction between the three repressors and delay in protein production, the genetic circuit produces oscillatory dynamics, observed by fluorescent protein expression under a microscope. Since then, several other oscillators have been constructed by teams including ones led by Drs. Alexander J. Ninfa, James C. Liao, and Jeff Hasty.[32, 33, 34]

Since these oscillators functioned only within a single cell, we engineered a synthetic clock circuit that is capable of generating synchronized oscillations in a growing population of bacteria (figure 7.3).[35] In this circuit, a combination of positive and negative feedback produces and degrades AHL, which can freely travel among the neighboring bacteria and binds to the promoter to express a green fluorescent protein (GFP). This mechanism established the communication between bacteria to synchronize these oscillations across a population, with spatiotemporal waves occurring at millimeter scales.
The aforementioned works demonstrate the multiple layers of genetic circuits that can enable diverse and complex bacterial responses to environmental cues. Utilizing this system, synthetic biologists are now engineering bacteria that utilize logic circuits and sensing in therapeutic contexts. Below, we will discuss recent works in cancer as a primary example.

V. CANCER THERAPY

Cancer is a heterogeneous disease caused by the uncontrolled and abnormal division of cells. It can originate anywhere in the human body and may metastasize if left untreated. The National Cancer Institute estimates that in 2016 there will be about 1.6 million new cases of cancer diagnosed in the United States and that almost 600,000 people will die from the disease.[36] With the growing number of cancer incidents per year, there is a pressing need for a clinically effective therapy. Current cancer treatments

a. 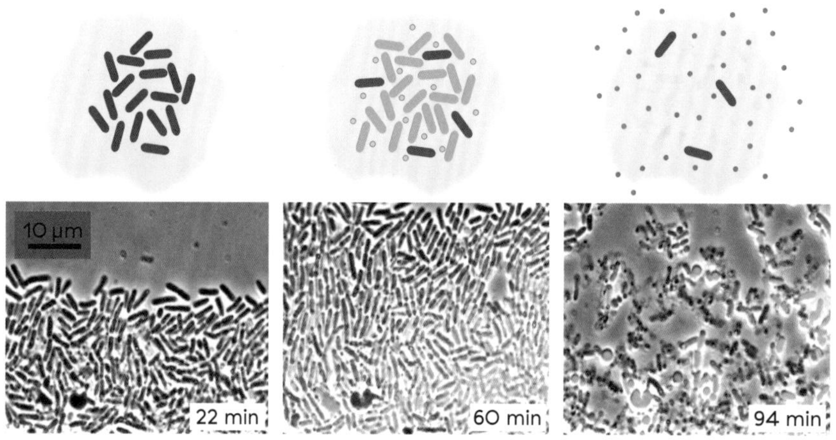 Growth Quorum threshold Synchronized lysis

10 µm

22 min 60 min 94 min

often include surgery, radiation therapy, or chemotherapy. A major limitation of these options is the treatment's inability to discriminate between benign and malignant cells, thus causing damaging side effects for patients. Thus, there is a need for a targeted therapeutic that homes to tumor cells, adequately penetrates through the tumor mass, and is cytotoxic to all of the cancer cells.

Over the past century, several bacteria have been explored for their potential in cancer therapy. Studies have specifically documented bacterial colonization and growth within the necrotic cores of tumors, microenvironments previously thought to be sterile.[37] Bacteria's role as a cancer therapeutic was first noticed in 1868 by physicians W. Busch and F. Fehleisen, who observed regressions of tumors in cancer patients with bacterial infections, showing minimal harm to the patient's overall health.[38] Then in 1891, bone surgeon William Coley intentionally infected his cancer patients with *Streptococcus* bacteria and reproduced similar results to those seen in 1868.[39] He hypothesized that the induced bacterial infection shrank the malignancy by provoking an immune response. This method was initially controversial because of inconsistent results and the use of toxic bacterial strains; however, the idea later resurfaced when more was known about the tumor microenvironment and attenuated bacterial strains were available. Later studies demonstrated that attenuated *Clostridium novyi* spores showed colonization and regression of tumors over time in mice, but the treatment was difficult to standardize, with effectiveness being highly dependent on tumor size and spore dose.[40, 41, 42] With less toxic bacterial strains available, bacteria were seen as an optimal therapeutic with their natural selectivity to tumors, motility to penetrate into the tumor mass, and cytotoxicity.

With the emergence of genetic engineering, there has been a shift toward bacterial hosts such as *Salmonella typhimurium* and *E. coli*, primarily due to their ease of genetic modification, allowing for strains with attenuated virulence properties such as the ability to infect macrophages, and reduced endotoxicity.[42, 43, 44, 45] In particular, a 2002 study demonstrated that chromosomal deletions of genes *purI* and *msbB* in *S. typhimurium* VNP20009 had the potential to facilitate bacterial targeting and colonization of tumors in patients with metastatic melanoma cancer. The modified strain was taken as far as Phase I human clinical trials showing that 10^8 bacteria can be injected safely in humans, though no efficacy was observed.[46, 47] Subsequently, a more efficacious *S. typhimurium* A1-R strain

Figure 7.4 (left page, right) Bacterial dynamics of the synchronized lysis circuit (SLC). (A) The main stages of each lysis cycle from seeding to quorum "firing." The schematic depictions are typical time series images of the circuit-harboring cells undergoing the three main stages of quorum firing. Shown below is a fluorescence profile of the bacteria in a chamber. (B) *In vivo* bacterial dynamics. Pictures shown above are *in vivo* imaging over time of a mouse bearing tumors injected with the bacteria. Below graph is a bacteria population over time tracked by luminescence. Data for each line represent tumor mass on the right (blue) and left (black) side of a mouse.

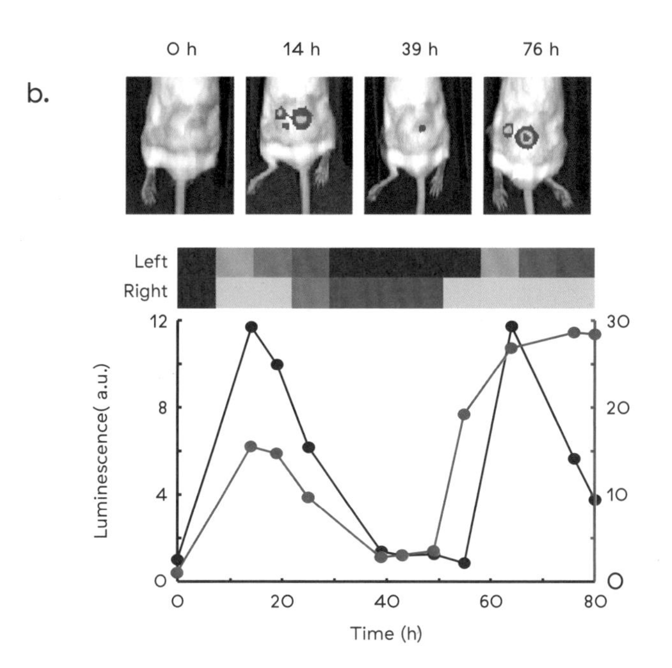

was tested in animal models and showed increased tumor targeting and virulence, with attenuated growth in normal tissue. Furthermore, mice are able to tolerate the A1-R strain at higher doses than with VNP20009, suggesting that A1-R has greater clinical potential.[43, 48, 49, 50]

In addition to engineering bacteria to more safely colonize tumors, anticancer agents have been engineered into bacterial genomes to increase their therapeutic effectiveness.[37] Two major mechanisms studied include the direct expression of proteins with antitumor activity and the transferring of expression vectors to infected cancer cells. In both mechanisms, bacteria express an anticancer agent that is either cytotoxic or provokes an immune response. Bacterial toxins, such as cytosolin A (ClyA), were early cytotoxic agents used because their genes are native to bacterial physiology.[37] In particular, ClyA functions by forming pores in cell membranes and inducing apoptosis.[51, 52, 53] Other cytotoxic agents induce apoptosis through mediators such as caspase 8 and 3.[54, 55] Additionally, bacteria can be engineered to deliver cytokines which stimulate immune cell activation, proliferation, and migration to the tumor site.[37] Examples of cytokines recombined into the bacterial genome include IL-2, which promotes lymphocyte proliferation, and IL-18, which suppresses angiogenesis, thereby cutting off the blood supply to the tumor.[56, 57, 58] Alternatively, bacteria can also be engineered to express markers upregulated by tumor cells, thus sensitizing the immune cells and destroying tumors presenting those markers.[37] Recent studies show that the production of Raf1, an upregulated transcription factor in tumor cells, by attenuated *S. typhimurium* using the *E. coli* hemolysin secretion system led to a regression of tumors in Raf-1-induced lung adenomas.[59] Utilizing bacteria as vehicles for therapeutics is beneficial in that it allows for the localized production of therapeutics at the tumor site, but the control of therapy often stops once the bacteria is injected, resulting in an "always on" cargo production leading to high dosages, off-target effects, and potential development of host resistance.

With rapid advances in the field of synthetic biology, gene circuits and logic gates have been incorporated into bacterial genomes to provide more control over these bacterial therapeutics. Early examples include regulatory circuits that allow *E. coli* to invade cancer cells in low-oxygen conditions, characteristic of tumor microenvironments.[9] Recently, we have expanded upon these initial oscillatory circuits and explored the use of bacteria for cyclic drug release.[60] We constructed a genetic circuit to mediate drug production and release in a synchronized mechanism (figure 7.4). The synchronized lysis circuit (SLC) relies on quorum-sensing to identify a critical density and threshold concentration of a signal molecule. When the threshold concentration is reached, the circuit is activated and the cells lyse in synchrony, releasing the therapeutic. The few bacteria that survive lysis repopulate, feeding the next cycle. This circuit not only allows temporal control of drug release, but also serves as a safety mechanism to ensure a controlled growth of the bacterial population. In a separate experiment, we engineered three different therapeutics into the SLC, allowing for a multicombination therapy in a single delivery vehicle. Haemolysin (hylE), an immune cell recruitment therapeutic (CCL21), and a trigger for tumor cell apoptosis (CDD-iRGD) were combined into an SLC triple strain (SLC-3) and transfected into *S. typhimurium*. Upon oral delivery of the SLC-3 combined with a common clinical chemotherapy, 5-FU, to colorectal tumor-bearing mice, a 50% increase in mean survival time for the animals was observed. We demonstrated that SLC could produce a significant therapeutic effect both on its own and in combination with chemotherapy in subcutaneous and liver metastasis models.

One challenge in exploiting synthetic biology for translational applications is to develop robust systems that function in *in vivo* contexts. One major hurdle is that culture-based selective pressures such as antibiotic resistance are absent *in vivo*, leading to strain-dependent instability of plasmid-based networks over time.[61] To minimize the extent of plasmid loss in the absence of antibiotic selection, stabilizing elements for plasmid retention can be incorporated into the strain.[62] For example, a toxin-antitoxin system can be employed in which the toxin and antitoxin are produced simultaneously and, in the event of plasmid loss, the cell will be killed by the long-lived toxin.[63, 64] Furthermore, *alp*, originating from the *Bacillus subtils* plasmid, ensures equal segregation of plasmids during cell division by producing filaments that push plasmids to the poles of cells.[62] Additionally, genes of interest can be integrated into bacterial genomic DNA, which differs from plasmid DNA in that it cannot be lost to the environment or transferred to other bacteria,[65, 66, 67] where it might produce less therapeutic results or mutate with higher frequency.

While plasmid stability strategies allow for the bacterial circuits to be more viable in animal models, they also may allow for transmission and survival of the engineered plasmids at low frequency, posing a biosafety issue. Moving forward, it will be critical to ensure that these engineered bacteria or their engineered plasmids do not escape from and survive outside of the laboratory environment. This can be done by attenuations that limit a bacteria's growth on a synthetic amino acid[68] or other nutrient not present in the environment. Alternatively, biocontainment circuits, like the previously mentioned "kill switch," have been developed to couple biosensing with circuit control of cell viability.[31] These circuits specify environmental "input" signals that lead to cellular destruction, ensuring the controlled growth of bacterial populations in the intended microenvironments. The challenge with these circuits will be to have them maintain functionality in the environment, given their strong selective pressure with expressed toxins.

VI. FUTURE DIRECTIONS

The treatment standards for cancer are being revolutionized by reprogramming biology. Today, the tools to engineer regulatory circuits and kill switches to control bacterial cancer therapeutic production and release are rapidly evolving. As safety mechanisms develop, it will be possible in the future to incorporate therapeutic bacteria into our daily diets and have them detect, target, and treat tumors unbeknownst to us.

Cancer therapy is just one of the many applications for programmable bacteria. The ability to engineer one of life's oldest forms as smart therapeutic agents has the potential to transform our current therapeutic capabilities across a range of diseases. Since bacteria are found almost everywhere in our bodies, this presents an opportunity to engineer bacteria for many applications such as prevention of infections in the upper respiratory of cystic fibrosis patients and engineering of skin microbes for UV protection, to name a few.

Over the past century, research has accelerated from the first discovery of the "hereditary units" dubbed "genes" to the intentional reprogramming of life. Parallel to the field of biology, we have seen the rapid advancement of biotechnologies such as improved computation, high-throughput assays, and clinically relevant animal models that have made these scientific strides possible. Despite the progress, synthetic biology has several challenges ahead before reaching its full clinical potential. With issues like public perception and integration into standard treatments, there is still room for the field to grow. Even so, with the fast-paced design of novel bacterial circuits and the concurrent development of innovative technologies, the diverse capabilities of engineered bacteria will soon be realized.

NOTES

1 Brian Charlesworth and Deborah Charlesworth, "Darwin and Genetics," *Genetics* 183, no. 3 (2009): 757–766.

2 Stanley N. Cohen, Annie C. Y. Chang, and Leslie Hsu, "Nonchromosomal Antibiotic Resistance in Bacteria: Genetic Transformation of Escherichia coli by R-Factor DNA," *Proceedings of the National Academy of Sciences* 69, no. 8 (1972): 2110–2114.

3 Stanley N. Cohen, Annie C. Y. Chang, and Leslie Hsu, *Genetics and Genomics Timeline* 2004, available from http://www.genomenewsnetwork.org/resources/timeline/1973_Boyer.php

4 Jacques Monod and François Jacob, "General Conclusions: Teleonomic Mechanisms in Cellular Metabolism, Growth, and Differentiation," paper read at Cold Spring Harbor symposium on quantitative biology, 1961.

5 D. Ewen Cameron, Caleb J. Bashor, and James J. Collins, "A Brief History of Synthetic Biology," Nature Reviews Microbiology 12, no. 5 (2014): 381–390.

6 Timothy S. Gardner, Charles R. Cantor, and James J. Collins, "Construction of a Genetic Toggle Switch in Escherichia coli," Nature 403, no. 6767 (2000): 339–342.

7 Michael B. Elowitz and Stanislas Leibler, "A Synthetic Oscillatory Network of Transcriptional Regulators," *Nature* 403, no. 6767 (2000): 335–338.

8 Anselm Levskaya, Aaron A. Chevalier, Jeffrey J. Tabor, Zachary Booth Simpson, Laura A. Lavery, Matthew Levy, Eric A. Davidson, Alexander Scouras, Andrew D. Ellington, and Edward M. Marcotte, "Synthetic Biology: Engineering Escherichia coli to See Light," *Nature* 438, no. 7067 (2005): 441–442.

9 J. Christopher Anderson, Elizabeth J. Clarke, Adam P. Arkin, and Christopher A. Voigt, "Environmentally Controlled Invasion of Cancer Cells by Engineered Bacteria," *Journal of Molecular Biology* 355, no. 4 (2006): 619–627.

10 John E. Dueber, Gabriel C. Wu, G. Reza Malmirchegini, Tae Seok Moon, Christopher J. Petzold, Adeeti V. Ullal, Kristala L. J. Prather, and Jay D. Keasling, "Synthetic Protein Scaffolds Provide Modular Control over Metabolic Flux," *Nature Biotechnology* 27, no. 8 (2009): 753–759.

11 Jeffrey J. Tabor, Howard M. Salis, Zachary Booth Simpson, Aaron A. Chevalier, Anselm Levskaya, Edward M. Marcotte, Christopher A. Voigt, and Andrew D. Ellington, "A Synthetic Genetic Edge Detection Program," *Cell* 137, no. 7 (2009): 1272–1281.

12 Ahmad S. Khalil and James J. Collins, "Synthetic Biology: Applications Come of Age," *Nature Reviews Genetics* 11, no. 5 (2010): 367–379.

13 Wilfried Weber, Ronald Schoenmakers, Bettina Keller, Marc Gitzinger, Thomas Grau, Marie Daoud-El Baba, Peter Sander, and Martin Fussenegger, "A Synthetic Mammalian Gene Circuit Reveals Antituberculosis Compounds," *Proceedings of the National Academy of Sciences* 105, no. 29 (2008): 9994–9998.

14 Dae-Kyun Ro, Eric M. Paradise, Mario Ouellet, Karl J. Fisher, Karyn L. Newman, John M. Ndungu, Kimberly A. Ho, Rachel A. Eachus, Timothy S. Ham, and James Kirby, "Production of the Antimalarial Drug Precursor Artemisinic Acid in Engineered Yeast," *Nature* 440, no. 7086 (2006): 940–943.

15 J. L. Fortman, Swapnil Chhabra, Aindrila Mukhopadhyay, Howard Chou, Taek Soon Lee, Eric Steen, and Jay D. Keasling, "Biofuel Alternatives to Ethanol: Pumping the Microbial Well," *Trends in Biotechnology* 26, no. 7 (2008): 375–381.

16 Clyde A. Hutchison, Ray-Yuan Chuang, Vladimir N. Noskov, Nacyra Assad-Garcia, Thomas J. Deerinck, Mark H. Ellisman, John Gill, Krishna Kannan, Bogumil J. Karas, and Li Ma, "Design and Synthesis of a Minimal Bacterial Genome," *Science* 351, no. 6280 (2016): aad6253.

17 Ingrid M. Keseler, Julio Collado-Vides, Socorro Gama-Castro, John Ingraham, Suzanne Paley, Ian T. Paulsen, Martín Peralta-Gil, and Peter D. Karp, "EcoCyc: A Comprehensive Database Resource for Escherichia coli," *Nucleic Acids Research* 33, suppl. 1 (2005): D334–D337.

18 W. W. Bromer, L. G. Sinn, A. Staub, and Otto K. Behrens, "The Amino Acid Sequence of Glucagon," *Journal of the American Chemical Society* 78, no. 15 (1956): 3858–3860.

19 Siegfried Labeit and Bernhard Kolmerer, "Titins: Giant Proteins in Charge of Muscle Ultrastructure and Elasticity," *Science* 270, no. 5234 (1995): 293.

20 Frederick Sanger, Steven Nicklen, and Alan R. Coulson, "DNA Sequencing with Chain-Terminating Inhibitors," *Proceedings of the National Academy of Sciences* 74, no. 12 (1977): 5463–5467.

21 Illumina. Illumina Inc. 2016. Available from http://www.illumina.com/content/dam/illumina-marketing/documents/products/illumina_sequencing_introduction.pdf).

22 Sara Goodwin, John D. McPherson, and W. Richard McCombie, "Coming of Age: Ten Years of Next-Generation Sequencing Technologies," *Nature Reviews Genetics* 17, no. 6 (2016): 333–351.

23 R. K. Saiki, D. H. Gelfand, S. Stoffel, S. T. Scharf, R. Higuchi, G. T. Horn, K. B. Mullis, and H. A. Ehrlich, "Primer-Directed Enzymatic Amplification of DNA," *Science* 239 (1988): 487–491.

24 Daniel G. Gibson, John I. Glass, Carole Lartigue, Vladimir N. Noskov, Ray-Yuan Chuang, Mikkel A. Algire, Gwynedd A. Benders, Michael G. Montague, Li Ma, and Monzia M. Moodie, "Creation of a Bacterial Cell Controlled by a Chemically Synthesized Genome," *Science* 329, no. 5987 (2010): 52–56.

25 John F. Morrow and Paul Berg, "Cleavage of Simian Virus 40 DNA at a Unique Site by a Bacterial Restriction Enzyme," *Proceedings of the National Academy of Sciences* 69, no. 11 (1972): 3365–3369.

26 David A. Jackson, Robert H. Symons, and Paul Berg, "Biochemical Method for Inserting New Genetic Information into DNA of Simian Virus 40: Circular SV40 DNA Molecules Containing Lambda Phage Genes and the Galactose Operon of Escherichia coli," *Proceedings of the National Academy of Sciences* 69, no. 10 (1972): 2904–2909.

27 Daniel G. Gibson, Lei Young, Ray-Yuan Chuang, J. Craig Venter, Clyde A. Hutchison, and Hamilton O. Smith, "Enzymatic Assembly of DNA Molecules up to Several Hundred Kilobases," *Nature Methods* 6, no. 5 (2009): 343–345.

28 Prashant Mali, Luhan Yang, Kevin M. Esvelt, John Aach, Marc Guell, James E. DiCarlo, Julie E. Norville, and George M. Church, "RNA-Guided Human Genome Engineering via Cas9," *Science* 339, no. 6121 (2013): 823–826.

29 Le Cong, F. Ann Ran, David Cox, Shuailiang Lin, Robert Barretto, Naomi Habib, Patrick D. Hsu, Xuebing Wu, Wenyan Jiang, and Luciano A. Marraffini, "Multiplex Genome Engineering Using CRISPR/Cas Systems," *Science* 339, no. 6121 (2013): 819—823.

30 Martin Jinek, Krzysztof Chylinski, Ines Fonfara, Michael Hauer, Jennifer A. Doudna, and Emmanuelle Charpentier, "A Programmable Dual-RNA—Guided DNA Endonuclease in Adaptive Bacterial Immunity," *Science* 337, no. 6096 (2012): 816—821.

31 Clement T. Y. Chan, Jeong Wook Lee, D. Ewen Cameron, Caleb J. Bashor, and James J. Collins, "'Deadman' and 'Passcode' Microbial Kill Switches for Bacterial Containment," *Nature Chemical Biology* 12, no. 2 (2016): 82—86.

32 Mariette R. Atkinson, Michael A. Savageau, Jesse T. Myers, and Alexander J. Ninfa, "Development of Genetic Circuitry Exhibiting Toggle Switch or Oscillatory Behavior in Escherichia coli," *Cell* 113, no. 5 (2003): 597—607.

33 Eileen Fung, Wilson W. Wong, Jason K. Suen, Thomas Bulter, Sun-gu Lee, and James C. Liao, "A Synthetic Gene—Metabolic Oscillator," *Nature* 435, no. 7038 (2005): 118—122.

34 Jesse Stricker, Scott Cookson, Matthew R. Bennett, William H. Mather, Lev S. Tsimring, and Jeff Hasty, "A Fast, Robust and Tunable Synthetic Gene Oscillator," *Nature* 456, no. 7221 (2008): 516—519.

35 Tal Danino, Octavio Mondragón-Palomino, Lev Tsimring, and Jeff Hasty, "A Synchronized Quorum of Genetic Clocks," *Nature* 463, no. 7279 (2010): 326—330.

36 National Institute of Health, *"Understanding Cancer"* (2015). Available from http://www.cancer.gov/about-cancer/understanding/what-is-cancer.

37 Neil S. Forbes, "Engineering the Perfect (Bacterial) Cancer Therapy," *Nature Reviews Cancer* 10, no. 11 (2010): 785—794.

38 Tobias A. Oelschlaeger, "Bacteria as Tumor Therapeutics?," *Bioengineered Bugs* 1 no. 2 (2010): 146—147.

39 Edward F. McCarthy, "The Toxins of William B. Coley and the Treatment of Bone and Soft-Tissue Sarcomas," *Iowa Orthopaedic Journal* no. 26 (2006): 154.

40 Long H. Dang, Chetan Bettegowda, David L. Huso, Kenneth W. Kinzler, and Bert Vogelstein, "Combination Bacteriolytic Therapy for the Treatment of Experimental Tumors," *Proceedings of the National Academy of Sciences* 98, no. 26 (2001): 15155—15160.

41 Richard A. Malmgren and Clyde C. Flanigan, "Localization of the Vegetative Form of Clostridium Tetani in Mouse Tumors Following Intravenous Spore Administration," *Cancer Research* 15, no. 7 (1955): 473—478.

42 Leschner, Sara, Kathrin Westphal, Nicole Dietrich, Nuno Viegas, Jadwiga Jablonska, Marcin Lyszkiewicz, Stefan Lienenklaus, Werner Falk, Nelson Gekara, and Holger Loessner, "Tumor Invasion of Salmonella enterica serovar Typhimurium Is Accompanied by Strong Hemorrhage Promoted by TNF-α," *PloS one* 4, no. 8 (2009): e6692.

43 Ming Zhao, Meng Yang, Xiao-Ming Li, Ping Jiang, Eugene Baranov, Shukuan Li, Mingxu Xu, Sheldon Penman, and Robert M. Hoffman, "Tumor-Targeting Bacterial Therapy with Amino Acid Auxotrophs of GFP-Expressing Salmonella typhimurium," *Proceedings of the National Academy of Sciences* 102, no. 3 (2005): 755—760.

44 Ling Zhang, Lifang Gao, Lijuan Zhao, Baofeng Guo, Kun Ji, Yong Tian, Jinguo Wang, Hao Yu, Jiadi Hu, and Dhananjaya V. Kalvakolanu, "Intratumoral Delivery and Suppression of Prostate Tumor Growth by Attenuated Salmonella enterica serovar Typhimurium Carrying Plasmid-Based Small Interfering RNAs," *Cancer Research* 67, no. 12 (2007): 5859—5864.

45 Qian Zhang, A. Yu Yong, Ena Wang, Nanhai Chen, Robert L. Danner, Peter J. Munson, Francesco M. Marincola, and Aladar A. Szalay, "Eradication of Solid Human Breast Tumors in Nude Mice with an Intravenously Injected Light-Emitting Oncolytic Vaccinia Virus," *Cancer Research* 67 no. 20 (2007): 10038—10046.

46 John F. Toso, Vee J. Gill, Patrick Hwu, Francesco M. Marincola, Nicholas P. Restifo, Douglas J. Schwartzentruber, Richard M. Sherry, Suzanne L. Topalian, James C. Yang, and Frida Stock, "Phase I Study of the Intravenous Administration of Attenuated Salmonella typhimurium to Patients with Metastatic Melanoma," *Journal of Clinical Oncology* 20, no. 1 (2002): 142—152.

47 David M. Heimann and Steven A. Rosenberg, "Continuous Intravenous Administration of Live Genetically Modified Salmonella typhimurium in Patients with Metastatic Melanoma," *Journal of Immunotherapy* 26, no. 2 (2003): 179.

48 Yong Zhang, Nan Zhang, Ming Zhao, and Robert M. Hoffman, "Comparison of the Selective Targeting Efficacy of Salmonella typhimurium A1-R and VNP20009 on the Lewis Lung Carcinoma in Nude Mice," *Oncotarget* 6, no. 16 (2015): 14625.

49 Ming Zhao, Meng Yang, Huaiyu Ma, Xiaoming Li, Xiuying Tan, Shukuan Li, Zhijian Yang, and Robert M. Hoffman, "Targeted Therapy with a Salmonella typhimurium Leucine-Arginine Auxotroph Cures Orthotopic Human Breast Tumors in Nude Mice," *Cancer Research* 66, no. 15 (2006): 7647—7652

50 Ming Zhao, Jack Geller, Huaiyu Ma, Meng Yang, Sheldon Penman, and Robert M. Hoffman, "Monotherapy with a Tumor-Targeting Mutant of Salmonella typhimurium Cures Orthotopic Metastatic Mouse Models of Human Prostate Cancer," *Proceedings of the National Academy of Sciences* 104, no. 24 (2007): 10170—10174.

51 Sheng-Nan Jiang, Thuy X. Phan, Taek-Keun Nam, Vu H. Nguyen, Hyung-Seok Kim, Hee-Seung Bom, Hyon E. Choy, Yeongjin Hong, and Jung-Joon Min, "Inhibition of Tumor Growth and Metastasis by a Combination of Escherichia coli—Mediated Cytolytic Therapy and Radiotherapy," *Molecular Therapy* 18, no. 3 (2010): 635—642.

52 R. M. Ryan, J. Green, P. J. Williams, S. Tazzyman, S. Hunt, J. H. Harmey, S. C. Kehoe, and C. E. Lewis, "Bacterial Delivery of a Novel Cytolysin to Hypoxic Areas of Solid Tumors," *Gene Therapy* 16, no. 3 (2009): 329—339.

53 Vu H. Nguyen, Hyung-Seok Kim, Jung-Min Ha, Yeongjin Hong, Hyon E. Choy, and Jung-Joon Min, "Genetically Engineered Salmonella typhimurium as an Imageable Therapeutic Probe for Cancer," *Cancer Research* 70, no. 1 (2010): 18—23.

54 Sabha Ganai, R. B. Arenas, and N. S. Forbes, "Tumour-Targeted Delivery of TRAIL Using Salmonella typhimurium Enhances Breast Cancer Survival in Mice," *British Journal of Cancer* 101, no. 10 (2009): 1683—1691.

55 Markus Loeffler, Gaelle Le'Negrate, Maryla Krajewska, and John C. Reed, "Inhibition of Tumor Growth Using Salmonella Expressing Fas Ligand," *Journal of the National Cancer Institute* 100, no. 15 (2008): 1113—1116.

56 Markus Loeffler, Gaelle Le'Negrate, Maryla Krajewska, and John C. Reed, "IL-18-Producing Salmonella Inhibit Tumor Growth," *Cancer Gene Therapy* 15, no. 12 (2008): 787—794.

57 Mark J. Micallef, Tadao Tanimoto, Keizo Kohno, Masao Ikeda, and Masashi Kurimoto, "Interleukin 18 Induces the Sequential Activation of Natural Killer Cells and Cytotoxic T Lymphocytes to Protect Syngeneic Mice from Transplantation with Meth A Sarcoma," *Cancer Research* 57, no. 20 (1997): 4557—4563.

58 Sofie Barbé, Lieve Van Mellaert, Jan Theys, Nick Geukens, Elke Lammertyn, Philippe Lambin, and Jozef Anné, "Secretory Production of Biologically Active Rat Interleukin-2 by Clostridium acetobutylicum DSM792 as a Tool for Anti-Tumor Treatment," *FEMS Microbiology Letters* 246, no. 1 (2005): 67—73.

59 Ivaylo Gentschev, Joachim Fensterle, Andreas Schmidt, Tamara Potapenko, Jakob Troppmair, Werner Goebel, and Ulf R. Rapp, "Use of a Recombinant Salmonella enterica serovar Typhimurium Strain Expressing C-Raf for Protection against C-Raf Induced Lung Adenoma in Mice," *BMC Cancer* 5, no. 1 (2005): 1.

60 M. Omar Din, Tal Danino, Arthur Prindle, Matt Skalak, Jangir Selimkhanov, Kaitlin Allen, Ellixis Julio, Eta Atolia, Lev S. Tsimring, and Sangeeta N. Bhatia, "Synchronized Cycles of Bacterial Lysis for In Vivo Delivery," *Nature* 536, no. 7614 (2016): 81—85.

61 Tal Danino, Justin Lo, Arthur Prindle, Jeff Hasty, and Sangeeta N. Bhatia, "In Vivo Gene Expression Dynamics of Tumor-Targeted Bacteria," *ACS Synthetic Biology* 1, no. 10 (2012): 465—470.

62 Tal Danino, Arthur Prindle, Gabriel A. Kwong, Matthew Skalak, Howard Li, Kaitlin Allen, Jeff Hasty, and Sangeeta N. Bhatia, "Programmable Probiotics for Detection of Cancer in Urine," *Science Translational Medicine* 7, no. 289 (2015): 289ra84—289ra84.

63 T. K. Wood, R. H. Kuhn, and S. W. Peretti, "Enhanced Plasmid Stability through Post-Segregational Killing of Plasmid-Free Cells," *Biotechnology Techniques* 4, no. 1 (1990): 39—44.

64 Kenn Gerdes, "The parB (hok/sok) Locus of Plasmid R1: A General Purpose Plasmid Stabilization System," *Nature Biotechnology* 6, no. 12 (1988): 1402—1405.

65 François St-Pierre, Lun Cui, David G. Priest, Drew Endy, Ian B. Dodd, and Keith E. Shearwin, "One-Step Cloning and Chromosomal Integration of DNA," *ACS Synthetic Biology* 2, no. 9 (2013): 537—541.

66 Andreas Haldimann and Barry L. Wanner, "Conditional-Replication, Integration, Excision, and Retrieval Plasmid-Host Systems for Gene Structure-Function Studies of Bacteria," *Journal of Bacteriology* 183, no. 21 (2001): 6384—6393.

67 Marion H. Sibley and Elisabeth A. Raleigh, "A Versatile Element for Gene Addition in Bacterial Chromosomes," *Nucleic Acids Research* 40, no. 3 (2012): e19—e19.

68 Daniel J. Mandell, Marc J. Lajoie, Michael T. Mee, Ryo Takeuchi, Gleb Kuznetsov, Julie E. Norville, Christopher J. Gregg, Barry L. Stoddard, and George M. Church, "Biocontainment of Genetically Modified Organisms by Synthetic Protein Design," *Nature* 518, no. 7537 (2015):55—60.

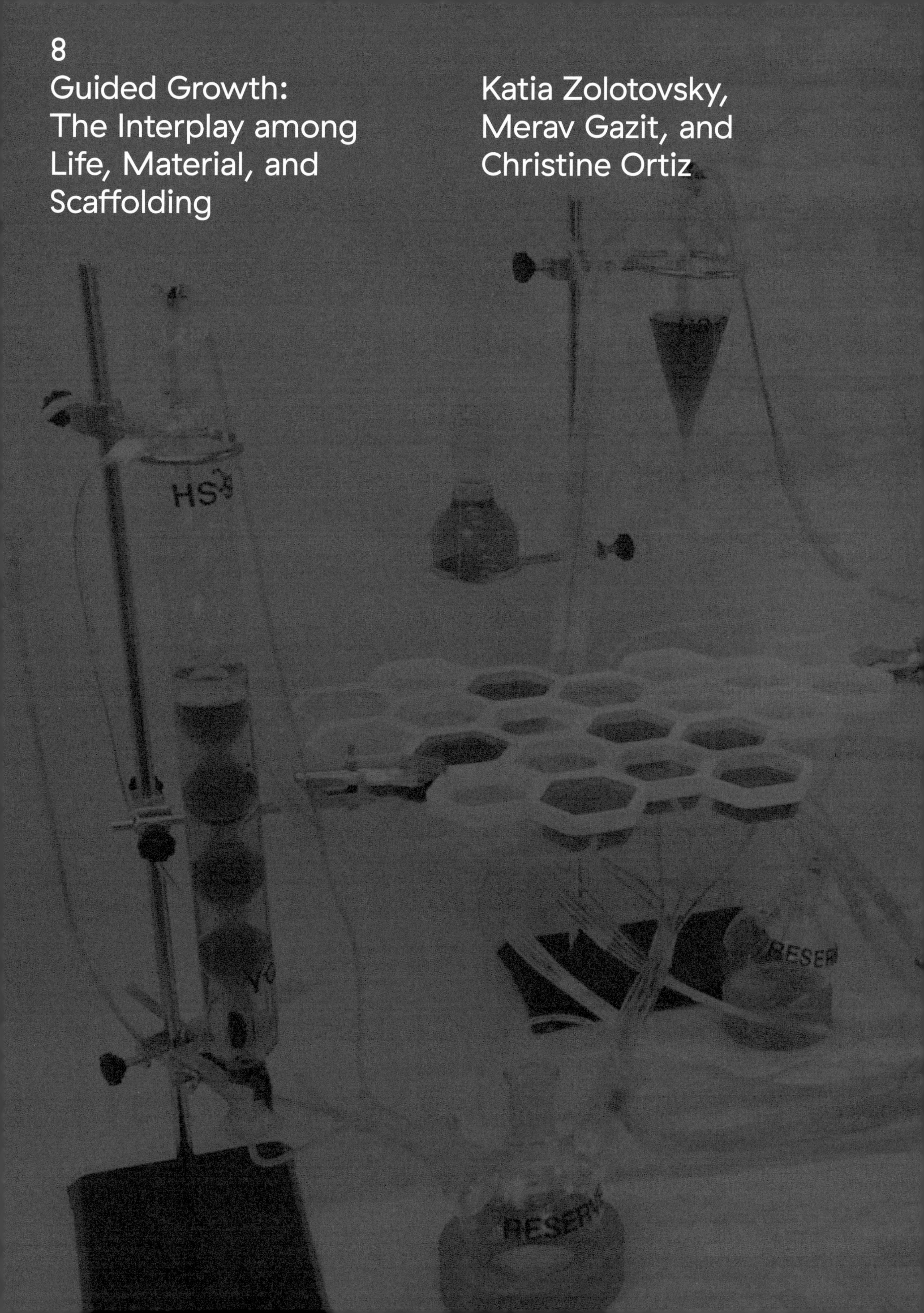

8 Guided Growth: The Interplay among Life, Material, and Scaffolding

Katia Zolotovsky, Merav Gazit, and Christine Ortiz

8
Guided Growth: The Interplay among Life, Material, and Scaffolding

Katia Zolotovsky,
Merav Gazit, and
Christine Ortiz

Cells are alive and responsive. With the tools of DNA design, made more accessible via synthetic biology, it is possible to program cells and their material production. This research explores synthetic biology as a computational method to program living materials and to grow active and adaptive building components for architecture, with the possibility of designing them for new functions such as air filtering and purification, self-repair, and photosynthesis.

To realize the potential for self-assembling, self-healing, adaptive, and functional materials, we propose a guided growth design process. Using the gram-negative bacterium *Gluconacetobacter xylinus* and a biofilm it produces, we grow active and adaptive hybrid materials and three-dimensional components. Biofilm here is a structural layer of cellulose produced by *G. xylinus* cells on a surface of nutrient-rich liquid where bacterial cells are embedded. In guided growth, using methods of synthetic biology, we modify bacterial cells' behavior in their growth environment and the structure and properties of biofilms they produce. We then design and modulate the growth environment as scaffolding for patterning, shaping, harvesting, and processing composite biofilms, while keeping the bacterial cells alive and responsive. These methods of combining genetic engineering with environmental regulation and scaffold design can be further generalized to other material systems where bacteria act as a matter-organizing agent.[1]

Recently, biomaterials have been used for architecture, grown instead of fabricated. For example, architects have developed mycelium, a mushroom grown on wood chips, into structurally sound material[2] and have aggregated minerals into hard blocks using bacteria.[3] Growing materials on site helps save nature's resources and energy, currently wasted on environmentally harmful manufacturing processes of natural materials and their transportation. However, after the growth process the above-mentioned materials are rendered dead and don't preserve the biological advantages of the organisms that created them, such as healing in response to damage. The vision of this research is hybrid materials composed of engineered living cells and nonliving scaffolds that give structure to and support their long-term viability.

In the guided growth design process, we employ three scales of resolution—life, material, and scaffolding—that work in parallel within the structural living material system (figure 8.1). The workflow of design on these three parallel levels of resolution aims to create engineered structural cellulose-based biofilms with viable living bacterial cells, capable of responding to the environment and tuning their material production. This work will open a new design space for building construction and other large-scale applications.

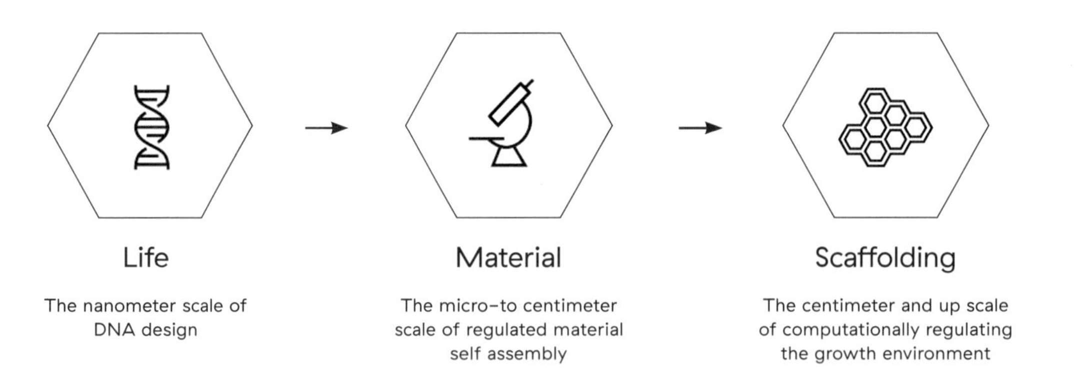

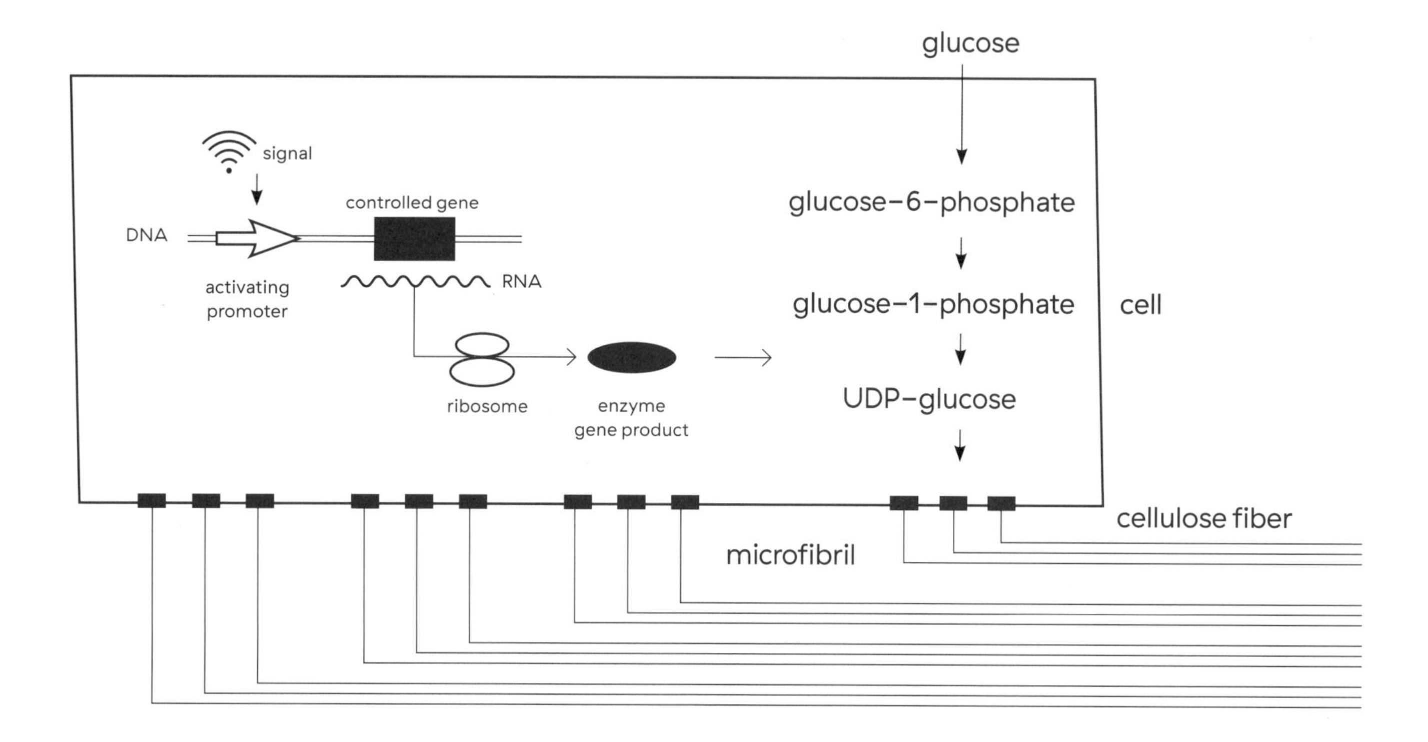

Figure 8.2a

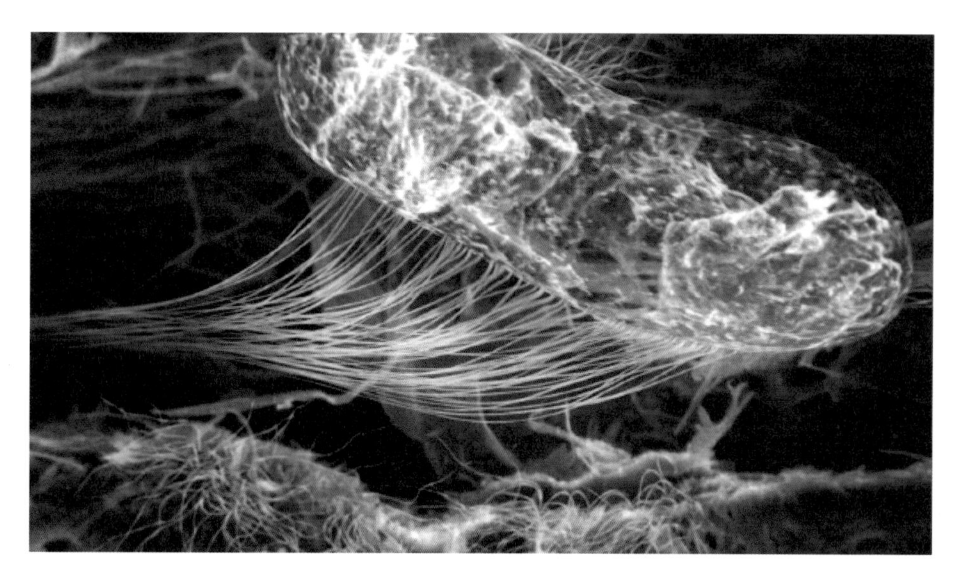

Figure 8.2b

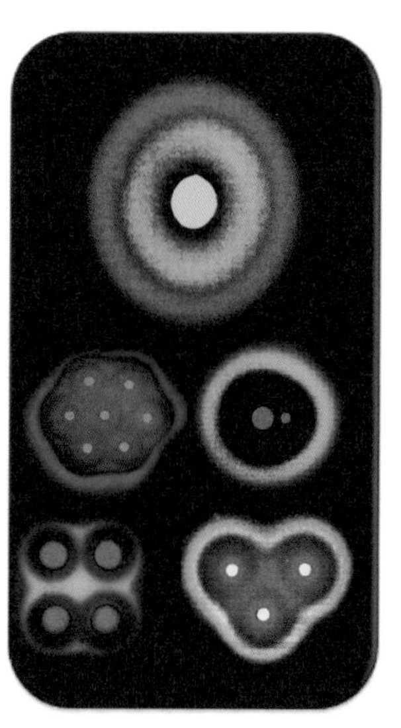

Figure 8.2c

Figure 8.1 (left page)
The three scales of resolution combined in the guided growth design process: life, material, and scaffolding

Figure 8.2 (top: a, bottom left: b, bottom right: c) Engineering bacterial cells to sense their environment, process information, and tune material production. (a) Schematic description of regulating the self-assembly of cellulose fibers via DNA design. (b) Rendering of *Gluconacetobacter xylinus* cell extruding cellulose fibers.[4]
(c) Using synthetic gene networks and fluorescent reporting proteins to design cell-to-cell communication and spatial patterning. Reprinted by permission from Macmillan Publishers Ltd on behalf of Cancer Research UK: *Nature* © 2005.[5]

Life: the nanometer scale of DNA design. We use tools of synthetic biology for genetic design of living components (cells) to sense their environment, process information, and actuate material production. Collaborating with synthetic biologists from the Weiss Lab, we introduce synthetic gene networks, engineered DNA circuits, to pattern cellulose fibers through the intake of substances in response to signals in their environment (such as small chemicals, UV light) (figure 8.2a–b). We further explore hierarchical material patterning through cell-to-cell communication within a biofilm that allows us to define distinct spatial situations such as center and edge (figure 8.2c).

Material: the micro- to centimeter scale of regulated material self-assembly. We regulate the physical and chemical parameters of the growth environment to tune the composition and properties of biofilms as they grow. Biofilm grows on the interface

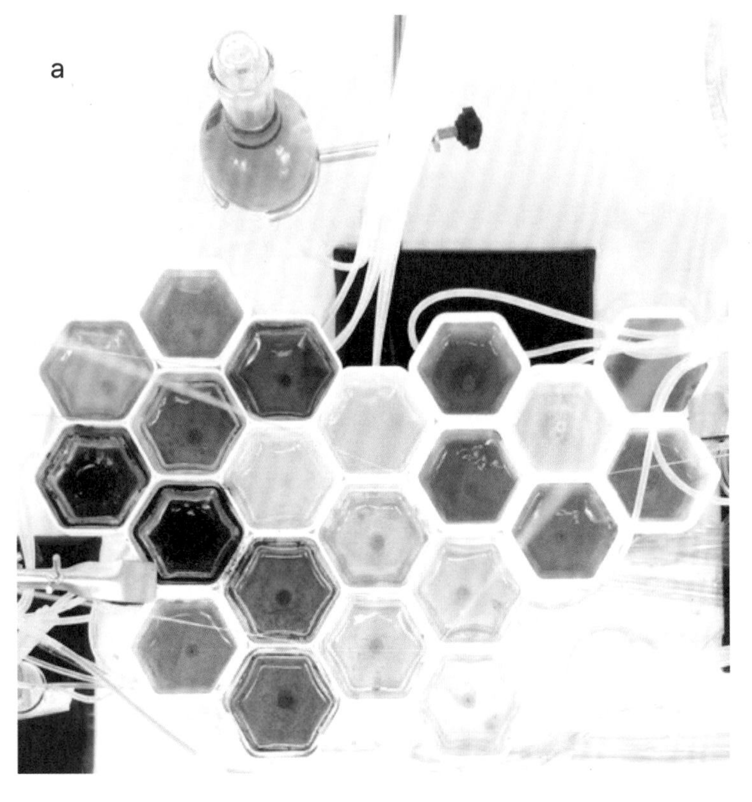

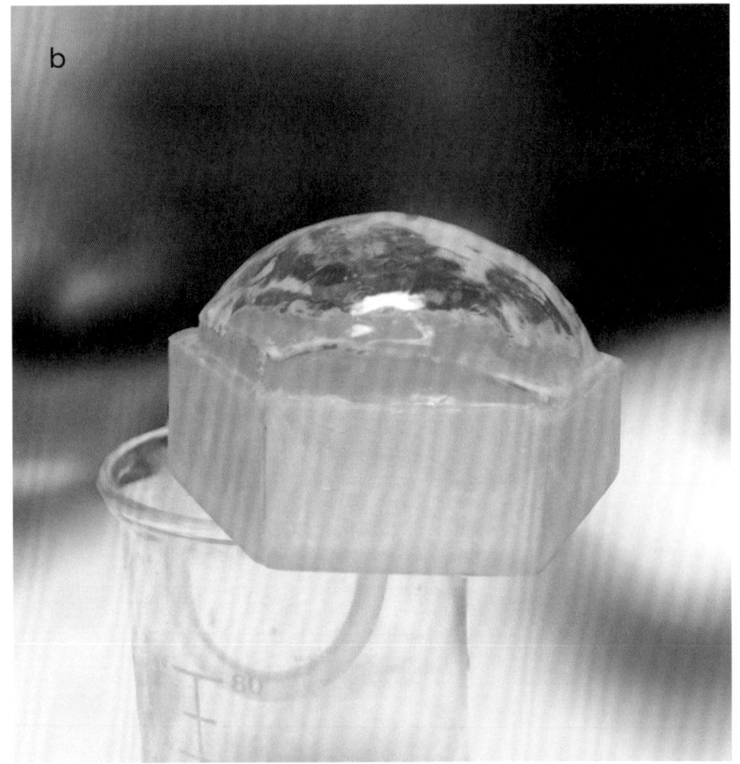

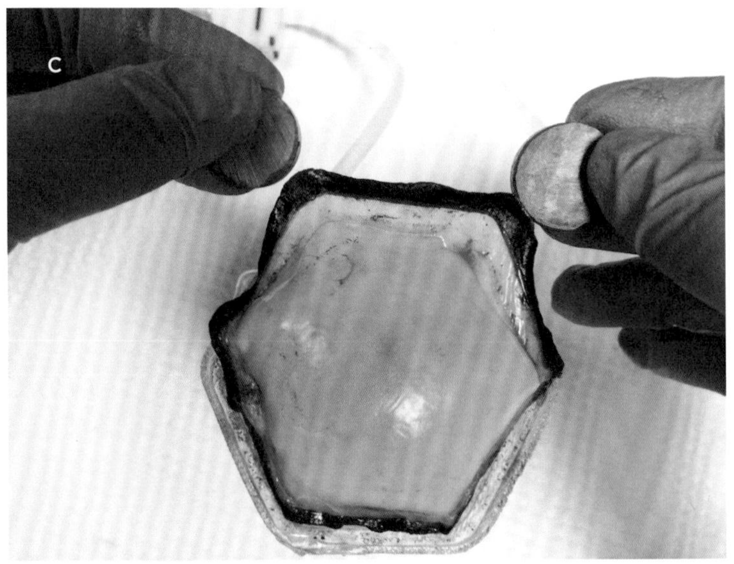

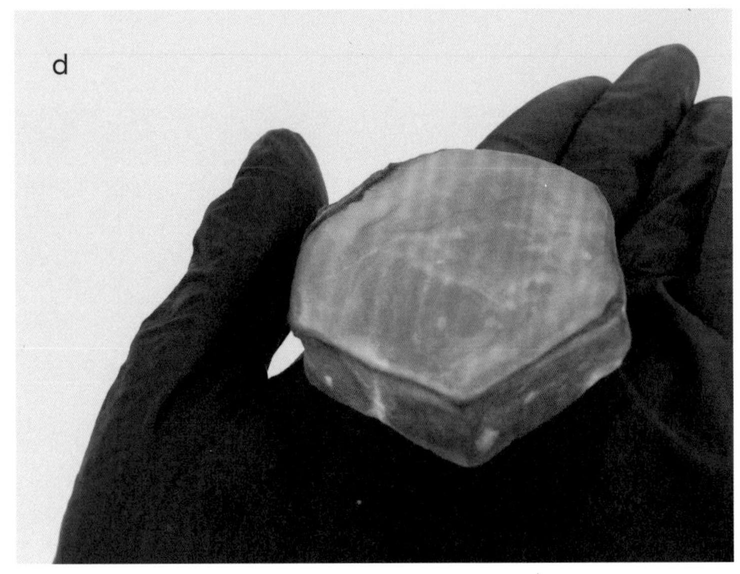

Figure 8.3 (top)
Design of biofilm shape, structure, and properties by the growth environment. (a) Biofilm grows on the interface of nutrients-rich liquid and air. (b) Pneumatic actuation post-growth. (c) In-situ biocomposite with magnetite to generate 3D shaped magnetic biofilms. (d) 3D biofilm grown from bacterial cellulose.

Figure 8.4a (right page)
We designed a macro-fluidic pneumatic scaffolding that computationally regulates the growth environment.

of nutrient-rich liquid and air, and we use silicone as a substrate that allows permeability to oxygen and not to liquid to shape the biofilm as it grows (figure 8.3a–b). We study the effect of timing, composition, and flow of the nutrient-rich liquid on biofilms' structure and properties. We create in-situ composites, including magnetite to produce magnetic biofilms and polyvinyl alcohol for tougher skins (figure 8.3c–d). To characterize the patterned and composite biofilms' structure and properties we use methods of materials science (electro and focused ion beam microscopy, tension and compression testing—data not shown here). Cells are kept alive via the use of the growth scaffolding.

Scaffolding: the centimeter-and-up scale of computationally regulating the growth environment. We design and fabricate a macrofluidic pneumatic scaffolding, as shown in figures 8.3a and 8.4. The scaffolding demonstrates the possibility of using active living components at architectural scale. It both structurally supports and computationally regulates the flow of nutrients, added substances, and air for biological growth and material processing. The scaffolding allows the adjustment of medium composition, in-situ substance concentration, solvent exchange in order to create biocomposites, and pneumatic actuation of the biofilms post-growth. We built custom-made microcontrollers and designed computational scripts to: (a) control physicochemical conditions in each growth vessel, (b) allow post-growth pneumatic actuation to stabilize the 3D cellulose structures by the pressure of compressed air.

The scaffold design presented here provides all of the following: four types of growth mediums; added substances for in-situ biocomposites (polyvinyl alcohol solution and magnetite (Fe_3O_4) powder); reserve containers for renewable waste; 10 miniature solenoid valves; 6 liquid and air pumps;

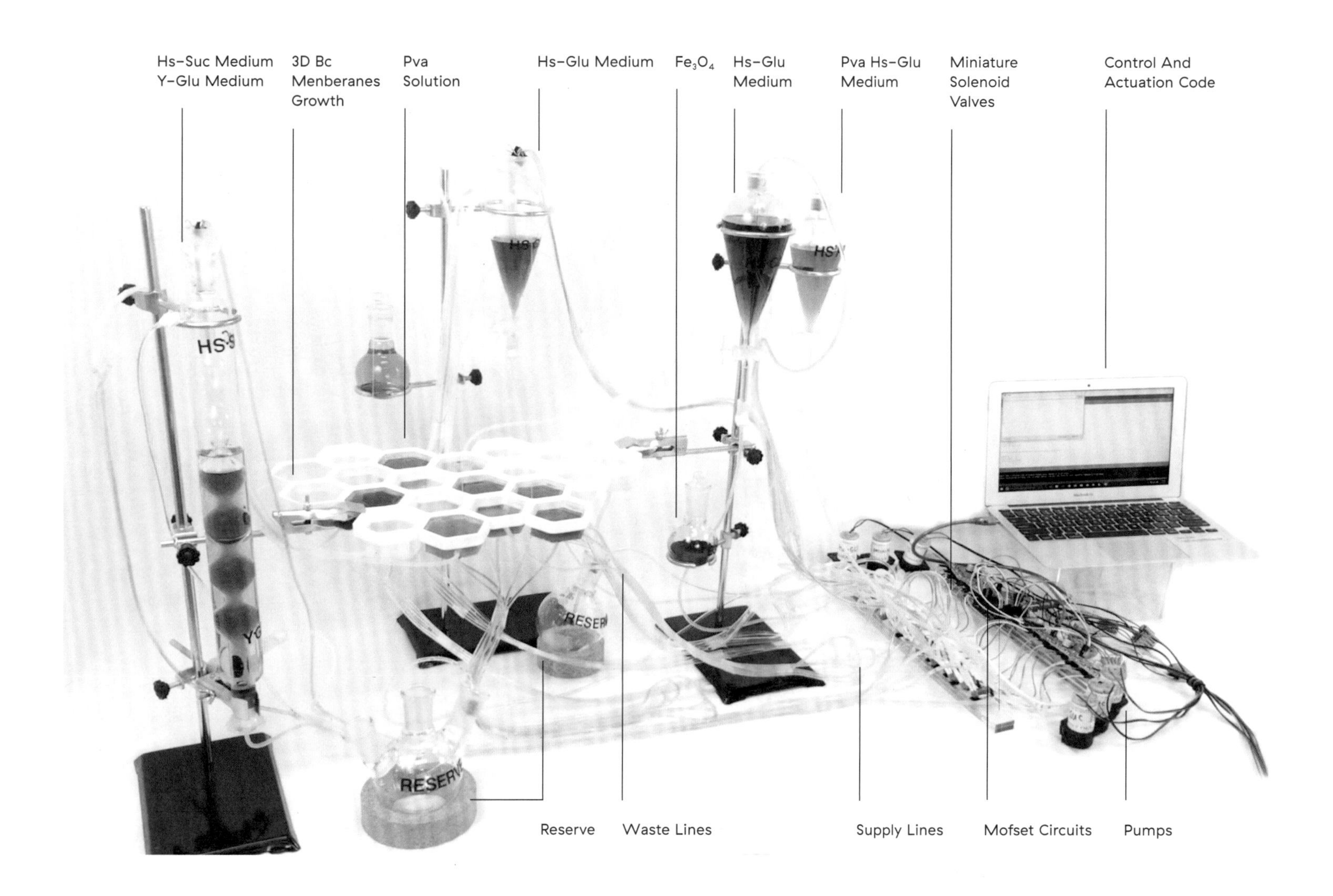

Figure 8.4b
We designed a macro-fluidic pneumatic scaffolding that computationally regulates the growth environment.

custom-made electric circuits; and Arduino microcontroller connected to a laptop that actuate, monitor, and receive feedback from the system. Each growth vessel has an inlet and outlet, and the entire system is routed by silicone tubes (figure 8.4).

In our latest work, working with the tools of synthetic biology, we aim to reduce the scaffolding needs and to grow materials bottom-up, exploring the relations between living and nonliving components of a self-supporting material system that is engineered to grow. Working with the tools of synthetic biology, we explore how these tools, commonly used for precision and predictability for biomedical applications, can be used differently in the domain of architectural design. This new design mindset will combine the rigor of engineering life with an openness of experimenting with materials and shapes.

This research was funded by the National Science Foundation Division of Materials Research (NSF DMR) under the grant #1508072 named "Material and morphometric control of bacterial cellulose via genetic engineering post-processing and 3D printed molding".

NOTES

1 M. Dade-Robertson, C. Ramirez Figueroa, and M. Zhang, "Material Ecologies for Synthetic Biology: Biomineralization and the State Space of Design," *Computer-Aided Design* 60 (March 2015): 28—39.

2 B. Brownell, "Transmaterial 3: A Catalog of Materials That Redefine Our Physical Environment," in *Transmaterial: A Catalog of Materials That Redefine Our Physical Environment* (New York: Princeton Architectural Press, 2010).

3 Biomason, Inc., "Production and Manufacture with Enzymes and Bacteria," U.S. Patent US9428418, filed November 12, 2015, and published August 30, 2016.

4 Stefan Schwabe with Emilia Fostreuter, *Bacteria Producing Cellulose Fibers, Animation by Emilia Forstreuter* (illustration), 2013, retrieved from http://www.stschwabe.com/work/XylinumCones/dublin.php

5 Basu, S., Gerchman, Y., Collins, C. H., Arnold, F. H., and Weiss, R. (2005). A synthetic multicellular system for programmed pattern formation. *Nature, 434*(7037), 1130-1134.

9
Mechanically Guided Deterministic Assembly of 3D Mesostructures in Advanced Materials

John A. Rogers

9 Mechanically Guided Deterministic Assembly of 3D Mesostructures in Advanced Materials

John A. Rogers

Artificial three-dimensional (3D) architectures that mimic complex 3D structures in nature provide compelling opportunities in electronics, energy, biomedical, and many other areas. Existing approaches to make 3D constructs involve limiting constraints in geometries, materials, and length scales. A newly developed deterministic route to 3D mesostructures in advanced materials exploits compressive buckling, as a mechanism to transform 2D precursors into programmed 3D shapes on elastomer substrates. Structural geometries enabled by this approach range from single- (figure 9.1) and multiple-level (figure 9.2) configurations, to releasable multilayers with fully separated (figures 9.3 and 9.4), entangled (figure 9.5), and coherently coupled (figure 9.6) configurations, to 3D sheets/membranes (figure 9.6) inspired by the concept of kirigami. Broad sets of materials, including but not limited to monocrystalline silicon and other inorganic semiconductors (figures 9.1 and 9.2), polymers (figure 9.3), their heterogeneous combinations (figure 9.4), and bilayers of metal and polymer (figures 9.5 and 9.6), are compatible with this mechanically guided assembly process. Another appealing aspect of this approach is its implementation over a broad span of length scales from sub-microscale (figure 9.1) and microscale (figures 9.2—9.4) to millimeter scale (figure 9.5) and centimeter scale (figure 9.6), which suggests a diverse collection of potential applications. Future opportunities lie in the integration of a variety of functional materials such as natural biomaterials and 2D materials, and applications in various fields such as active scaffolds for cells/tissues engineering and micro/nano robotics.

The figures show mechanical assemblies of complex, 3D mesostructures from 2D precursors of advanced materials at different length scales:

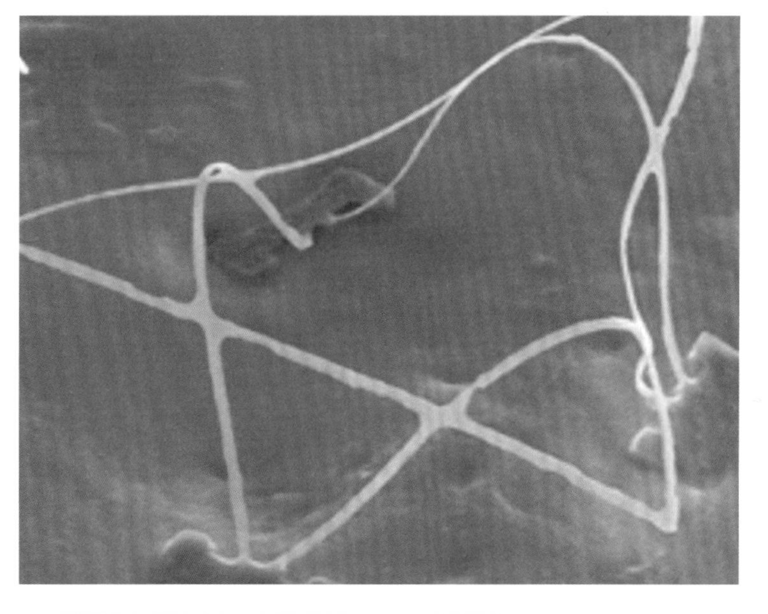

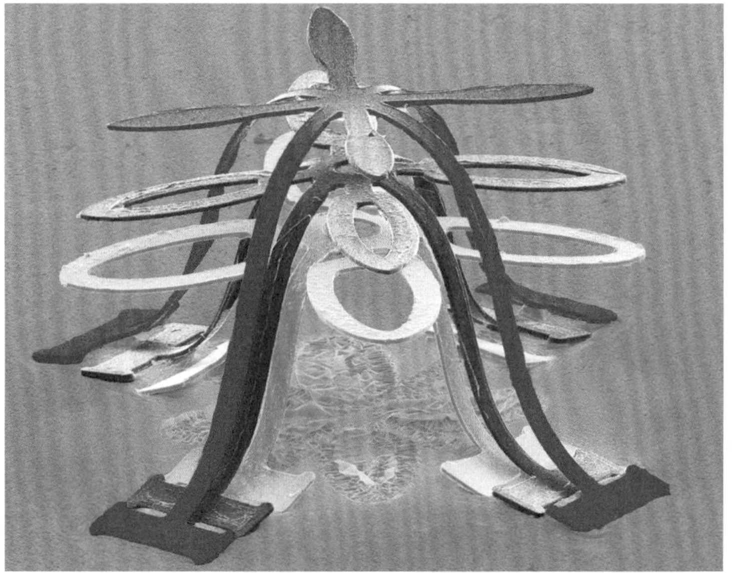

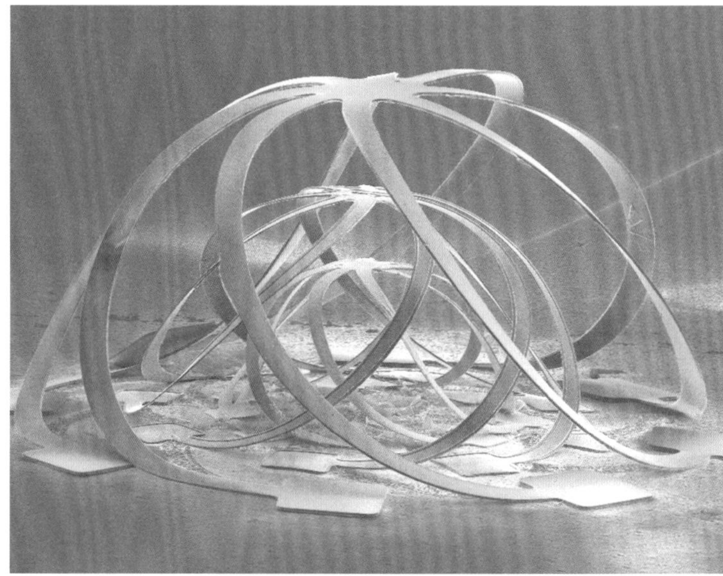

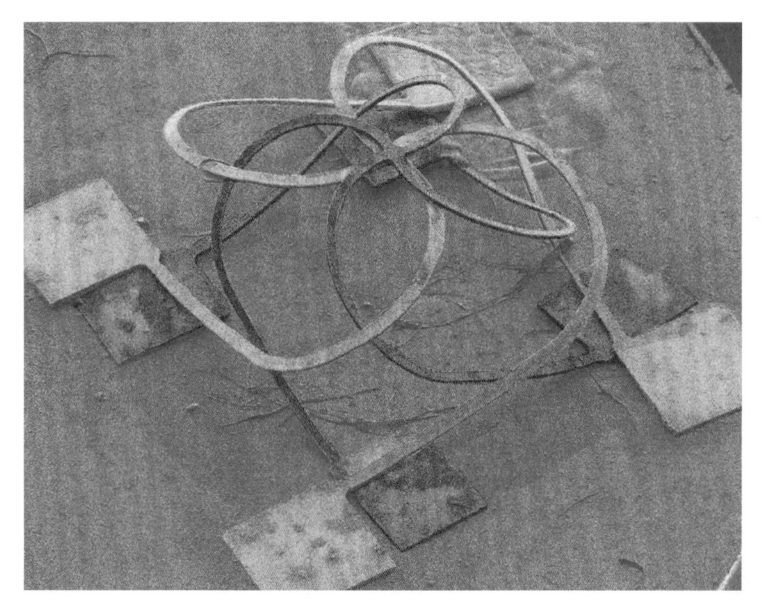

(From top left to bottom right)

Figure 9.1
3D "starfish" framework that incorporates silicon nanoribbons with widths of 800 nm and thicknesses of 100 nm.

Figure 9.2
3D bilayer networks of silicon microribbons with widths of 50 μm and thicknesses of 2 μm.

Figure 9.3
3D trilayer microstructure of epoxy that resembles a tree.

Figure 9.4
3D hybrid nested cages of epoxy (green) and silicon (gray).

Figure 9.5
3D entangled structure made from flexible printed circuit board.

Figure 9.6
3D kirigami structure made of bilayers of copper and polyethylene terephthalate.

10

Biomimetic 4D Printing

A. Sydney Gladman,
Elisabetta A. Matsumoto,
L. Mahadevan, and
Jennifer A. Lewis

10
Biomimetic 4D Printing

A. Sydney Gladman,
Elisabetta A. Matsumoto,
L. Mahadevan, and
Jennifer A. Lewis

We developed a new 4D printing method inspired by the movements of natural plants. Specifically, we encode swelling and elastic anisotropy in printed hydrogel composites through the alignment of stiff cellulose fibrils on the fly during printing. Filler alignment parallel to the print path leads to enhanced stiffness in that direction; hence, upon immersion in water, the printed filaments expand preferentially in the direction orthogonal to the printing path. When structures are patterned with broken symmetry, i.e., as bilayers, their anisotropic swelling leads to programmable out-of-plane deformation, determined by the orientation of printed filaments. We have demonstrated complex changes in curvature including bending, twisting, ruffling, conical defects, and more, all using a single hydrogel-based ink printed in a single step. We have demonstrated the ability to precisely control curvature by varying the actual and the effective thickness, the latter of which is governed by the interfilament spacing within the printed architectures. Our filled hydrogel ink is modular, allowing a broad range of hydrogel chemistries and anisotropic filler compositions to be explored. This biomimetic 4D printing platform enables the design and fabrication of complex, reversible shape-changing architectures printed with one composite hydrogel ink in a single step. These biocompatible shape-shifting architectures with interesting mechanical and photothermal properties may find applications in smart textiles, tissue micrograppers or scaffolds, or as actuators and sensors in soft machines.

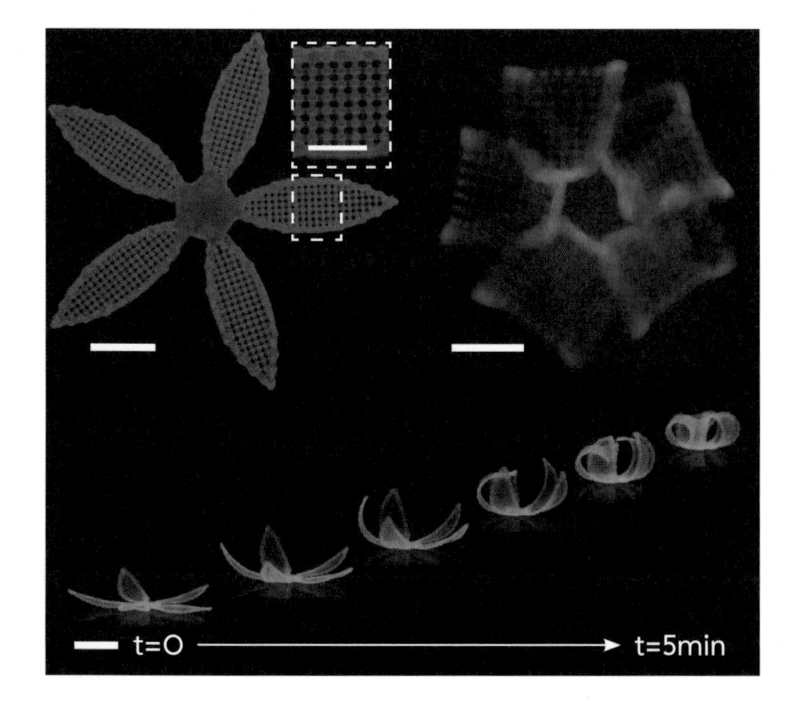

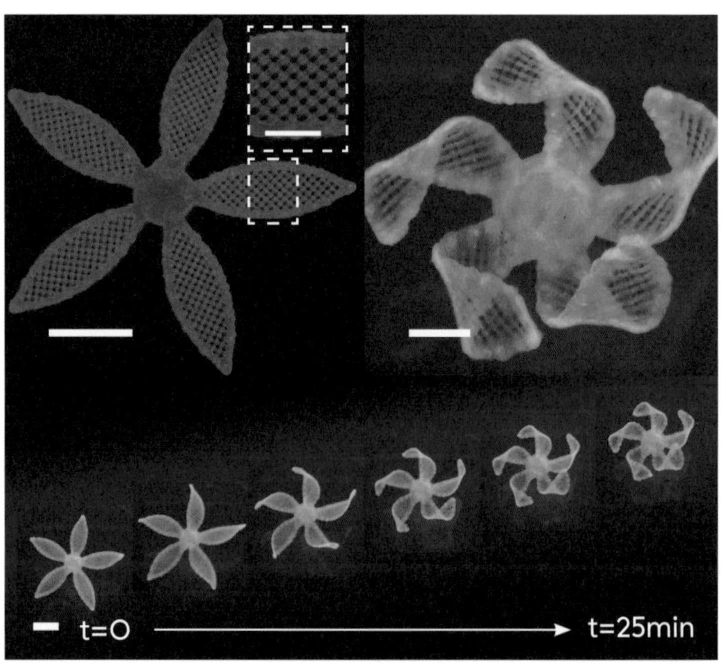

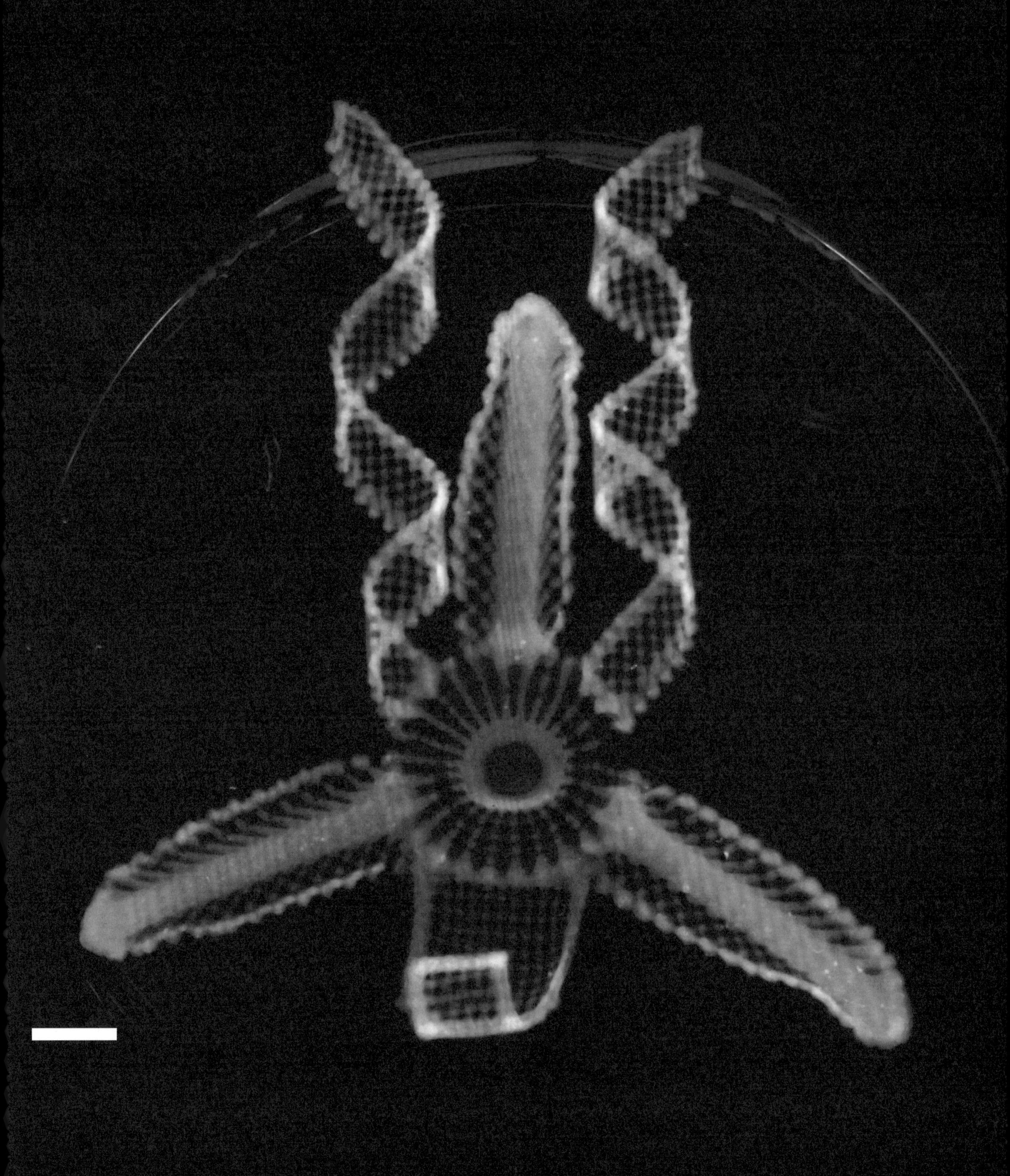

Figure 10.1 (left page, left)
4D-printed bending flower. A 4D-printed flower is programmed to bend its petals upon immersion in water due to anisotropic swelling. Scale bar = 5 mm (2.5 mm inset).

Figure 10.2 (left page, right)
4D-printed twisting flower. A 4D printed flower is programmed to twist its petals upon immersion in water due to anisotropic swelling. Scale bar = 5mm (2.5 mm inset)

Figure 10.3 (top)
4D-printed orchid (1). A 4D-printed orchid displays a variety of unique curvatures after transformation in water, programmed from the print path.

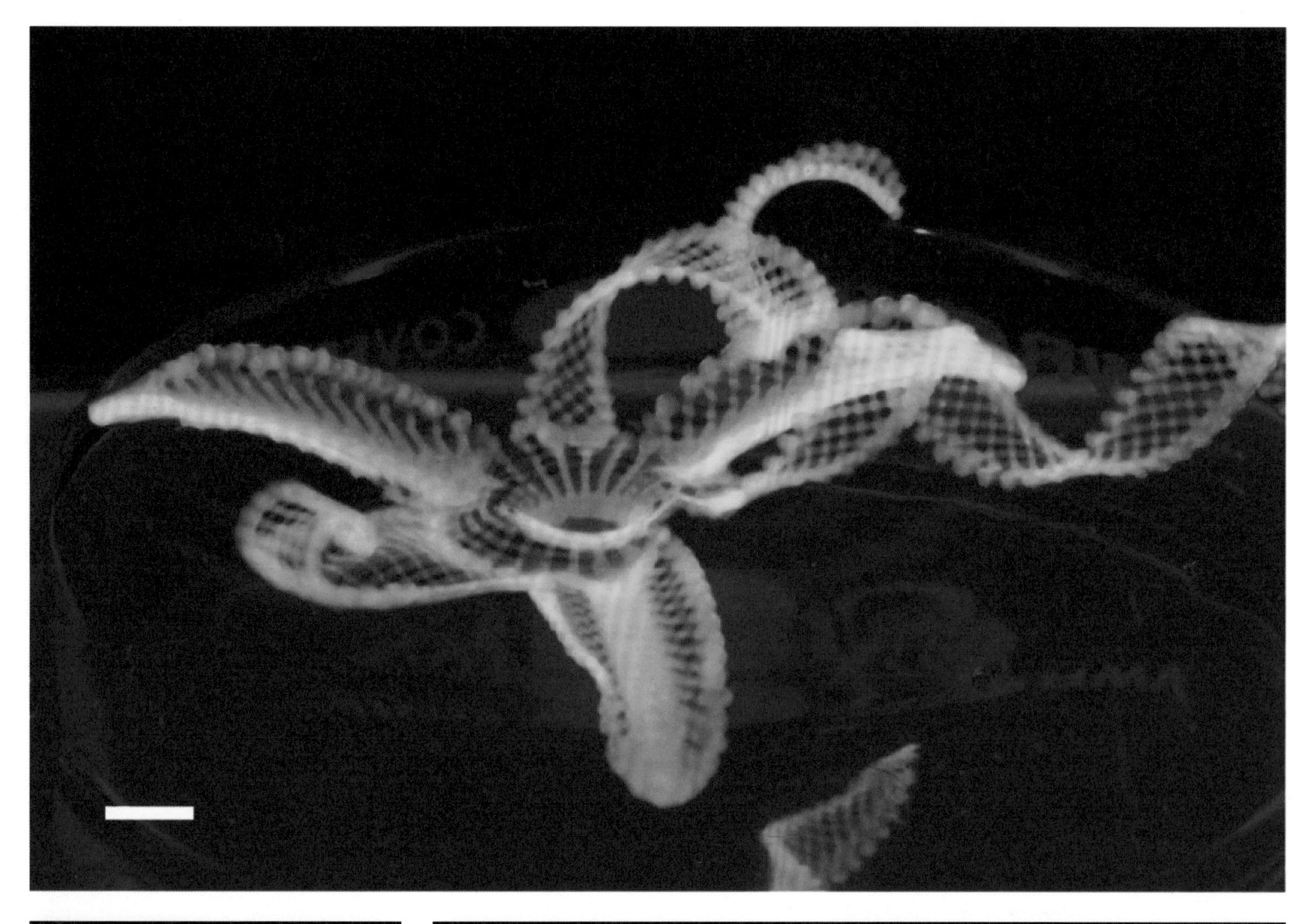

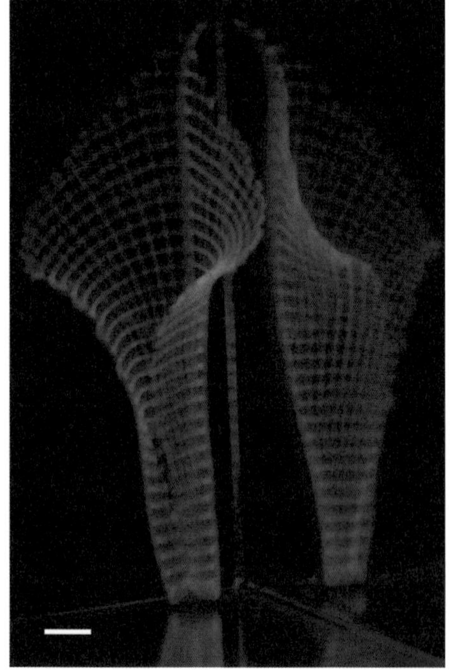

Figure 10.4 (top)
4D-printed orchid (2). A 4D-printed orchid displays a variety of unique curvatures after transformation in water, programmed from the print path. Scale bar = 5 mm.

Figure 10.5 (bottom left)
4D-printed calla lily. The inverse design problem is solved by predicting the print path necessary to achieve the 3D curvature of a calla lily flower upon transformation in water.
Scale bar = 5 mm.

Figure 10.6 (bottom right)
4D printing of calla lily. The programmed print path that will evolve into a calla lily architecture upon transformation in water is printed using our composite hydrogel ink.

11
Hydrogel Devices and Machines

Xuanhe Zhao

11
Hydrogel Devices and Machines

Xuanhe Zhao

Hydrogels are polymer networks infiltrated with large amounts of water, often comprising more than 90% of their weight. Hydrogels' unique properties such as softness, wetness, and biocompatibility make them ideal material candidates for a wide range of advanced applications including biomedicine, wearable devices, and soft robotics. In this research, we cover a number of topics on hydrogel devices and machines from the mechanics of hydrogels to their advanced applications. Our research ranges from studies on the fundamental physics and mechanics of hydrogels, including robust adhesion of hydrogels on solid surfaces, swelling, and instabilities of hydrogels, to advanced applications such as smart wound dressing, hydrogel soft robots, microfluidics, and stretchable electronics.

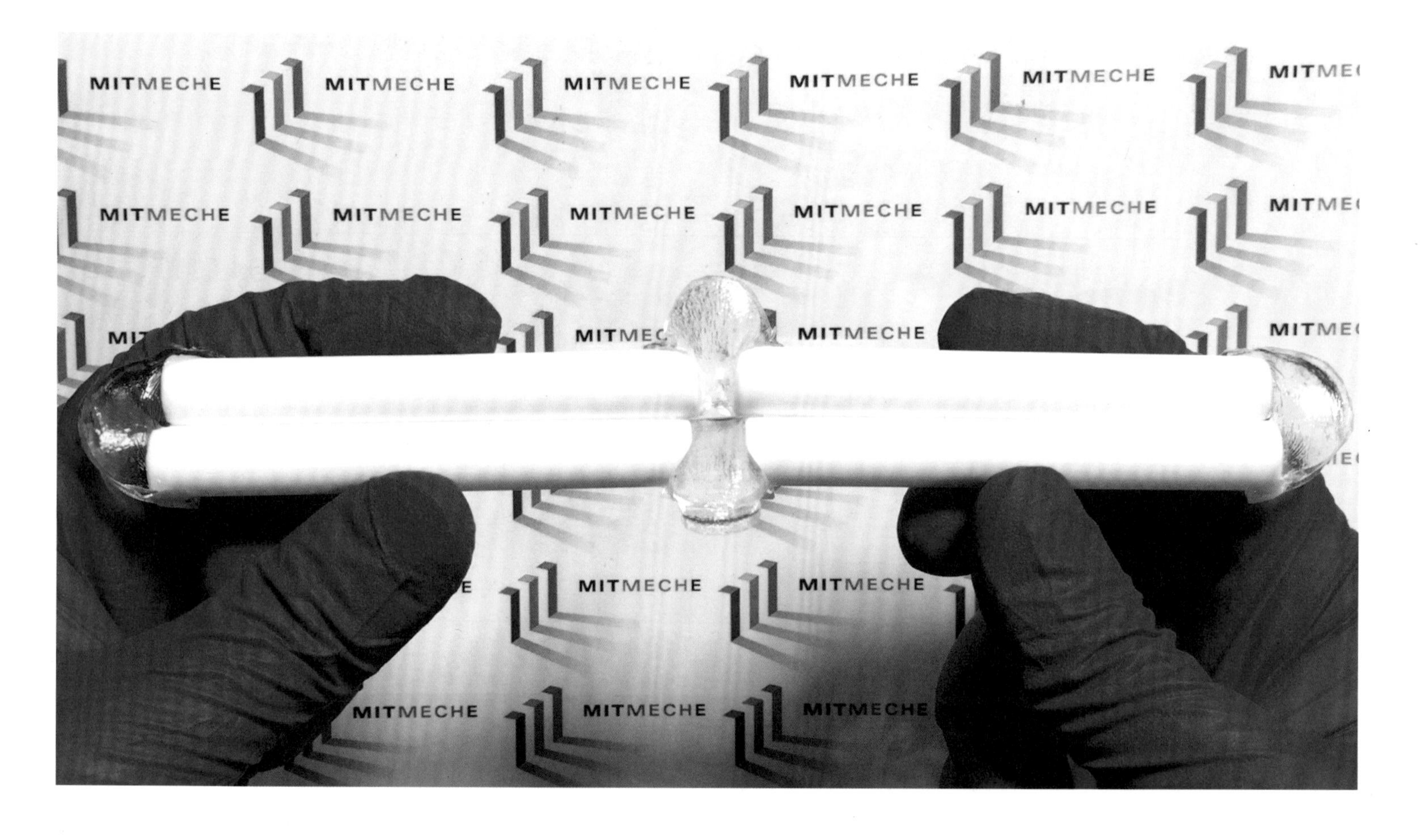

Figure 11.1 (left page)
Soft hydrogel joints. Tough hydrogels forming soft joints between ceramic rods. Soft and tough hydrogel joints robustly adhered onto ceramic rods are an engineered replica of biological joints made of tendon and bone. Such artificial soft hydrogel joints may find applications like soft robotic joints.

Figure 11.2
Tough wet adhesion of hydrogel. Engineered hydrogel being pulled away from a glass surface. The material shows a property called "tough wet adhesion," comparable to tendon-and-bone interface. The wavy-edge instability at the interface is a hallmark of strongly adhered soft material on a rigid surface

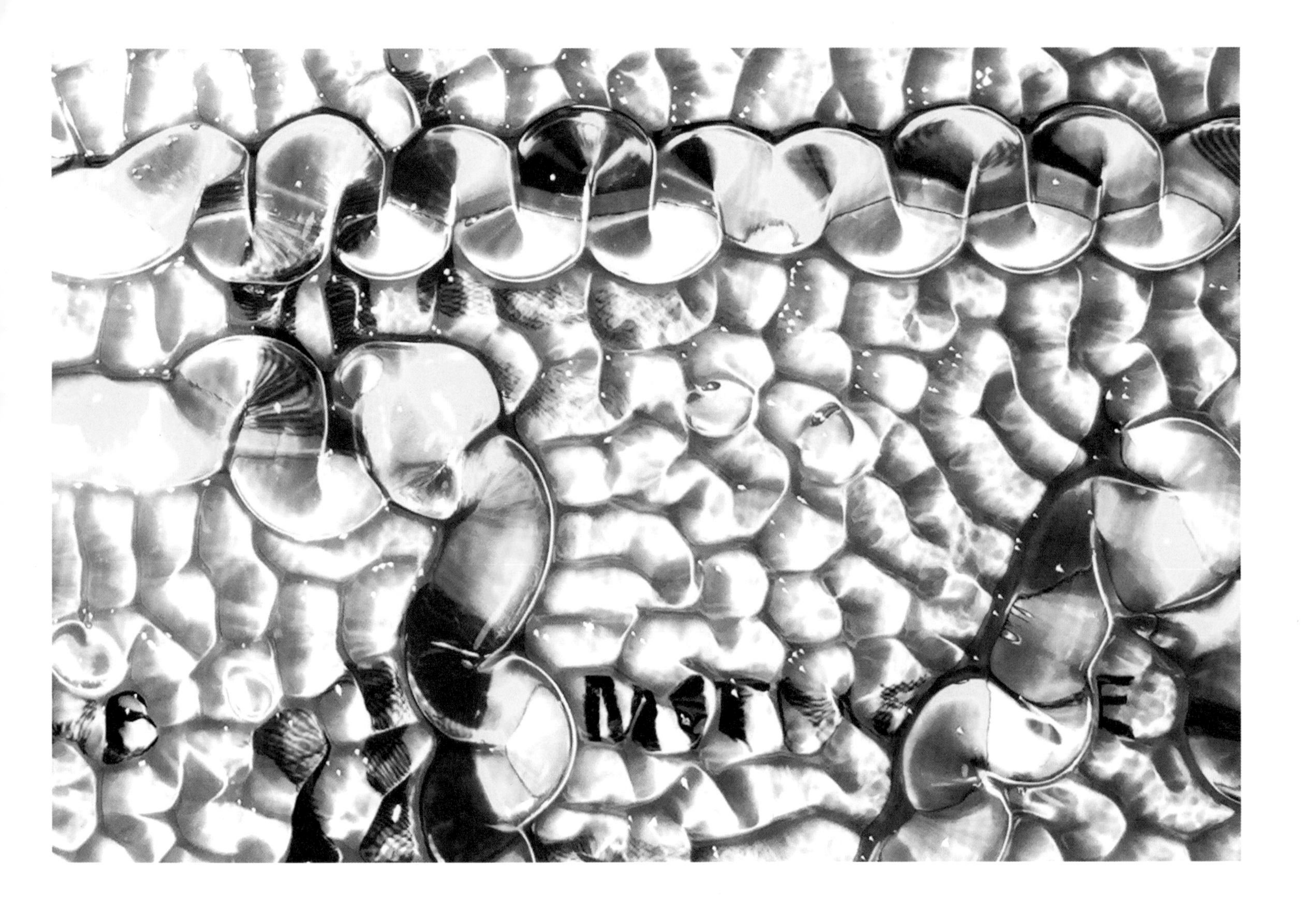

Figure 11.3 (top)
Swelling pattern of hydrogel. Intricate creasing and buckling patterns being formed by swollen hydrogels adhered on a glass surface. Swelling of hydrogels in water can form complex "instability" patterns when hydrogels are strongly adhered onto rigid surfaces.

Figure 11.4
Stretchable hydrogel fluidic channels. Stretchable hydrogel fluidics with multiple channels filled with different color dyes. Hydrogel fluidic channels can be stretched to more than 10 times their original length without losing fluidic functionality while allowing diffusion of liquid through hydrogels.

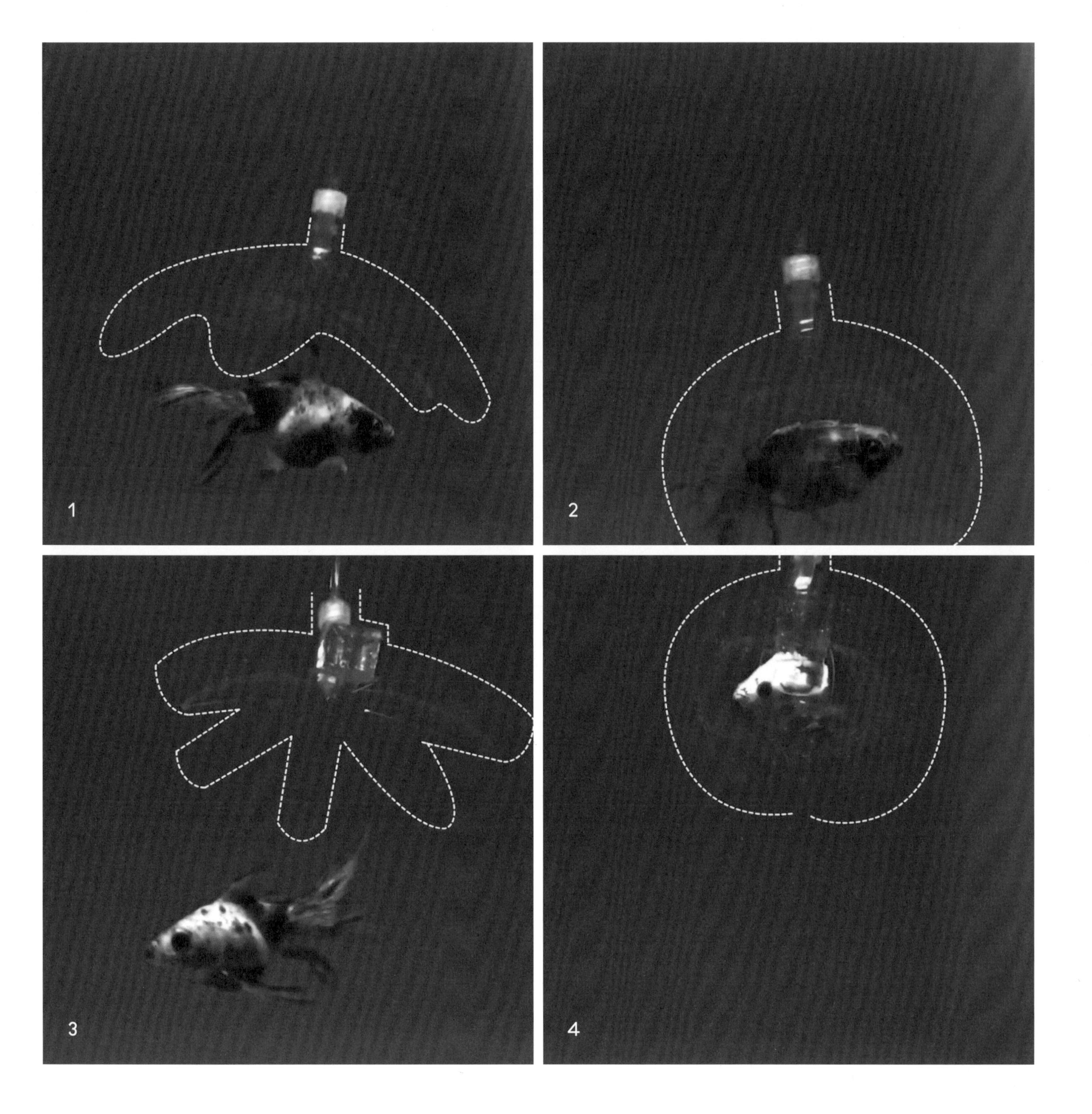

Figure 11.5
Transparent hydrogel robotic gripper catching goldfish. Tough hydrogels can be fabricated into rapidly actuating soft robotic structures with optical and acoustic invisibility in water due to their very high water contents. The transparent hydrogel robotic gripper can catch a live goldfish and release it without harming it owing to its soft and biocompatible nature.

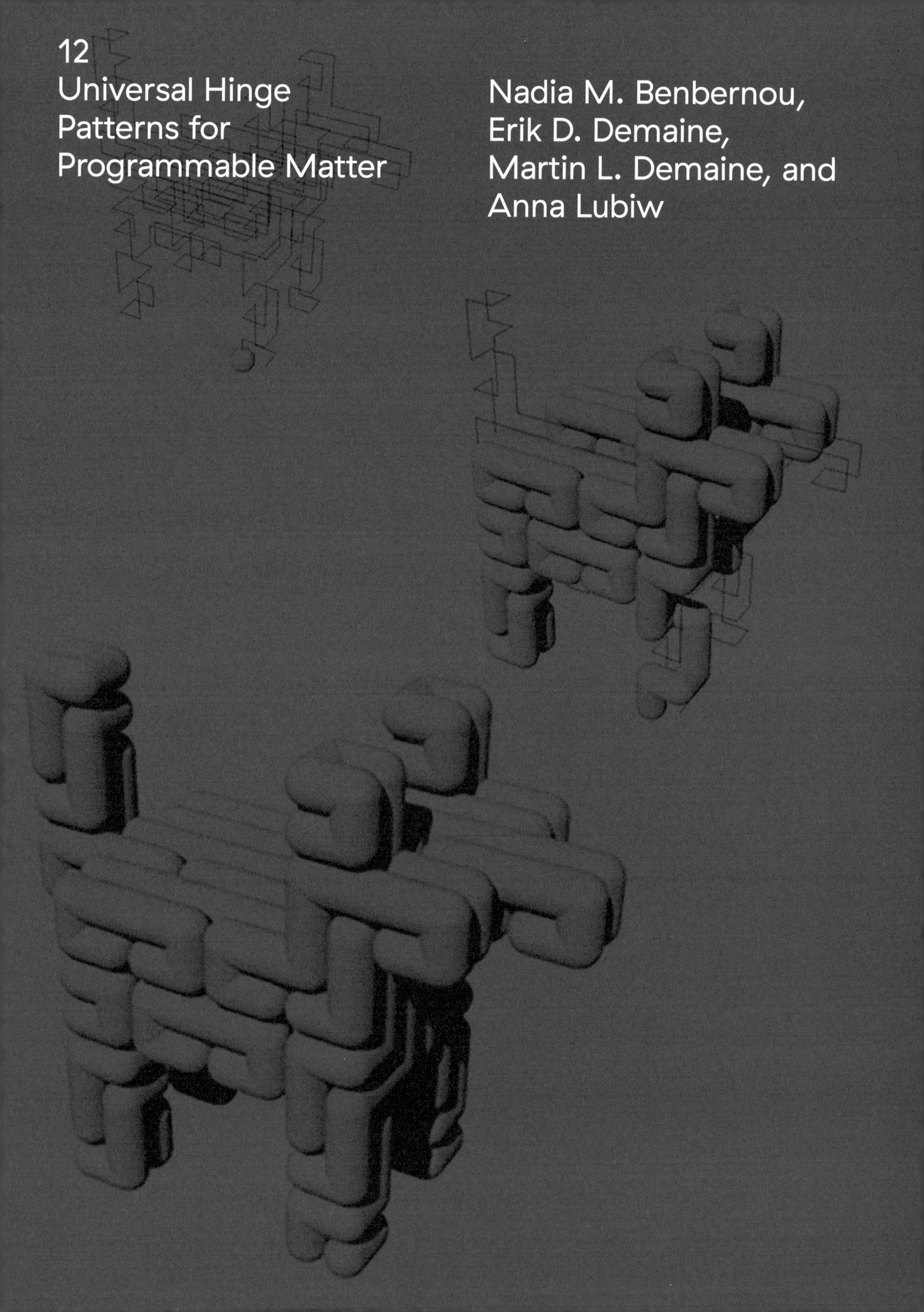

12

Universal Hinge Patterns for Programmable Matter

Nadia M. Benbernou, Erik D. Demaine, Martin L. Demaine, and Anna Lubiw

12
Universal Hinge Patterns for Programmable Matter

Nadia M. Benbernou,
Erik D. Demaine,
Martin L. Demaine, and
Anna Lubiw

Folding offers a powerful way to transform the shape of an object, often turning a one- or two-dimensional construction into a complex three-dimensional shape. Here we survey different ways in which folding has been used to implement *programmable matter*—material whose shape can be externally programmed.

1. SELF-FOLDING SHEETS

One approach to programmable matter, developed in an MIT-Harvard collaboration, is a self-folding sheet—a sheet of material that can fold itself into several different origami designs, without manipulation by a human origamist.[1, 2] (See figure 12.1.) For practicality, the sheet must consist of a fixed pattern of hinges, each with an embedded actuator that can be programmed to fold or not. Thus for the programmable matter to be able to form a universal set of shapes, we need a universal hinge pattern.

1.1. UNIVERSAL HINGE PATTERNS AND BOX PLEATING

The idea of a universal hinge pattern[3] is that a finite set of *hinges* (possible creases) suffices to make exponentially many different shapes. The main theoretical result underpinning the self-folding sheet presented above is that an $N \times N$ box-pleat grid suffices to make any polycube made of $O(N)$ cubes.[3] The *box-pleat grid* is a square grid plus alternating diagonals in the squares, also known as the tetrakis tiling. For each target polycube, a subset of the hinges in the grid serves as the crease pattern for that shape. Polycubes form a *universal* set of shapes in that they can arbitrarily closely approximate (in the Hausdorff sense) any desired volume.

Universal hinge patterns contrast with the usual goal in computational origami design: given a desired shape or property, find a crease pattern that folds into an origami with that shape or property. Examples include folding any shape,[4] folding approximately any shape while being watertight,[5] and optimally folding a shape whose projection is a

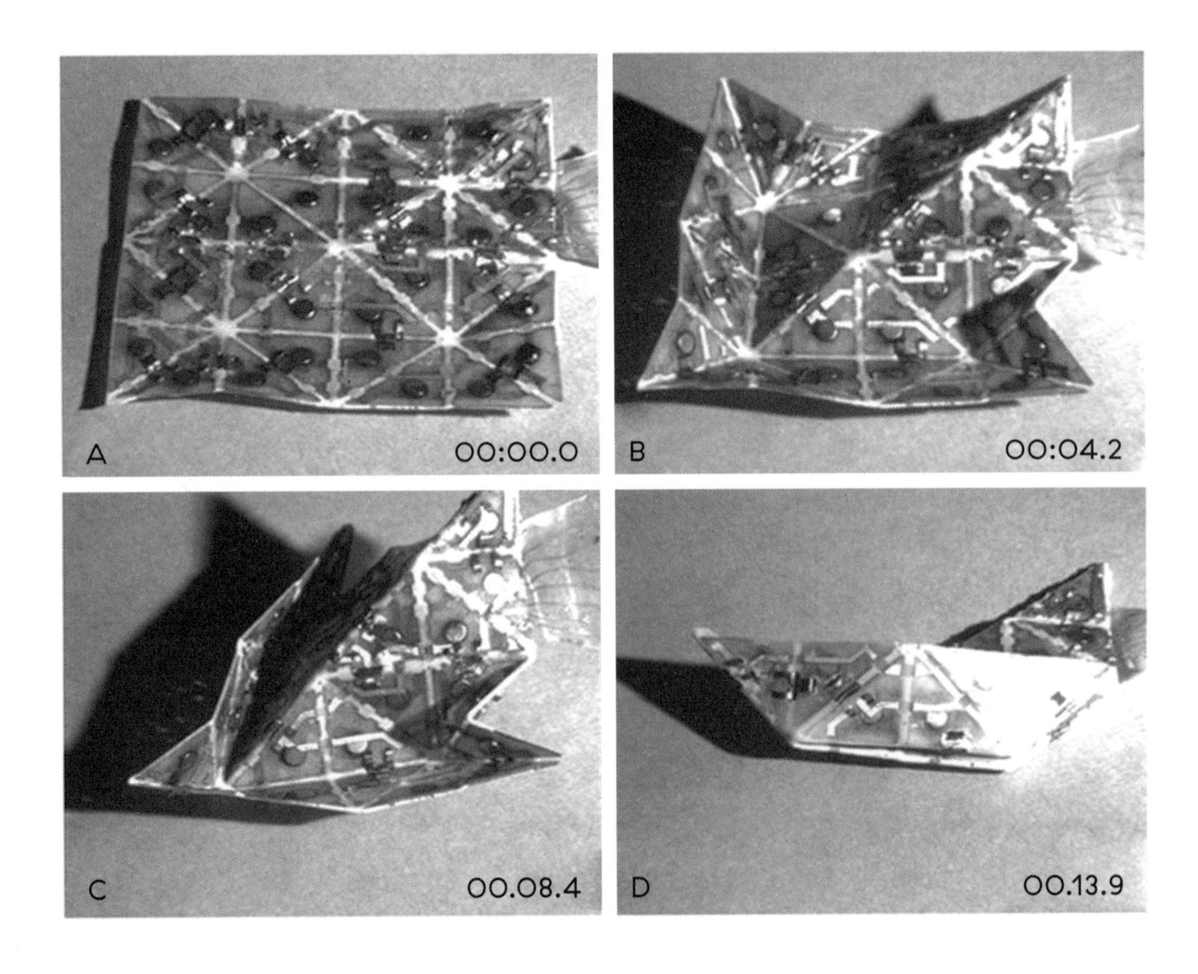

Figure 12.1 (a: left page, b: top) Programmable matter via sheet folding: universal 44-hinge pattern can fold in particular into a boat or paper airplane. Referenced from Hawkes et al. (see note 1), figs. 4 and 5.

desired metric tree.[6, 7] In all of these results, every different shape or tree results in a completely different crease pattern; two shapes rarely share many or even any creases. While useful for manufacturing, these results are difficult to adapt to the case of one sheet self-folding into many different shapes.

The box-pleated universal hinge pattern,[3] however, has some practical limitations that reduce practical application to programmable matter. Specifically, using a sheet of area $\Theta(N^2)$ to fold N cubes means that all but a $\Theta(1/N)$ fraction of the surface area is wasted. Unfortunately, this reduction in surface area is necessary for a roughly square sheet, as folding a $1 \times 1 \times N$ tube requires a sheet of diameter $\Omega(N)$. Furthermore, a polycube made from N cubes can have surface area as low as $\Theta(N^{2/3})$, resulting in further wastage of surface area in the worst case. Given the factor $\Omega(N)$ reduction in surface area, an average of $\Omega(N)$ layers of material come together on the polycube surface. Indeed, the current approach can have up to $\Theta(N^2)$ layers coming together at a single point.[3] Real-world robotic materials have significant thickness, given the embedded actuation and electronics, meaning that only a few overlapping layers are really practical.[1]

1.2. UNIVERSAL HINGE PATTERNS IN STRIPS

Recently, we introduced two new universal hinge patterns that avoid these inefficiencies, by using sheets of material that are long only in one dimension (strips).[8] Figure 12.2 shows the two hinge patterns: the *canonical strip* is a $1 \times N$ strip with hinges at integer grid lines and same-oriented diagonals, while the *zigzag strip* is an N-square zigzag with hinges at just integer grid lines. We show that any *grid surface*—any connected surface made up of unit squares on the 3D cube grid—can be folded from either strip. The strip length only needs to be a constant factor larger than the surface area, and the number of layers is at most a constant throughout the folding. Most of our analysis concerns (genus 0) *grid polyhedra*, that is, when the surface is topologically equivalent to a sphere (a manifold without boundary, so that every edge is incident to exactly two grid squares, and without handles, unlike a torus). We show that a grid polyhedron of surface area N can be folded from a canonical strip of length $2N$ with at most two layers everywhere, or from a zigzag strip of length $4N$ with at most four layers everywhere.

Figure 12.2
Two universal hinge patterns in strips. (A) A canonical strip of length 5. (B) A zigzag strip of length 6. The dashed lines are hinges. Referenced from Benbernou et al. (see note 8), fig. 1.

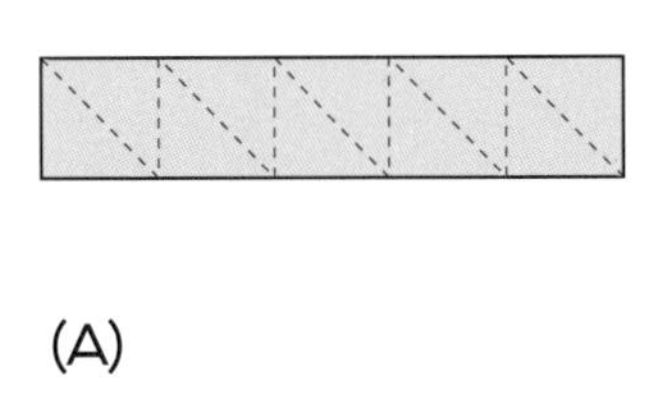

(A)

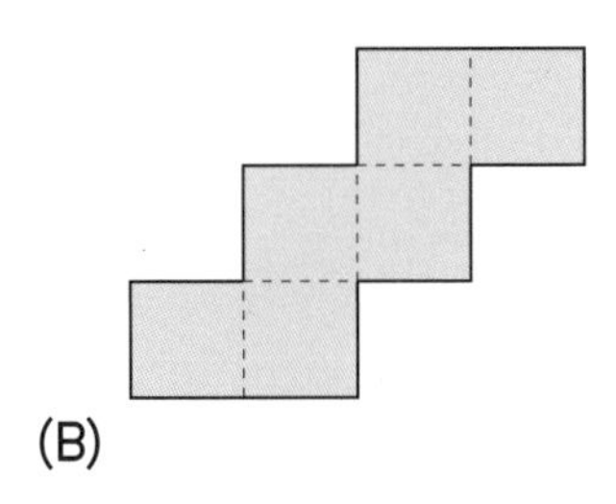

(B)

It is relatively easy to prove that these hinge patterns are universal, by showing that the strip can be navigated to go straight or turn left/right as desired. For example, figure 12.3 shows how a canonical strip can turn left or right; it goes straight without any folding. But a naive application of these gadgets would lead to less-efficient foldings. We show how to achieve the efficient bounds mentioned above by viewing the problem as a type of milling (similar to rapid-fabrication CNC milling/cutting tools) applied to the surface of the grid polyhedron.[8] In our situation, both length and number of turns in the milling tour are important, as both influence the required length of a strip to cover the surface. Thus we develop one algorithm that simultaneously approximates both measures, specifically achieving both a 2-approximation in length and an 8/3-approximation in number of turns. Such results have also been achieved for 2D pockets;[9] our results are the first we know for surfaces in 3D.

The improved surface efficiency and reduced layering of these strip results seem more practical than the previous approaches for programmable matter, though we have yet to put this to the test. In addition, the panels of either strip (the facets delineated by hinges) are connected acyclically into a path, making them potentially easier to control. One potential drawback is that the reduced connectivity makes for a flimsier device; this issue can be mitigated by adding tabs to the edges of the strips to make full two-dimensional contacts across seams and thereby increase strength.

We also show an important practical result for our strip foldings:[8] under a small assumption about feature size,[10] we give an algorithm for actually folding the strip into the desired shape, while keeping the panels rigid (flat) and avoiding self-intersection throughout the motion. Such a rigid folding process is important given current fabrication materials, which allow flexibility only in the creases between panels.[1] Our approach is to keep the yet-unused portion of the strip folded up into an accordion and then to unfold only what is needed for each move: straight, left, or right. Figure 12.4 shows two key cases for unrolling part of the *accordion* of a canonical strip. This approach may still suffer from practical issues, as it requires a large (temporary) accumulation of many layers in an accordion form.

1.3. FONTS

To further illustrate the power of strip folding, we designed a typeface, representing each letter of the alphabet and each of the ten digits by a folding of a 1 × X strip for some X, as shown in figure 12.5. The typeface consists of two fonts: with the unfolded font it is a puzzle to figure out each letter or number, while the folded font is easy to read. These crease patterns adhere to an integer grid with orthogonal and/or diagonal creases, but are not necessarily subpatterns of the canonical hinge pattern. This extra flexibility gives us control to produce folded half-squares as desired, increasing the font's fidelity.

We have developed a web app that visualizes the font.[11] Currently in development is the ability to chain letters together into one long strip folding; figure 12.7 shows one example.

2. 1D CHAIN ROBOTS

Another approach to programmable matter through folding, developed by the MIT Center for Bits and Atoms, is based on 1D chains.[12] Our inspiration here is nature's folding of proteins or DNA (as exploited in DNA origami).[13] 1D chains are also quite similar to long strips of 2D paper as considered above.

The theoretical underpinning is a theorem about *hinged dissections*: every polycube (or in 2D every polyomino) can be divided into a hinged chain of identical units, whose overall geometry is dependent only on the number of voxels/pixels in the desired shape.[14, 15, 16] Therefore one chain of blocks can fold universally into any polycube shape of the same size. By representing each cube in the polycube by a 2 × 2 × 2 grid of subcube pieces in the chain, we make the number of pieces in the chain just a factor of $2^3 = 8$ larger than the number of cubes in the polycube; and by using noncubical pieces, we can reduce this factor to 6.[15] (These factors are still substantially larger than the factor of 2 from strip folding, showing the power of working with thin material and allowing just two overlapping layers.)

This approach has been applied to building 1D self-folding chain robots at a variety of scales, specifically units of diameter 10 cm[12] and 1 cm.[17] These robots implement the hinges by pivoting about the center of the previous unit. Again it can be shown that the chain can rigidly deform into the desired shape, without collisions, essentially by pulling the chain into the folded form.[12]

3. CONCLUSION

Self-folding offers many exciting possibilities for realizing the dream of programming matter and making hardware as easy to program as software. An important tradeoff that deserves further study is the tradeoff between long/thin folding structures (such as the 1D chains of section 2 and the flat strips of section 1.2) and more uniformly two-dimensional folding structures (such as the square sheets of section 1.1). While the theory has explored this tradeoff at the extremes, with two dimensions perhaps winning in structural stiffness and one dimension winning in efficiency and number of layers, perhaps the optimal tradeoff in practice is some point in the middle. Alternatively, can we get the best of both worlds by achieving structural stiffness with a folding structure that is optimally efficient and uses few layers? We are excited to see how physical challenges will lead to further codevelopment of the theory and practice of folding structures.

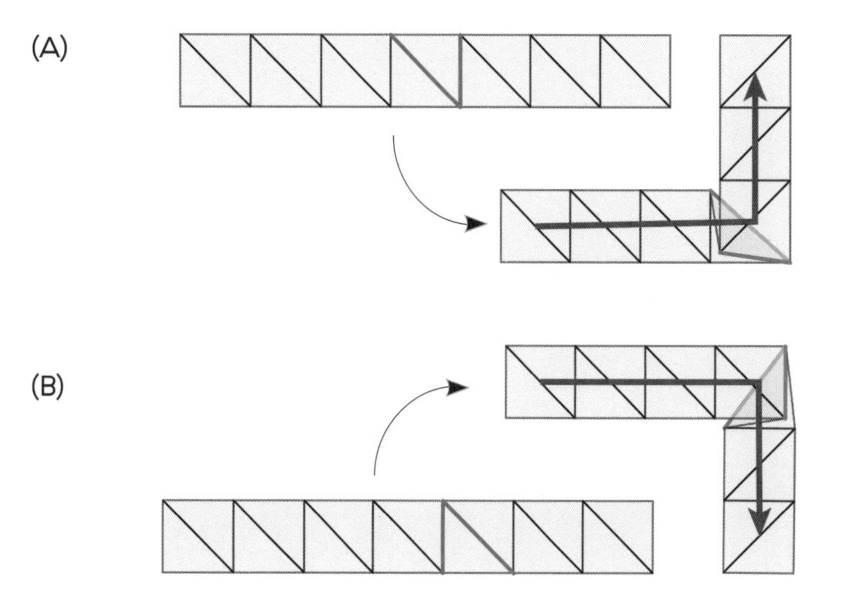

Figure 12.3
Left turn (A) and right turn (B) with a canonical strip. Referenced from Benbernou et al. (see note 8), fig. 3.

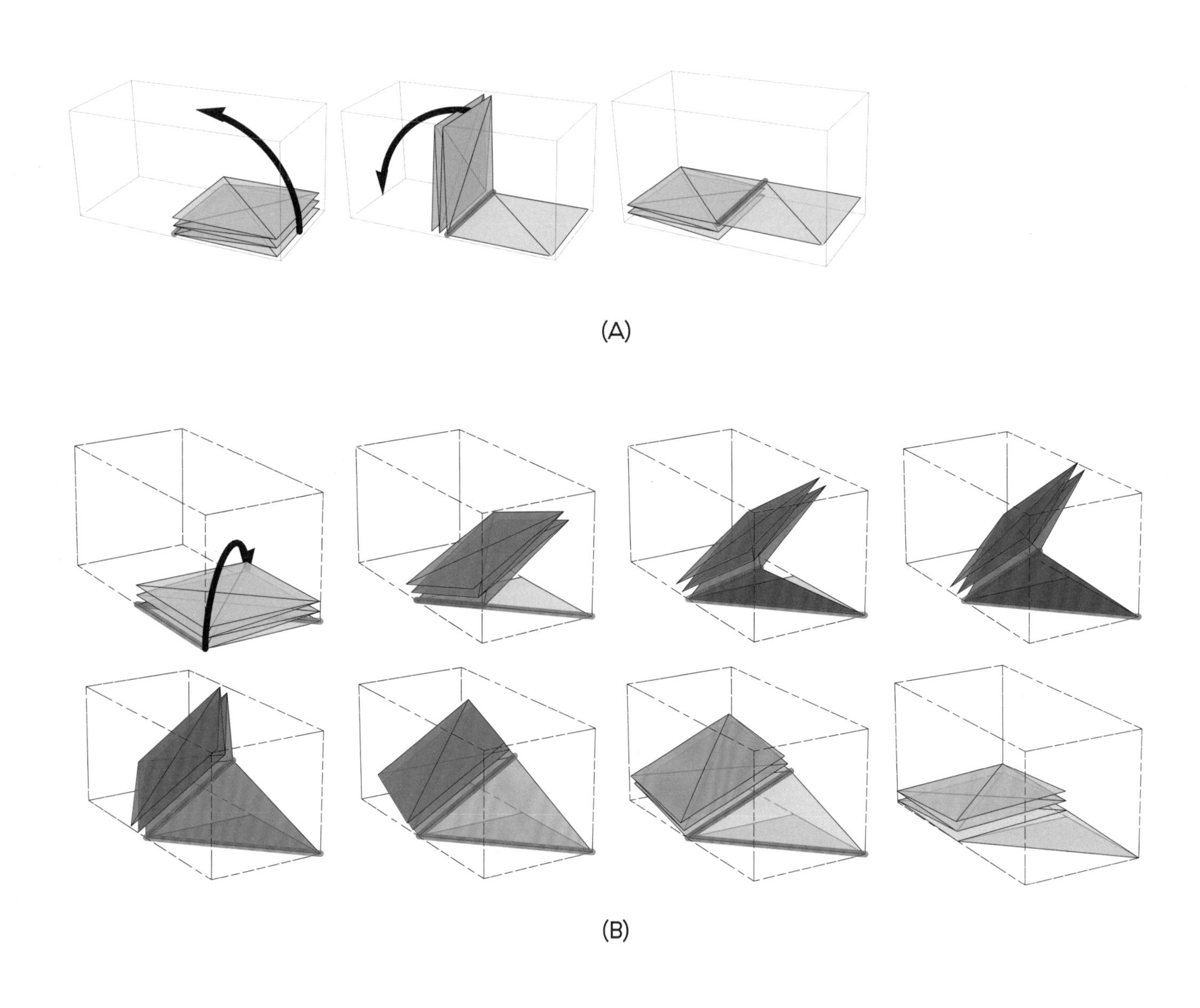

Figure 12.4
Rigid folding of canonical-strip accordion, face-up cases. (A) Straight. (B) Turn where *e23* is flat. Referenced from Benbernou et al. (see note 8), fig. 10.

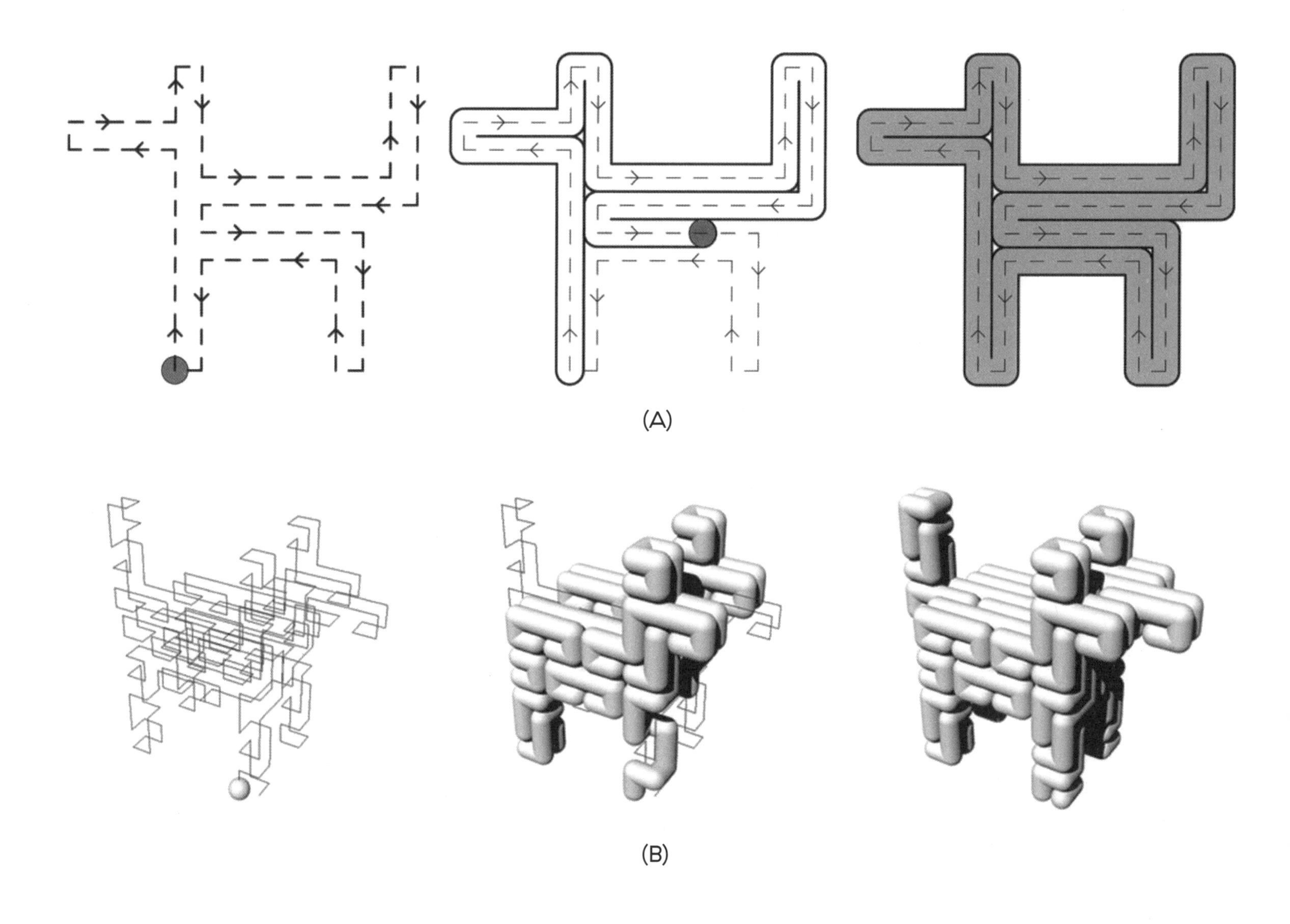

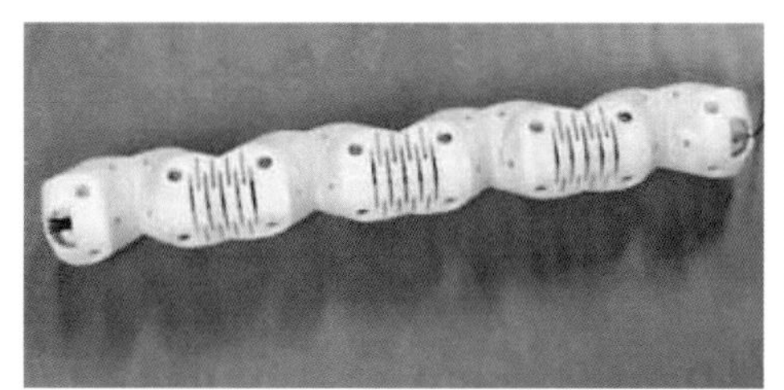
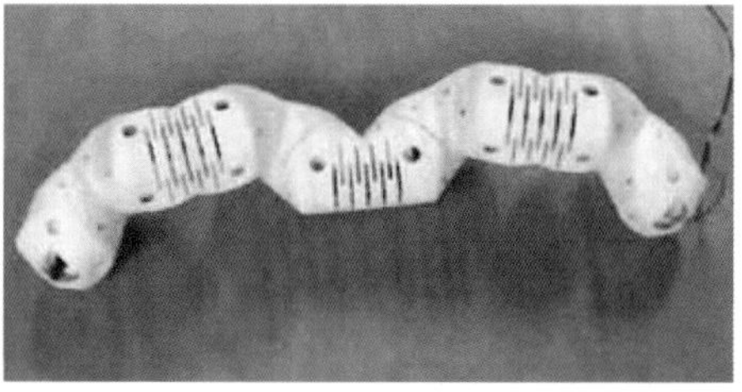
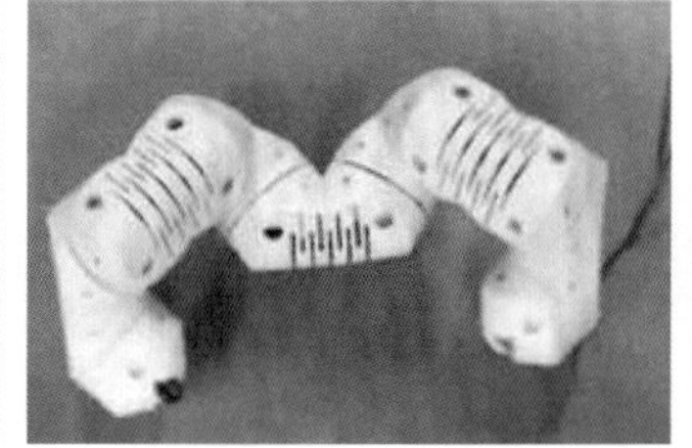
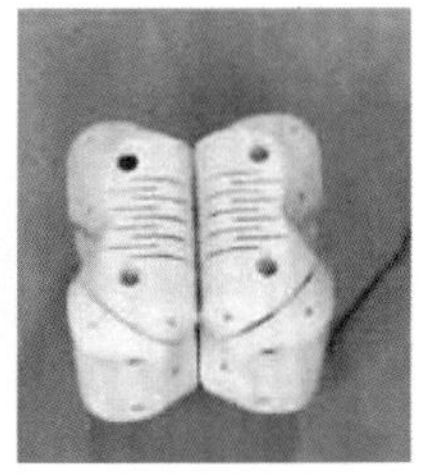

Figure 12.6
Hinged 1D chain robots (moteins): (A) designing any 2D shape; (B) designing any 3D shape; (C) folding a physical robot with units of diameter 10 cm. Referenced from Cheung et al. (see note 12), figs. 4 and 8.

Figure 12.5 (right)
Strip folding of a typeface, letters A–Z and numbers 0–9: unfolded font (top) and folded font (bottom), where the face incident to the bottom edge remains face-up. Referenced from Benbernou et al. (see note 8), fig. 2.

Figure 12.7 (botom)
An example of joining together a few letters from our typeface in figure 5. Unfolding (bottom), not to scale with folding (top). Referenced from Benbernou et al. (see note 8), fig. 15.

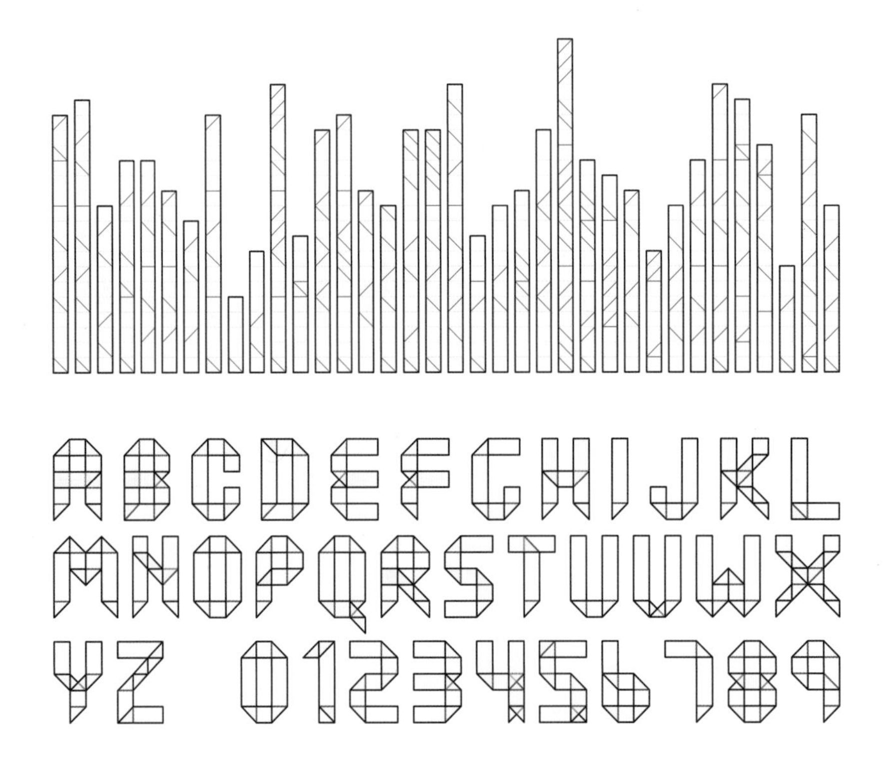

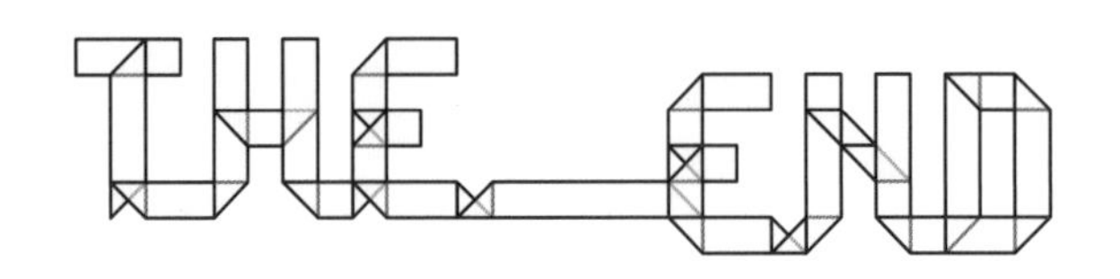

NOTES

1 E. Hawkes, B. An, N. M. Benbernou, H. Tanaka, S. Kim, E. D. Demaine, D. Rus, and R. J. Wood, "Programmable Matter by Folding," *Proceedings of the National Academy of Sciences* 107, no. 28 (2010): 12441—12445.

2 Byoungkwon An, Nadia Benbernou, Erik D. Demaine, and Daniela Rus, "Planning to fold multiple objects from a single self-folding sheet," *Robotica* 29, no. 1 (2011): 87—102. Special issue on Robotic Self-X Systems.

3 Nadia M. Benbernou, Erik D. Demaine, Martin L. Demaine, and Aviv Ovadya, "Universal Hinge Patterns to Fold Orthogonal Shapes," in Origami 5: Fifth International Meeting of *Origami in Science, Mathematics and Education*, ed. Patsy Wang-Iverson, Robert J. Lang, and Mark Yim (Boca Raton, FL: CRC Press, 2011).

4 Erik D. Demaine, Martin L. Demaine, and Joseph S. B. Mitchell, "Folding at Silhouettes and Wrapping Polyhedral Packages: New Results in Computational Origami," *Computational Geometry: Theory and Applications* 16, no. 1 (2000): 3—21.

5 Erik D. Demaine and Tomohiro Tachi, "Origamizer: A Practical Algorithm for Folding Any Polyhedron," manuscript, 2016.

6 Robert J. Lang, "A Computational Algorithm for Origami Design," in Proceedings of the 12th Annual ACM Symposium on Computational Geometry, Philadelphia, May 1996, 98—105.

7 Robert J. Lang and Erik D. Demaine, "Facet Ordering and Crease Assignment in Uniaxial Bases," in *Origami 4: Fourth International Meeting of Origami in Science, Mathematics, and Education*, ed. Robert J. Lang (Natick, MA: A. K. Peters, 2009),, 189—205.

8 Nadia M. Benbernou, Erik D. Demaine, Martin L. Demaine, and Anna Lubiw, "Universal Hinge Patterns for Folding Strips Efficiently into Any Grid Polyhedron," 2016, https://arXiv.org/abs/1611.03187

9 Esther M. Arkin, Michael A. Bender, Erik D. Demaine, Sandor P. Fekete, Joseph S. B. Mitchell, and Saurabh Sethia, "Optimal Covering Tours with Turn Costs," *SIAM Journal on Computing* 35, no. 3 (2005): 531—566.

10 Specifically, we need that every exterior voxel is contained in some empty 2 × 2 × 2 box.

11 http://erikdemaine.org/fonts/strip/

12 Kenneth C. Cheung, Erik D. Demaine, Jonathan Bachrach, and Saul Grith, "Programmable Assembly with Universally Foldable Strings (Moteins)," *IEEE Transactions on Robotics* 27, no. 4 (2011): 718—729.

13 Paul W. K. Rothemund, "Folding DNA to Create Nanoscale Shapes and Patterns," *Nature* 440 (March 2006): 297—302.

14 Erik D. Demaine, Martin L. Demaine, David Eppstein, Greg N. Frederickson, and Erich Friedman, "Hinged Dissection of Polyominoes and Polyforms," *Computational Geometry: Theory and Applications* 31, no. 3 (June 2005): 237—262.

15 Erik D. Demaine, Martin L. Demaine, Jeffrey F. Lindy, and Diane L. Souvaine, "Hinged Dissection of Polypolyhedra," in Proceedings of the 9th Workshop on Algorithms and Data Structures, volume 3608 of *Lecture Notes in Computer Science*, Waterloo, Canada, 2005, 205—217.

16 Timothy G. Abbott, Zachary Abel, David Charlton, Erik D. Demaine, Martin L. Demaine, and Scott Duke Kominers, "Hinged Dissections Exist," *Discrete and Computational Geometry* 47, no. 1 (2012): 150—186.

17 Ara N. Knaian, Kenneth C. Cheung, Maxim B. Lobovsky, Asa J. Oines, Peter Schmidt-Neilsen, and Neil A. Gershenfeld, "The Milli-Motein: A Self-Folding Chain of Programmable Matter with a One Centimeter Module Pitch," in Proceedings of the 2012 IEEE/RSJ International Conference on Intelligent Robots and Systems, 2012.

13

Microrobotics

Rob Wood

13
Microrobotics

Rob Wood

Artificial intelligence is typically focused on algorithms for perception and control of autonomous robots able to make and act on decisions in real environments. Our research is focused instead on the design, mechanics, materials, and manufacturing of novel robot platforms that make the perception, control, or action easier or more robust for natural, unstructured, and often unpredictable

environments. Key principles in this pursuit include bioinspired designs, smart materials for novel sensors and actuators, and the development of multiscale, multimaterial manufacturing methods. Representative examples of this philosophy include the creation of two unique classes of robots: soft-bodied autonomous robots and highly agile aerial and terrestrial robotic insects.

Figure 13.1
Prototype robotic insects created through a popup-book-inspired self-assembly process.

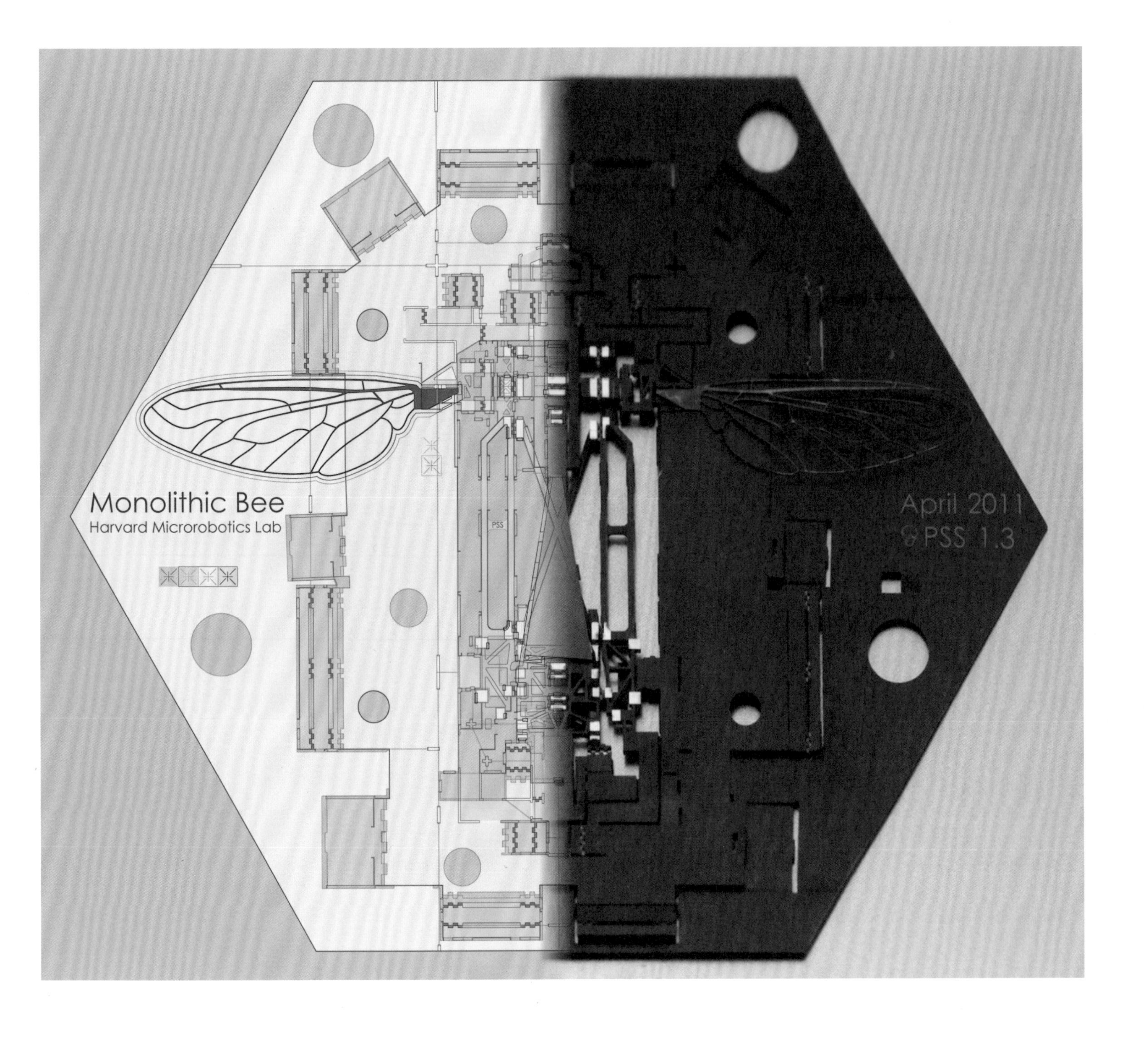

Figure 13.2 (left page)
A composite image of a CAD drawing (left) and preassembly image (right) of the popup assembly mechanism for insect-like robots.

14
Self-Reconfiguring Robot Pebbles

Kyle Gilpin and Daniela Rus

Self-Reconfiguring Robot Pebbles

Kyle Gilpin and
Daniela Rus

ABSTRACT

In this paper, we present the Pebbles, a modular self-reconfiguring system of 1 cm cubes. Each Pebble node contains a microprocessor and four switchable electropermanent magnets that enable bonding, point-to-point communication, and power transfer between nearest neighbors in a cubic lattice. These electropermanent magnets provide a robust communication link between nodes in contact with one another, and they allow rapid, on-the-fly topology changes.

We detail the communication hardware and protocols, and we describe robust routing algorithms that operate independent of the topology of a multi-Pebble system, adapting to topological changes. The resulting hardware-software infrastructure enables desktop mobility experiments using this plug-and-play platform.

1. INTRODUCTION

The ability to form diverse objects with specific form and material properties out of a set of standardized modules would be the ultimate universal toolkit. Such a system would be immensely useful for an astronaut on an interplanetary mission or a scientist isolated at the South Pole. Even for the average mechanic or surgeon, the ability to form arbitrary, task-specific tools would be immensely valuable while inspecting and/or working in tight spaces. We envision a bag of these intelligent particles that one can shake in order to form a goal object. The modules contained within the bag would first crystallize into a regular structure and then self-disassemble in an organized fashion to form the goal shape. In the same way a sculptor removes the extra stone from a block of marble to reveal a statue, our system subtracts modules from the crystallized block to form the goal structure. After it is sculpted, the object can be retrieved by reaching into the bag and brushing off the extra modules. When the user is done with the object, he returns it to the bag where it disintegrates back to its component modules. The free modules are incorporated into the universal structure of the bag and can be reused at a later time.

The goal shape may be a robot built for a specific task (a snake to pass through a tunnel or a rolling belt to quickly cover open ground), or an object designed for a particular job (such as a wrench or surgical instrument). In all cases, the resulting structure, due to the intelligent nature of its component parts, is imbued with unique properties. An object formed from programmable matter could provide real-time, in-situ feedback on internal stresses; disintegrate at a controlled rate; or dynamically change its rigidity to correspond with the task at hand.

In the quest to create a universal programmable matter system, there are many challenges which must still be overcome:

- due to their size, grain-size modules must be able to self-assemble;

- due to power and storage-size constraints, shape description information cannot be broadcast in million-module systems;

- fabrication of 3D structures that are electrically and mechanically active is difficult;

- small, easily switchable, high-strength connectors do not exist;

- supplying power to millions of grain-sized modules is challenging;

- debugging massively parallel systems is not supported by current development tools.

In this chapter we present a case study for self-reconfiguration by "sculpting." Self-reconfiguring robots have the ability to modify their geometry without human intervention. They take on the body shape most suited for the task at hand. Imagine a hardware voxel that can connect with other voxels and then

Figure 14.1
The Pebble nodes are built by wrapping a flex circuit (a) around a brass frame (b). A capacitor (d) provides energy to four electropermanent magnets (c) which enable mechanical bonding, communication, and power transfer between neighboring nodes.

move to connect with different voxels. In this way, they can turn themselves into desired robot shapes. Say you want to create a car with a changeable trunk. The trunk could be large for shopping or small for city parking. By making the trunk self-reconfiguring, we can achieve the goal. Alternatively, suppose a factory robot needs a screwdriver from a shelf, and the screwdriver is too high to reach. With self-reconfiguration capabilities, the robot could grow an extra-long arm to get that screwdriver.

Self-reconfiguration by sculpting, which we call self-disassembly, is a subtraction approach to shape formation. Given a block of modules, the desired shape is achieved by identifying which modules in the aggregate are part of the structure and then removing the extra modules. We focus on programmable matter systems consisting of collections of physically connected robotic modules that have the ability to communicate with and bond to their immediate neighbors. Using this functionality, we show that a system of these modules is capable of autonomously creating user-specified goal shapes through a process of self-assembly followed by self-disassembly.

To create self-reconfiguring robots, we need to design hardware modules capable of connecting with each other and achieving internal geometry changes autonomously. We also need supporting communication and control algorithms for achieving the desired shape and behavior. Here we describe the Pebbles, a hardware platform that enables self-reconfiguration by self-disassembly. Each Pebble is an autonomous 1 cm cube (see figure 14.1) that contains a microprocessor and four network interfaces. The Pebbles are held together

using switchable electropermanent (EP) magnets that can be turned on or off under software control. These magnets provide secure mechanical and electrical connections between neighboring nodes, but they are not strong enough to prevent the user from breaking their mechanical bonds in order to force a topology change. In addition, these magnetic connectors enable communication and power transfer between nodes, eliminating the need for batteries. Another advantage of the Robot Pebbles system is the mechanical simplicity. The Pebbles contain no moving parts, making them simpler to manufacture, less expensive, more reliable, and easier to miniaturize than more traditional modular robotic systems, which often rely on complex, gendered mechanical latches to connect neighboring modules. The simplicity of their shape also ensures that the modules passively self-align, eliminating the need for precision sensing and motion control.

The Robot Pebbles system employs a set of distributed algorithms to perform two discrete steps: (1) rely on stochastic forces to self-assemble a close-packed crystalline lattice of modules, and (2) use the process of self-disassembly to remove the extra material from this block, leaving behind the goal structure. By approaching shape formation in this manner, we hope to speed up the entire process, eliminate any global information that must be distributed throughout the system, and simplify the computing requirements of each module.

As the individual modules in self-reconfiguring and programmable matter systems continue to shrink in size, it will become increasingly difficult to actuate and precisely control the assembly process. In particular, designing modules capable of exerting the forces necessary to attract their neighbors from significant distances will be challenging. Instead, these systems may find assembly and disassembly much simpler when driven by stochastic environmental forces. The Pebble modules presented in section 3, which are able to latch together from distances approximately 2035% of the module dimensions, could easily take advantage of these stochastic assembly mechanisms to form an initial structure. Our particular system also relies on external forces to carry the unused modules away from the final shape. In our system, this force is often gravity, but it could also be vibration, fluid flow, or the user reaching into the bag of smart sand particles to extract the finished object.

1.1. ADVANTAGES OF SELF-DISASSEMBLY

Designing an electromagnetic module capable of exerting the force necessary to attract or repel other modules from a distance greater than the size of a module has proven challenging. Shape formation with electrostatic or magnetic modules is more feasible when driven by stochastic forces, so that the actuators only need to operate over short distances.

Traditional self-assembling systems aim to form complex shapes in a direct manner. As these structures grow from a single module, new modules are only allowed to attach to the structure in specific locations. By carefully controlling these locations and waiting for a sufficiently lengthy period of time, the desired structure grows in an organic manner. In contrast, our system greatly simplifies the assembly process by initially aiming to form a regular crystalline block of fully connected modules. We make only limited attempts to restrict which modules or faces are allowed to bond with the growing structure. These restrictions are only to ensure that we achieve a regular structure. After we form this initial block of material, we complete the shape formation process through self-disassembly and subtraction of the unwanted modules.

Subtraction has one distinct advantage over existing self-assembly techniques. Subtraction does not rely on complicated attachment mechanisms that require precise alignment or careful planning. Subtraction excels at shape formation because it is relatively easy, quick, and robust. One caveat is that

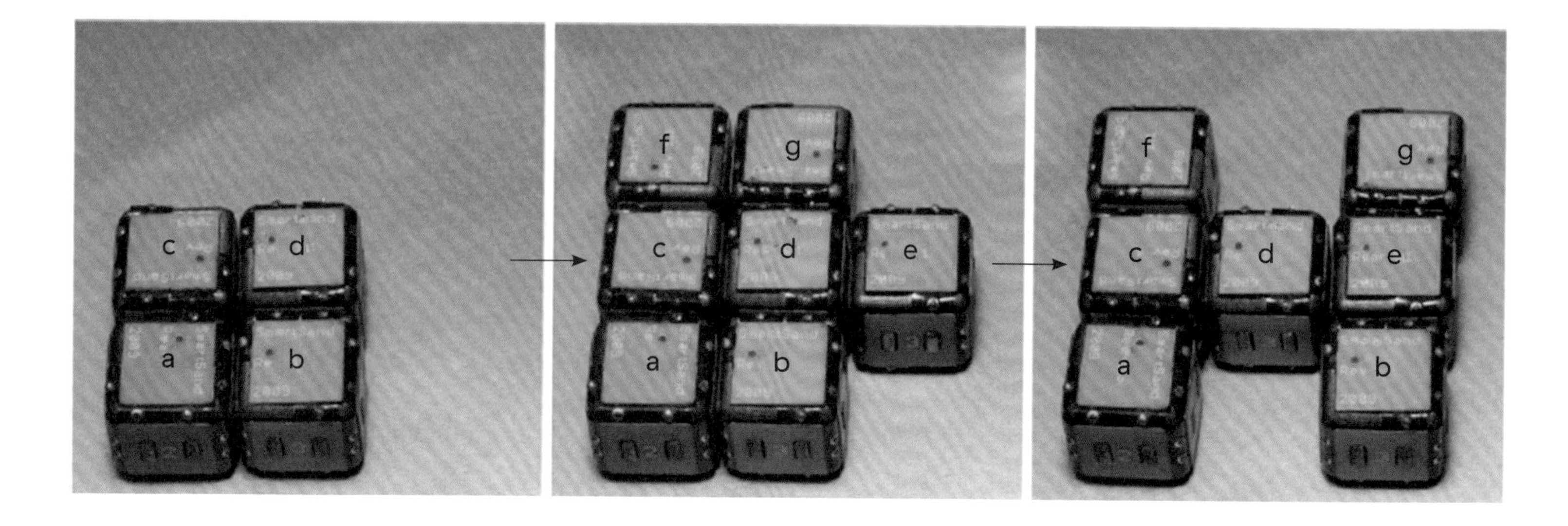

Figure 14.2
A system of Pebbles can be easily and quickly reconfigured on the fly by simply adding Pebbles or moving a Pebble from one location to another.

the initial mass of material must be preassembled, but this can be done in an automated way. Our modules, due to symmetry in their magnet-endowed faces, are rotation-invariant within a plane, so an inclined vibration table should function well to rapidly assemble large, regular sheets of modules. In the future, we envision deploying modules, such as spheres, which naturally pack 3D space efficiently.

It is also worth noting that as the individual modules in self-reconfiguring and programmable matter systems continue to shrink, it will become increasingly difficult to actuate and precisely control the assembly process. In particular, designing modules capable of exerting the forces necessary to attract their neighbors from significant distances will be challenging. Instead, these systems may find assembly and disassembly much simpler when driven by stochastic environmental forces. The modules addressed in this paper, which are able to latch together from distances approximately 2035% of the module dimensions, could easily take advantage of these stochastic assembly mechanisms to form an initial structure. Our particular system also relies on external forces to carry the unused modules away from the final shape. (The EP magnets that we employ cannot both attract and repel.) In our system, this force is often gravity, but it could also be vibration, fluid flow, or the user reaching into the bag of smart sand particles to extract the finished object.

1.2. RELATED WORK

Our research builds on previous work in programmable matter, self-assembly, and self-reconfiguring robotics that grew out of the modular robotics field. The first paper in this field was presented at the International Conference on Robotics and Automation in 1988 by Toshio Fukuda and S. Nakagawa titled "Dynamically Reconfigurable Robotic System."[1] In their paper, Fukuda and Nakagawa describe the abstract concept of a reconfigurable robotic system that can assume different shapes. They envisioned a robot system composed of different types of modules that can combine to accomplish a variety of tasks. Over the past twenty years, modular robotics research has developed many facets: hardware design; planning and control algorithms; the tradeoff between hardware and algorithmic complexity; efficient simulation; and system integration.

Modular robotic systems can be described and classified using several axes and properties. In what follows, we choose the traditional route of classifying modular robotic systems by the geometry of the system: chain, lattice, truss, or free-form. For a more detailed history of the modular robotics field, see Yim et al.[2] and Gilpin and Rus.[3]

Our research builds on work in programmable matter, self-assembly, and from the self-reconfiguring robotics community. Several interesting program-driven stochastic self-assembly systems are under active development.[4, 5, 6] Like the Robot Pebbles we propose, these systems rely on rigid particles for shape formation. Other approaches to shape formation rely on modules with internal degrees of freedom that are able to modify their topology in some way.[7, 8, 9, 10, 11] There are also systems[12, 13, 14] that fall between these two extremes in which neighboring modules join to accomplish relative actuation.

Other research has focused more directly on the concept of programmable matter—systems composed of small, intelligent modules able to form a variety of macroscale objects in response to external commands or stimuli. One particular system[15] uses rigid cylindrical modules adorned with electromagnets to achieve 2D shape formation. Some theoretical research has previously investigated the use of submillimeter intelligent particles as 3D sensing and replication devices.[16] More recent developments are utilizing deformable modules[17] as a way to realize programmable matter. Others have investigated using biology as the basis for programmable sheets able to fold into a different structure.[18] Finally, a system has been proposed[19] that has no actuation ability, but demonstrates what may be termed "virtual" programmable matter through the use of 1000 distributed modules to form an intelligent paintable display capable of forming text and images.

Some past research has focused specifically on self-disassembling systems as a basis for shape formation.[20] This past work was based on large modules (45 cm cubes) with internal moving parts. Additionally, the modules lacked symmetry, so they had to be assembled in a particular orientation. The work presented in this paper is an outgrowth of the Miche system,[20] but we have reduced the module size, eliminated all moving parts, and added symmetry to allow for arbitrary orientations. Finally, the system presented in this paper shows promise as both a self-disassembling and self-assembling system.

2. PEBBLE MODULE HARDWARE

The Pebble modules are a system of rapidly reconfigurable networked microcomputers. Unlike most wireless communication systems, the Pebbles communicate using magnetic induction, which ensures high directivity and minimal interference. Modules can only communicate if they are nearly touching, and a module always knows from which of its neighbors a message originated. The modules are magnetically bonded to their neighbors, providing a positive lock that remains easy to break when the user wishes reconfigure the system. The modules share power with their neighbors, so only a single module needs to be connected to an external power supply. This eliminates the need for internal batteries that need to be replaced or recharged on a regular basis. We have previously provided a general overview of the Pebble hardware,[21] but this work provides a significantly more detailed explanation of the communication hardware, protocols, and algorithms.

The Pebble modules (see figure 14.1) are 12 mm cubes that each contain a microprocessor, four EP magnets, power regulation circuitry, and an LED. The EP magnets give the system its unique functionality. Like a traditional electric magnet, they can be turned on and off, but like a permanent magnet, they retain their state without consuming any power. Power is only required to switch an EP magnet from off to on and vice versa. The EP magnets are used for intermodule bonding, power transfer, and communication. The four EP magnets all reside in the same horizontal plane. Consequently, the current instantiation of the system is limited to forming planar cubic lattices. Space and symmetry constraints prohibit 3D lattices for the current 1 cm cube. Doubling the size to 2 cm cubes would create space for axially symmetric EP magnets on all six faces, thereby enabling a 3D system.

2.1. ELECTROPERMANENT MAGNET BONDING

The EP magnets are composed of two magnetic cylinders laid side by side and wrapped in an actuation coil. One of these cylinders is composed of NdFeB, which is a high coercivity magnetic material making it difficult to repolarize. The other cylinder is alnico, a low-coercivity, easy-to-repolarize magnet. When the coil is temporarily energized by a pulse of current, the polarization of the alnico is permanently flipped, but the NdFeB is unaffected. After the pulse has subsided, if the alnico's polarity aligns with that of the NdFeB magnet, the two fields reinforce each other and the device attracts other ferromagnetic materials. If instead the current pulse is reversed, the alnico's polarization ends up opposite that of the NdFeB magnet, and the two fields cancel each other. With the two fields canceling each other, the device does not attract other materials. When two adjacent EP magnets are mated together, they can withstand over 3N of normal force before being pulled apart.[22, 23, 24] By reducing the current pulse width or magnitude, it should be possible to reduce the bond strength.[25]

2.2. POWER TRANSFER

The Pebbles form a power distribution network in order to eliminate the need for batteries. The hardware testbed supplies one module with power, which is then transferred to that module's neighbor through the EP magnet pole pieces. Power is passed from module to module though DC conduction, not induction. Within a single EP magnet assembly, the north and south poles are electrically isolated from one another by a thin Parylene coating that is applied during the manufacturing process. This insulation allows 20V to be applied across the poles. The external faces of the EP magnets are uncoated so that when two EP magnets come into contact they form a low-resistance (0.3 ohm) path between neighboring modules.

2.3. COMMUNICATION HARDWARE LAYER

The EP magnets are also used for intermodule communication. When two EP magnets in neighboring modules are mated north-to-south, they form a 1:1 transformer whose core is composed of the four NdFeB and alnico rods. A voltage pulse across the transmitting coil induces a corresponding pulse in the opposite coil. For communication, we use 1μs pulses with a polarity that reinforces the magnetic bond strength.

3. PEBBLE SYSTEM CAPABILITIES

The Robot Pebbles can form arbitrary shapes through a two-step process. First, a loose collection of modules self-assemble into a regular crystalline structure. Once this initial block of material is formed, the system sculpts itself into an arbitrary shape using self-disassembly. In contrast with self-assembly structure that grow from a single module, self-disassembly greatly simplifies the assembly process by initially aiming to form a regular crystalline block of fully connected modules. We make only limited attempts to restrict which modules or faces are allowed to bond with the growing structure. These restrictions ensure that we achieve a regular structure. After we form this initial block of material, we complete the shape formation process through self-disassembly and removal of the unwanted modules.

3.1. SELF-ASSEMBLY

Pebble self-assembly is realized stochastically, for example on a vibration table. During the self-assembly process, we want to ensure that no gaps are formed in the growing structure. If we allow new modules to be added at any location on the growing structure, it will be easy to create gaps in the structure that are theoretically difficult and practically impossible to fill. To avoid holes in the self-assembled structure, all modules know the location of a special module called the root module. Second, once each module is added to the structure, it can determine its (x,y) position.

The self-assembly process begins as the free module receives power when it comes into contact with a module that is already a part of the crystallized structure. Immediately, the free module queries its neighbor to determine its location. Based on this location, the module then constructs a root vector pointing back to the root module. The vector may have *x* and *y* components. The new module permanently bonds with the structure if it detects that it has neighbors in both the *x* and *y* directions of the root vector, if they exist. If the new module does not detect neighbors in the appropriate locations, it informs whatever neighbors it is contacting, and they deactivate their connectors, allowing the pebble to continue moving to a new location. The module will lose power so the next time it contacts the structure, the self-assembly process will restart.

3.2. SHAPE FORMATION

Once an initial block of sculptable host material has been assembled, we must convey the desired final shape to the system. There are two broad approaches to this problem. The user may either convey the desired shape to the system manually, or employ the intelligence embedded in each of the modules to automate the shape distribution process. Pebbles support three approaches: inclusion chains, magnification, and replication.

One approach to shape distribution is to specifically inform each module whether it should be included in the goal structure. This is the default approach to shape formation. The approach can be used to form multiple shapes that are contiguous or separated by any number of unused modules. This flexibility allows the sculpting of objects with interlocking subparts and internal degrees of freedom. The shape distribution algorithm operates by transmitting a single inclusion message to each module in the initial structure that is destined to be a part of a goal shape. Modules not included in any goal shape do not receive an inclusion message. Modules assume, by default, that they are not included in the final structure. Inclusion messages originate from the sculptor's PC and, once in the structure, create and follow a dynamic path constructed from a constant amount of information per message. The algorithm avoids encoding the detailed path that each inclusion message must follow, and it avoids flooding the system with inclusion messages.

A second approach to shape formation that we wish to explore is magnification. Given the description of an initial shape to be formed and some magnification factor, the modules in the initial host material should create a magnified version of the original. The initial shape description may be distributed using the approach described above or using a sensing-based approach.

There are several unique problems to be solved when designing the magnification algorithms. Primarily, we must determine how the modules in the original structure go about informing other modules in the host material that they should be a part of the magnified shape. In our system, there is no centralized controller to inform all of the modules composing the magnified shape of their status. Instead, this information must be generated and dispersed in a distributed manner. During the process, we must ensure that a minimal number of messages are exchanged. The magnification algorithms must also be able to operate in initial host material that contains holes or broken communication links between neighboring modules. In the case of broken communication channels, the modules in the structure should still successfully distribute a description of the magnified shape to all relevant modules. In the case of holes in the initial host material, the magnification algorithm should approximate the finished shape as accurately as possible. We also need the capability to inform some modules that are a part of the original structure that they are not a part of the magnified structure. This scenario arises when the original structure contains holes or concavities. As the original structure is magnified, these voids shift position within the initial host material and may land on modules that were a part of the initial structure.

A third approach, shape formation by replication, works as follows. Given an initial description of a shape (created either by the inclusion approach or by magnification), we intend to create as many replicas of the shape as possible. The total number of replicas should only be limited by the amount of initial host material available for sculpting. We intend to perform the replication process with distributed algorithms that minimize the number of messages between neighboring modules. In implementing the replication algorithm, we must deal with several challenges. First, we must ensure that the algorithm can handle the case in which a complete replica of the initial shape does not completely fit inside the confines of the initial block of host material. Second, the replication algorithm must, assuming it places replicated shapes in contact with their neighbors, ensure that neighboring replicas separate from one another. Third, as with the magnification algorithm, the replication algorithm should be robust to missing intermodule communication links. As before, these could be caused by hardware malfunctions or modules that are missing from the initial close-packed structure.

3.3. DISCONNECTION

One of the most challenging aspects of shape formation with the Robot Pebbles is the intermodule disconnection process that must occur after all modules know whether to remain as part of a finished object or to disconnect completely. The complexity in this disconnection process is caused by the fact that a module loses its ability to function once it breaks its connection with the neighbor supplying it with power. Furthermore, all modules that are dependent on that module for power will also lose power and will not be able to break additional magnetic bonds.

A tree can be used to represent how power is transmitted through an initial block of modules. The module connected to the user's PC, because it is also connected to an external power supply, is the root of this power transfer tree. Every other module in the tree has one parent that supplies the module with power. Conversely, every module to which a module supplies power is a child. The key concept is that, although different neighbors could also supply it with power, a module will never lose power as long as it is connected to its parent. We have designed and implemented an algorithm which ensures that an initial block of material can disassemble correctly, disconnecting bonds that should be broken and keeping those that should be preserved. Intuitively, the disconnection algorithm operates by ensuring that a module has no children before disconnecting from its parent.

4. DISCUSSION

Imagine a future in which robots are so integrated into the fabric of human life that they become as common as smart phones are today. The field of robotics has the potential to greatly improve the quality of our lives at work, at home, and at play. But there are significant gaps between where robots are today and the promise of pervasive integration of robots in everyday life. Some of the gaps concern the creation of robots. How do we design and fabricate new robots quickly and efficiently? Other gaps concern the computation and capabilities of robots to reason, change, and adapt for increasingly complex tasks in increasingly complex environments. Other gaps pertain to interactions between robots and between robots and people.

Today it takes too much time to make robots. One idea for speeding up the fabrication of robots is to modularize them: to make a universal particle that can be composed in many ways to create robots with different shapes—in other words to create self-reconfiguring robots. An alternative is to create software tools and fabrication processes to enable the creation of one-of-a-kind custom robots really fast; in other words, to create a robot compiler. We have described a modular robot system that is composed of a set of 1 cm modules called Pebbles that can communicate with their neighbors in a cubic lattice, form connections to these neighbors, and transmit power. These local operations enable a system of Pebbles to self-assemble as "robot blocks," program desired shapes, and self-disassemble to realize the desired shape. We hope that by uniting the ideas for self-assembly and 3D shape formation we can move the robotic Pebbles one step closer to smart sand. While there is still work required to further miniaturize the Pebbles, the current system is a useful testbed that allows us to quickly test the concepts and algorithms that drive the self-assembly and self-disassembly processes.

NOTES

1 T. Fukuda and S. Nakagawa, "Dynamically Reconfigurable Robotic System," *IEEE International Conference on Robotics and Automation*, April 1988, 1581—1586.

2 M. Yim, W.-M. Shen, B. Salemi, D. Rus, M. Moll, H. Lipson, E. Klavins, and G. S. Chirikjian, "Modular Self-Reconfigurable Robot Systems: Challenges and Opportunities for the Future," *IEEE Robotics and Automation Magazine* 14, no. 1 (March 2007): 43—52.

3 K. Gilpin and D. Rus, "Modular Robot Systems: From Self-Assembly to Self-Disassembly," *IEEE Robotics and Automation Magazine* 17, no. 3 (September 2010): 38—53.

4 P. White, V. Zykov, J. Bongard, and H. Lipson, "Three Dimensional Stochastic Reconfiguration of Modular Robots," *Robotics Science and Systems* (June 2005).

5 M. Tolley, J. Hiller, and H. Lipson, "Evolutionary Design and Assembly Planning for Stochastic Modular Robots," in *IEEE Conference on Intelligent Robotics and Systems* (IROS), October 2009, 73—78.

6 S. Griffith, D. Goldwater, and J. M. Jacobson, "Robotics: Self-Replication from Random Parts," *Nature* 437 (September 28, 2005): 636.

7 M. Yim, D. G. Duff, and K. D. Roufas, "Polybot: A Modular Reconfigurable Robot," in *IEEE International Conference on Robotics and Automation* (ICRA), April 2000, 514—520.

8 M. Yim, Y. Zhang, K. Roufas, D. Duff, and C. Eldershaw, "Connecting and Disconnecting for Self-Reconfiguration with Polybot," *IEEE/ASME Transaction on Mechatronics*, special issue on Information Technology in Mechatronics (2003).

9 A. Castano, A. Behar, and P. Will, "The Conro Modules for Reconfigurable Robots," *IEEE Transactions on Mechatronics* 7, no. 4 (December 2002): 403—409.

10 A. Kamimura, H. Kurokawa, E. Yoshida, S. Murata, K. Tomita, and S. Kokaji, "Automatic Locomotion Design and Experiments for a Modular Robotic System," *IEEE/ASME Transactions on Mechatronics* 10, no. 3 (June 2005): 314—325.

11 K. Kotay, D. Rus, M. Vona, and C. McGray, "The Self-Reconfiguring Robotic Molecule," *IEEE International Conference on Robotics and Automation* (ICRA), 1998, 424—431.

12 E. Yoshida, S. Murata, S. Kokaji, A. Kamimura, K. Tomita, and H. Kurokawa, "Get Back in Shape! A Hardware Prototype Self-Reconfigurable Modular Microrobot That Uses Shape Memory Alloy," *IEEE Robotics and Automation Magazine* 9, no. 4 (2002): 54—60.

13 M. Koseki, K. Minami, and N. Inou, "Cellular Robots Forming a Mechanical Structure (Evaluation of Structural Formation and Hardware Design of 'chobie ii')," *Proceedings of 7th International Symposium on Distributed Autonomous Robotic Systems* (DARS04), June 2004, 131—140.

14 D. Rus and M. Vona, "Crystalline Robots: Self-Reconfiguration with Compressible Unit Modules," *International Journal of Robotics Research* 22, no. 9 (2003): 699—715.

15 S. Goldstein, J. Campbell, and T. Mowry, "Programmable Matter," *IEEE Computer* 38, no. 6 (2005): 99—101.

16 P. Pillai, J. Campbell, G. Kedia, S. Moudgal, and K. Sheth, "A 3d Fax Machine Based on Claytronics," In *International Conference on Intelligent Robots and Systems* (IROS), October 2006, 4728—4735.

17 J. John Amend and H. Lipson, "Shape-Shifting Materials for Programmable Structures," International Conference on Ubiquitous Computing: Workshop on Architectural Robotics, September 2009.

18 R. Nagpal, "Programmable Self-Assembly Using Biologically-Inspired Multiagent Control," International Conference on Autonomous Agents and Multiagent Systems (AAMAS), July 2002.

19 W. Butera, "Text Display and Graphics Control on a Paintable Computer," in *Proceedings of the First International Conference on Self-Adaptive and Self-Organizing Systems (SASO)* (IEEE Computer Society Press, 2007), 45—54.

20 K. Gilpin, K. Kotay, D. Rus, and I. Vasilescu, "Miche: Modular Shape Formation by Self-Disassembly," *International Journal of Robotics Research* 27, no. 3—4 (2008): 345—372.

21 Withheld for blind review.

22 K. Gilpin, "Shape formation by self-disassembly in programmable matter systems," PhD. thesis, MIT, 2012.

23 K. Gilpin, K. Koyanagi, and D. Rus, "Making Self-Disassembling Objects with Multiple Components in the Robot Pebbles System," *IEEE International Conference on Robotics and Automation* (ICRA), 2011, 3614—3621.

24 K. Gilpin and D. Rus, "A Distributed Algorithm for 2d Shape Duplication with Smart Pebble Robots," in IEEE International Conference on Robotics and Automation (ICRA), 2012.

25 A. D. Marchese, H. Asada, and D. Rus, "Controlling the Locomotion of a Separated Inner Robot from an Outer Robot Using Electropermanent Magnets," IEEE International Conference on Robotics and Automation (ICRA), 2012.

15

General Principles for Programming Material

Athina Papadopoulou,
Jared Laucks, and
Skylar Tibbits

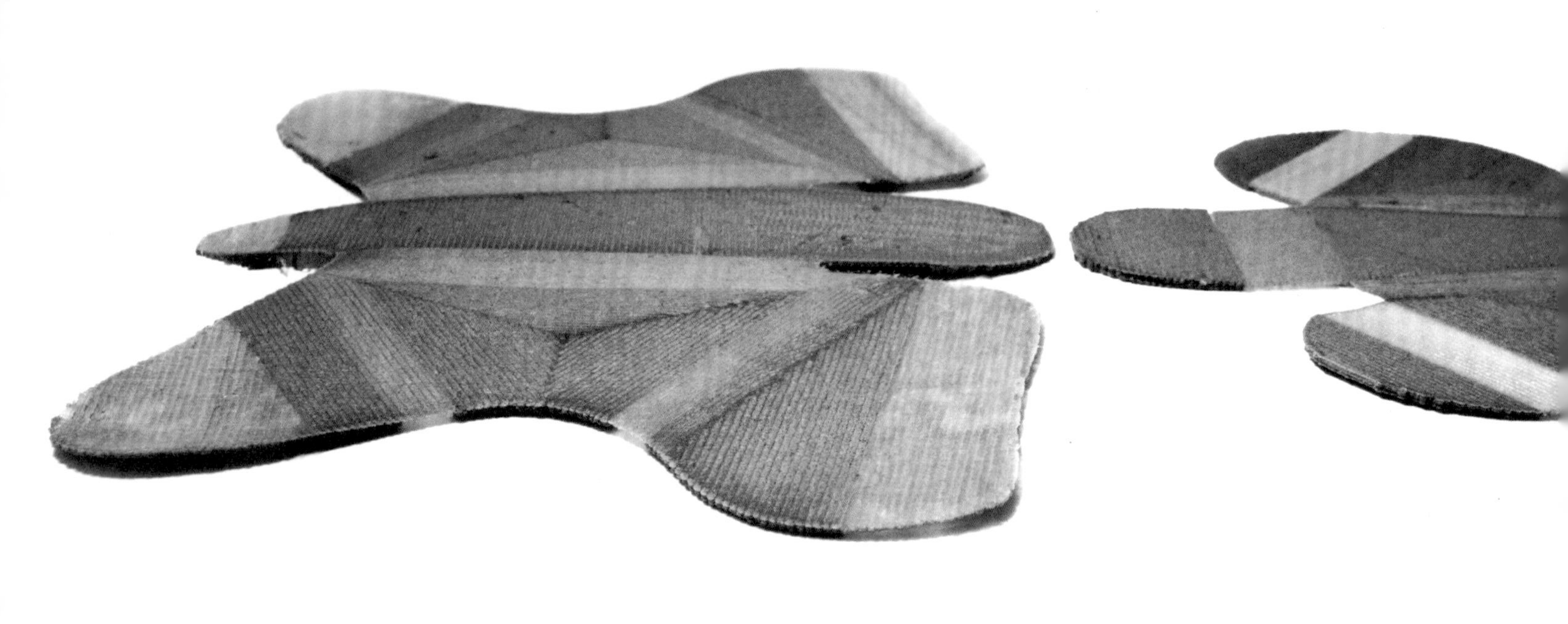

Figure 15.1
Printed wood Eames elephant that self-transforms from a flat sheet when exposed to moisture and heat.

15 General Principles for Programming Material

Athina Papadopoulou, Jared Laucks, and Skylar Tibbits

Over the past few decades, advances in software and hardware technologies have significantly shifted design and fabrication processes. For example, 3D printing has closed the gap between conceptual and material practices, offering designers greater design freedom and control over the fabrication process while making fabrication accessible to a wider range of users.[1, 2] In addition, advances in synthetic biology, bioengineering, materials science, and soft robotics have pushed materials research even further, leading to materials that grow, adapt, and transform.[3, 4, 5]

Although substantial research is being conducted on the controlled transformation of material systems, research on the general principles of materials' transformation across different scales remains limited. In this paper we propose a general methodology for programming materials by describing the principles for transformation and fabrication techniques of various off-the-shelf and custom-made materials. The wide applicability of the developed methods is being demonstrated through a growing series of programmable materials, including, wood, carbon fiber, polymers, textiles, foams, and rubbers, that can self-transform in a precise manner on demand.[6, 7]

The MIT Self-Assembly Lab's research on programmable materials investigates the design of composite materials, the development of methods for triggering material transformations, and the design and development of new tools that make this research possible. The developed methods demonstrate that a range of materials, from durable industrial materials such as carbon fiber to more commonly used materials such as fabrics, wood, and rubbers, can be rendered transformable through the design of their overall geometry and material composition. This research points toward a future in which smart materials will be as accessible, functional, and universally programmable as today's computers and machines. We envision a future in which any material at any scale can be programmed to dynamically change properties over time.

In programmable materials, the material and shape transformations are triggered by different types of energy such as heat, moisture, and light. Thus, natural environmental changes can lead to desired material transformations allowing for smart, energy-saving, and sustainable solutions in manufactured products designed to respond dynamically and adapt to their environment. Programmable materials have a wide range of applications in manufacturing, medical devices, sportswear, automotive, aviation, and fashion industries.

1. OVERVIEW OF METHODS

Although each programmable material has a unique set of properties, we can identify common principles and similarities in the methods for making and programming materials. Our aim here is to provide the framework, overall principles, and common challenges that characterize all programmable materials tested to date and demonstrate these general principles through highlighted applications.

Programmable materials have three relevant characteristics: (1) their *material composition*—what they are composed of and how the materials are organized; (2) the *activation energy*—what triggers the transformation; and (3) the *transformation mechanics*—the material's functional behavior when subjected to an activation energy. These three aspects can be further defined as follows:

Material Composition

Programmable materials are composed of one or more active material components. We define "active material" as a material whose properties change significantly when subjected to an activation energy: a programmable material could, for example, be composed of an active and a second "nonactive material" that transform based on an activation energy.

Figure 15.2
2D drawing of extrusion paths for 3D printing the Eames elephant that self-transforms from a flat sheet when exposed to moisture and heat. The black color represents the wood grain. The yellow color represents the active polymer grain placed in the bottom layer. The red color represents the active polymer grain placed at the top layer.

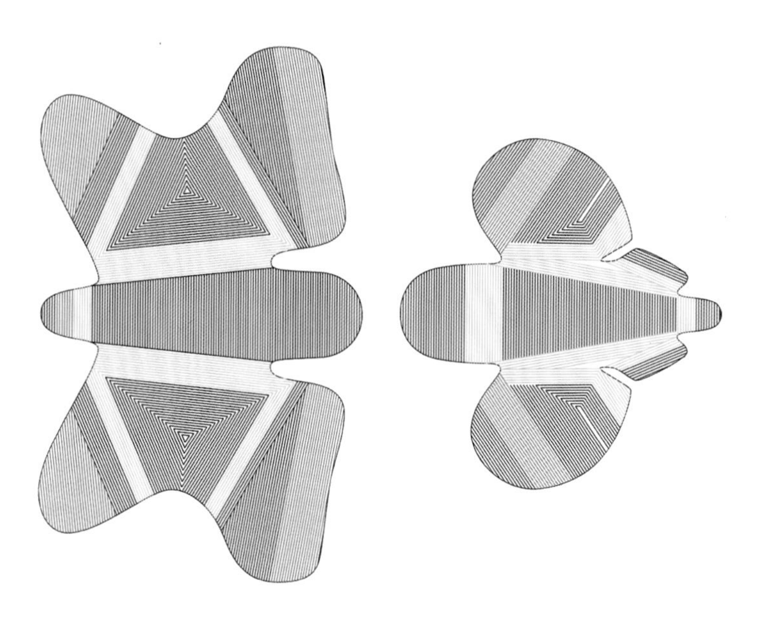

Alternatively, two or more materials could become highly active when bonded together as a single composite based on the differential properties of each material (differential thermal expansion, contraction, etc.). The spatial configuration of each of the materials can be produced through 3D printing, lamination, weaving, or other assembly techniques; or it might even be an off-the-shelf material cut into a desired shape. The orientation of the active material and the nonactive material is important for biasing the direction of transformation. Similarly, the geometry and placement of the materials can dramatically influence the final behavior in terms of their mechanical transformation (folding, curling, stretching, shrinking) and physical properties (folding force, flexibility, stiffness, etc.).

Activation Energy

Programmable materials are designed to respond to one or more types of energy. Possible types of energy include heat, light, sound, moisture, air pressure, and many others. A programmable material can transform while exposed to the activation energy or afterward. The resulting transformation may be reversible, ideally allowing for infinite repetition, or permanent, resulting in altered material properties and a final stable state. Different activation energies can be coupled with different materials, depending on the property of the material. For example, wood may be coupled with moisture, while heat is more frequently coupled with metal or plastics. Thus in programmable materials research, the material composition may be selected based on the desired activation energy, or conversely the activation energy may be selected based on the utilized material.

Transformation Mechanics

To be able to transform in a precise way, a programmable material needs to be composed of a specific combination of materials and receive a specific amount of exposure to the activation energy. Different proportions of nonactive and active material in the composition might affect the speed, duration, or force of transformation. Similarly, the position of each of the materials, especially the position of the active material in relation to the nonactive material, can affect the direction of transformation. In a layered programmable material, if a swelling active material is on the top surface of a nonactive material, the configuration will result in a downward curling of the material. Alternatively, if a shrinking active material is on the top surface of a nonactive material, the configuration will result in an upward curling transformation.

Different amounts of exposure to the activation energy might also result in a change of speed, force, permanent transformation, or even material failure. For example, with a heat-active bimaterial composite, less heat exposure might lead to less curvature or a longer transformation time. The grain direction of each of the materials in the composition also plays a significant role in the transformation mechanics of the final assembly. Most programmable materials will fold in a different manner depending on whether their grain is parallel or oblique in relation to the boundary shape of the material. The relation between the grain direction of the active and the nonactive materials in the material composition also plays an important role in the resultant transformation, as different grain directions may lead to different folding directions, slower or faster transformation, or no transformation at all.

Complex transformations can be encoded in the materials if we precisely design the grain direction of each material composite and the specification of joints within the main material component. For example, to encode a folding transformation of a "star-shaped" flat piece of fabric, wood, carbon, or other material into a three-dimensional pyramid, a second material with different expansion or contraction rate needs to be embedded in the composite along each of the folding axes, thus creating four joints along the center of the star-shaped piece that will be transformed into the pyramid's base. To fully control the transformation, the thickness and length of the joints are specified through the number of layers and the total area of the secondary material.

2. PROGRAMMABLE MATERIALS

2.1. PROGRAMMABLE WOOD

Wood is one of the best-known active materials traditionally used in architecture, naval architecture, furniture, and other products.[8] The techniques used to manipulate wood today are similar to the techniques used for centuries: steam-bending, molding, and lamination.[9] Such traditional woodworking techniques, when used in industrial processes for mass production, can be costly, as they require intensive manual labor and craftsmanship in production. One of the main challenges in wood-bending and other woodworking techniques is the anisotropic behavior of wood caused by the irregular patterns of the material's natural grain.[10] By taking advantage of 3D printing technologies, we can take advantage of wood's anisotropic behavior by designing custom wood grain patterns, and consequently gain control of wood's transformation and bending behavior. Offering a novel wood-folding method, 3D printed wood sheets can be fully customized and programmed to fold when subjected to moisture (figures 15.1 and 15.2).

Material Composition

By taking advantage of 3D printing, we are able to control the patterns of the printed wood grain by designing the machine's toolpath which deposits the wood and polymer filament. The grain patterns can be specifically designed to allow particular material transformations, guiding the material to fold in one or more specified directions when exposed to the activation energy. We have explored two different methods of programmable wood composition: wood-only printing and multimaterial printing. Wood-only printing is the custom deposition of wood filament through 3D printing (figures 15.3—15.6). Multimaterial wood printing is the custom deposition of wood filament and a second material filament with low moisture absorption (usually polymers such as ABS, PLA, etc.), resulting in a composite material consisting of wood as its main active component and a secondary material to guide or enhance the transformation (figures 15.1 and 15.2).

Activation Energy

The main activation energy used to trigger the transformation of programmable wood is moisture. Three moisture activating techniques have been explored so far with successful results: *water submersion*, *water vapor*, and *water with thermal radiation*.

The water submersion technique utilizes a wood composite that is fully submerged in the water for a specific amount of time, depending on density, thickness, and other parameters of the print (figure 15.6). After the submersion stage, which results in the wood swelling, the wood composite is placed in a dry environment, where it transforms. In the water vapor technique, the wood print absorbs moisture in the form of water vapor, and a closed chamber is used to control the amount of moisture. The setup functions at both high and low humidity levels and can also be used to create a dry environment to accelerate drying and speed up the material's transformation. The thermal radiation technique is used in combination with moisture intake (either the water submersion or the water vapor technique). After being exposed to moisture, the wood composites are exposed to thermal radiation during their drying process. Thermal radiation accelerates the transformation process and can lead to more controllable results through the localized application of the heat. These three techniques have been partially developed in collaboration with the Institute of Computational Design at the University of Stuttgart and have been more extensively discussed by Correa et al.[7]

Transformation Mechanics

The folding angle of the wood composites depends mainly on the direction of the printed wood grain. It has been demonstrated that in square samples of wood-only composites, if the wood grain is printed parallel to the boundaries of the sample, then the wood will fold in the axis perpendicular to the printed grain. Following the same principles, a wood-only composite with its grain printed diagonally in relation to its boundary shape will result in a diagonal fold perpendicular to the wood grain direction.

The transformation mechanics of the multimaterial wood composites are slightly more complex than those of the wood-only composites because the parameters of both the wood and the added polymer need to be addressed. The driving force of the multimaterial print's transformation is the differential behavior of the two materials in response to the activating energy, leading to a nonuniform deformation of the print. For example, a three-layer composite consisting of two layers of printed wood on the bottom and a layer of polymer on top, after exposure to the activation energy (immersion to moisture followed by exposure to heat), will fold upward. Similarly, an inversion of the order of the layers will lead to a downward fold, as the key factor is the differential expansion of the active and nonactive materials. In both cases, a greater amount of polymer on the corresponding layer will increase the folding angle as the wood expansion will be further constrained.

2.2. PROGRAMMABLE CARBON FIBER

Carbon fiber has been commonly used in the automotive, aviation, and apparel industries because of its high stiffness, tensile strength, and low weight.[11] Yet common forming processes for carbon fiber products require expensive and labor-intensive procedures with long lead times, complex lay-up,

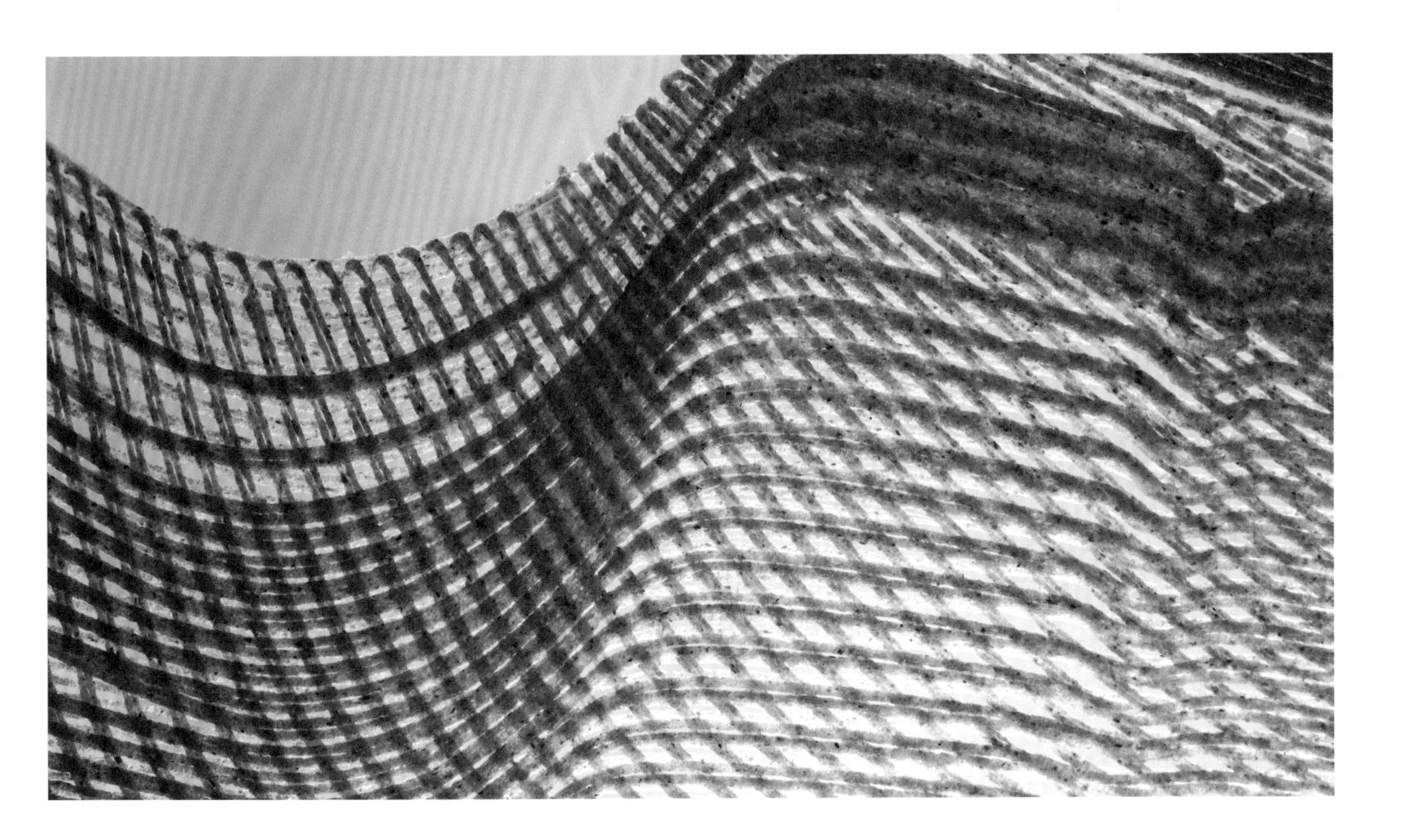

Figure 15.3
Printed wood grain.

and complex cure cycles, resulting in substantial energy requirements and cost in carbon-fiber-based manufacturing.[12, 13] Moreover, adaptive or responsive carbon fiber systems used for actuation or other means traditionally require electronics or mechanical parts that are more prone to failure than passive systems that require only heat radiation or other environmental activation energy.[14] To provide a solution to the limitations of current carbon fiber manufacturing, MIT's Self-Assembly Lab has collaborated with Carbitex LLC to develop a method for programming carbon to self-transform. Fully formed and functional products can be produced on demand and on site through flat carbon fiber sheets that are programmed to fold into predefined shapes when exposed to energy.

Material Composition

To program carbon fiber for controlled sensing and actuation, we fabricate composites made of flexible carbon fiber combined with heat-, moisture-, or light-active polymers. The flexible carbon fiber technology, CX6® developed by Carbitex LLC, is made of low-modulus matrix materials and processing techniques to harness carbon fiber's performance in a fully flexible form (figure 15.7).

One of the main challenges in making the programmable carbon fiber composite is the bonding of the flexible carbon fiber material with the polymer to allow for optimal transformation as well as ease of production. Various bonding methods have been tested, including 3D printing, lamination, and selective bonding. Each method demonstrates different advantages and constraints. For example, 3D printing permits customized and precise deposition of the active polymer on the carbon fiber sheet (figure 15.8). However, 3D printing is time-consuming and may impose size restrictions introduced by the print bed size. On the other hand, lamination tests with a heated platen press have demonstrated faster results than 3D printing. Compared to 3D printing, however, lamination does not offer the advantage of customized patterns. Finally, in the selective bonding technique, the two material components are only bonded in specific areas, using lamination, stitching, or other means. Selective bonding allows each material to transform with greater design freedom, taking advantage of the differential expansion or contraction properties of the composite material.

Activation Energy

When subjected to the corresponding energy, the added polymer layer within the programmable carbon composite acts as the active component of the composite, resulting in material transformation caused by differential expansion or contraction. For example, to program carbon fiber using heat, a material composite consisting of flexible carbon fiber and polymers with high expansion rate, such as nylon, polyethylene, or polyvinylidene fluoride, can be fabricated through either 3D printing, lamination, or selective bonding. In this case, the polymer used as the active material will demonstrate significant expansion with heat, whereas the carbon fiber, used as the nonactive material, will have negligible expansion. If flexible carbon fiber is used as the bottom surface of the composite and nylon as the top surface, when subjected to heat the composite will fold downward because of the differential thermal expansion of the two materials (figures 15.7 and 15.9).

An important parameter that needs to be taken into account when predicting the resultant transformation is the deformation of the material happening during the fabrication process: the high temperature at which the polymer is being extruded or laminated causes the polymer to bond with the carbon fiber while in a thermally expanded state and to shrink back to its natural state once it cools down, causing the composite to curl upward. Thus, after fabrication, the carbon composite has a natural curved state with the polymer on the inside. The transformation induced during the postfabrication activation process (using light, heat, moisture, etc.) is reversible, allowing the composite to alternate between the curved state and a flat or inversely curved state.

Three types of heat activation method have been tested to trigger the transformation of programmable carbon fiber: full-contact heating (for example using a heated plate), electroactive heating (using embedded wires), and noncontact heating (ambient heat radiation). To induce a desired transformation in a carbon composite, the temperature for contact heating needs to be within the range of 30 to 100 degrees Celsius, while for the same composite and transformation the temperature for ambient heating needs to be much higher, ranging from 200 to 400 Celsius.

Transformation Mechanics

Three main parameters determine the folding direction(s) of the programmable carbon fiber: the pattern of the carbon fiber weave, the grain or extrusion direction of the added active material, and the geometry (boundary shape) of the composite. The resultant transformation depends on (1) the relationship between the three main parameters, (2) various material properties such as thickness, stiffness, and the quantity and placement of the active material, and (3) the intensity and duration of exposure to the activation energy. The correlation among the three main parameters (fiber weave pattern, active material orientation, and boundary shape) mainly defines the folding direction(s) of the composite, whereas the other parameters (material properties and activation energy) mainly define the intensity of the transformation: folding angle and speed of transformation. The weave of the flexible carbon fiber can be either uniaxial, biaxial, or triaxial. To arrive at a controlled transformation, the polymer's grain or extrusion direction needs to be either aligned or perpendicular to one of the axes of the fiber's weave. Different combinations of the main parameters can lead to various types of geometric transformations: curve, fold, twist, spiral, or wave (figure 15.10).

2.3. PROGRAMMABLE TEXTILES

Textiles have a wide range of applications in architecture, furniture, apparel, sportswear, product design, automobiles, and aviation. In recent years various active and passive methods have been developed to render textiles responsive, interactive, and customizable according to users' needs.[15, 16, 17] However, most of the methods currently in use rely on complex sensing and actuation systems (requiring expensive electronics or biomaterial systems). Taking advantage of 3D printing and other fabrication technologies, the Self-Assembly Lab has developed methods to transform everyday textiles into desired 3D shapes, solely by designing the textiles' composition and the interaction with their environments. Two different methods have been developed to program textile transformation: prestressed textiles and active textiles. In prestressed textiles, patterns of deposited polymers form structural frames on top of or within the fabric, leading to controlled folding upon release from the prestressed state. In active textiles, composite textiles, consisting of textiles and added active polymers, transform when subjected to specific activation energies.

Material Composition Prestressed Textiles

In prestressed textiles, textiles with a high degree of elasticity, such as neoprene, jersey, and polyester, can be used as the active material of the programmable composite. Less elastic material, such as plastic, can be used as the nonactive material to structure the textile and constrain its stretched state. The textile is first stretched into the stressed state and then the polymer material is applied onto or within the textile composition through 3D printing, lamination, or other means of bonding (figures 15.11 and 15.12).

Active Textiles

Programmable active textiles are bimaterial composites consisting of textile as the nonactive component and an added active material that significantly changes properties (either expands or contracts) when subjected to an activation energy. The bimaterial composite is created through 3D printing, lamination, or other means of bonding that allow the deposition of the active component on or within the textile composition. The base textile can be either stretchable or nonstretchable and can be made of various materials, such as cotton, neoprene, jersey, vinyl, velvet, silk, and polyester-polyurethane copolymers (figure 15.13).

Activation Energy Prestressed Textiles

In programmable prestressed textiles, the textile is stretched in one or two dimensions with a specified force or until it reaches its limit. Once it is fixed in a flat stretched state, a nonstretchable material is added to the prestressed textile to prime the composite for transformation. The act of prestressing embeds activation energy in the programmable prestressed composite. Once the composite is released from its constrained flat state, the programmable prestressed composite jumps into a new state, transforming itself into the preprogrammed 3D shape.

Active Textiles

When subjected to an activation energy, the textile-based bimaterial composite will transform due to differential expansion or contraction of the various material components. For example, to program an active textile using heat, bimaterial composites can be used consisting of nonstretchable textile and an active polymer with a high expansion rate. The polymer can be selected from polyethylene (PE), polyethylene terephthalate (PET), polyvinylidene fluoride (PVDF), thermoplastics, or many others. If the textile is used as the

Figure 15.4 (right, top)
Printed wood basket that self-folds from a flat sheet when exposed to moisture and gravity.

Figure 15.5 (bottom)
Flat-printed wood pattern that will self-transform into a 3D basket.

Figure 15.6
Printed wood material under water.

bottom surface of the composite, and the active material as the top surface, then, when subjected to heat, the composite will fold downward because of the differential thermal expansion of the two materials. The transformation of programmable active textiles can be reversible: in a heat activation process, while the textile-based composite is being heated it gradually transforms into the desired shape, whereas while it cools down it gradually returns to its initial shape.

Transformation Mechanics Prestressed Textiles

The physical shape of the material deposited onto the stretched textile is the main agent that will determine the 3D shape transformation upon release of the textile. The textile properties (thickness, elasticity, weave, or knit pattern) and the properties of the added nonstretchable material (stiffness, elasticity, and thickness) also impact the resultant shape, as they affect the intensity (degrees of folding) and speed of transformation. In addition, in the case of a nonuniform stretching of the prestressed material, the directions of forces will bias the resultant transformation over specific folding directions. Finally, the position and amount of the nonstretchable material added onto or within the textile will also affect the resultant shape, in a desired and repeatable manner.

Active Textiles

Similarly to programmable carbon fiber, three main parameters define the transformation of programmable active textiles: the pattern of the textile weave or knit, the grain or extrusion direction of the added active material, and the geometry (boundary shape) of the textile-based composite. Other material properties are mostly responsible for the intensity of the transformation (thickness, elasticity, the placement and amount of the active material, and the intensity and duration of the exposure to the activation energy). The deposition of the active material can be either uniform or nonuniform in thickness and direction, resulting in controlled transformation of the textile-based composite in a selective manner, by allowing the intensity of the fold to vary locally. Similarly, the weave or knit density and thickness can also vary locally, making for a more customizable and controlled behavior. Thus, controlling the different parameters of the textile and additional components as well as their response to the activation energy allows for precise programmability of complex textile geometries.

3. CONCLUSIONS

Advances in materials capabilities combined with novel fabrication technologies have brought forth new perspectives on design and product performance through programmable materials. Still, the complexity and variety of material properties continue to limit the process of designing and fabricating programmable materials. We have shown a general approach to creating programmable materials based on three principles: material properties, activation energy, and transformation mechanics.

MIT's Self-Assembly Lab has developed methods to program materials and demonstrated their functionality, relying on the design of the material composition, shape, and interaction with the materials' physical environments. No complex electronic, hydraulic, or pneumatic systems are required (although electrical energy may be used if desired). We have shown that easily fabricated and ubiquitous materials, such as textiles, wood, plastics, and composites, can be used to create programmable materials.

To trigger the material transformation, an ambient energy source is used for activation energy, such as moisture, heat, light, air pressure, sound, or a combination of these. Adjusting the amount of exposure to the activation energy modifies the material transformation. When a programmable

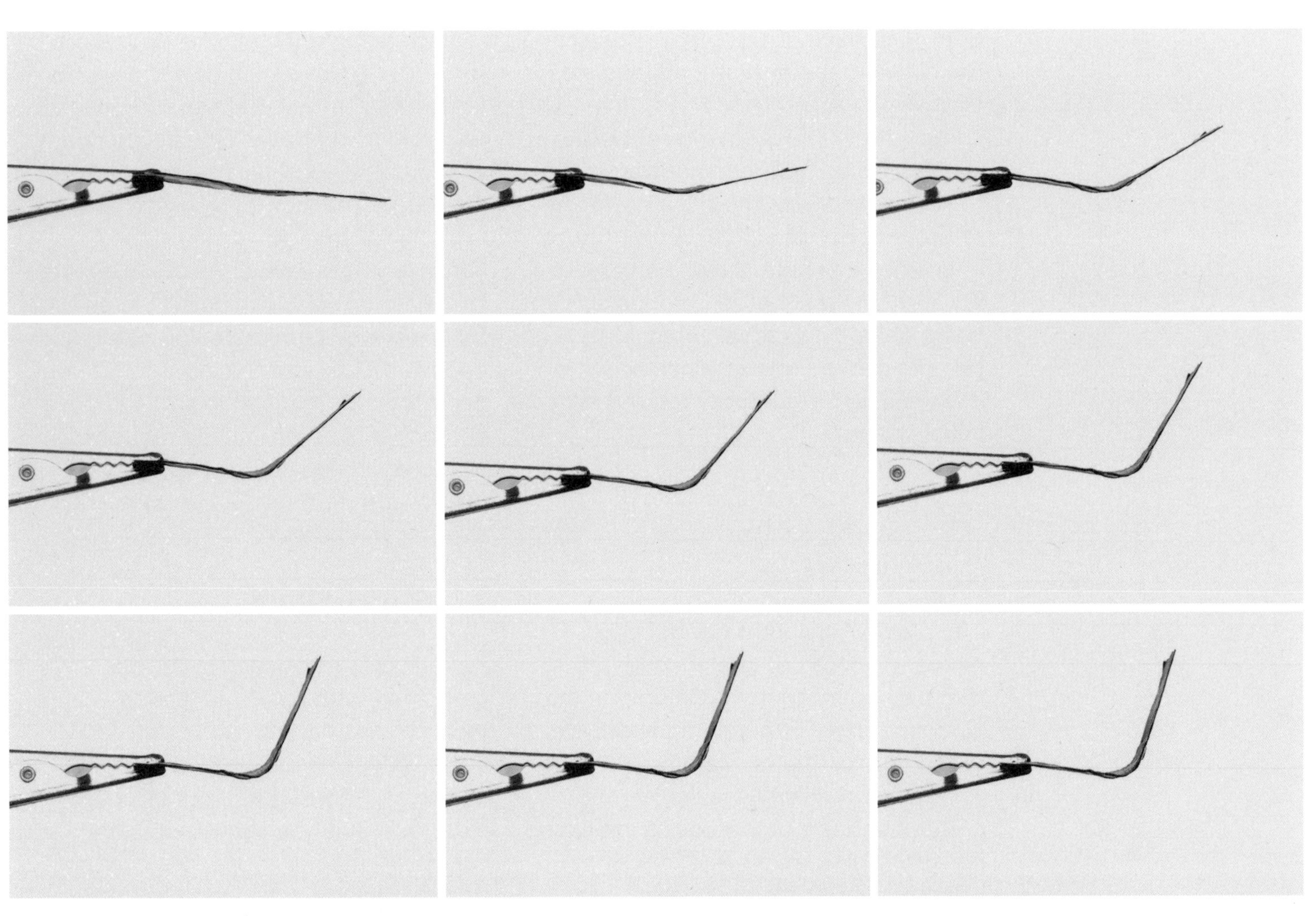

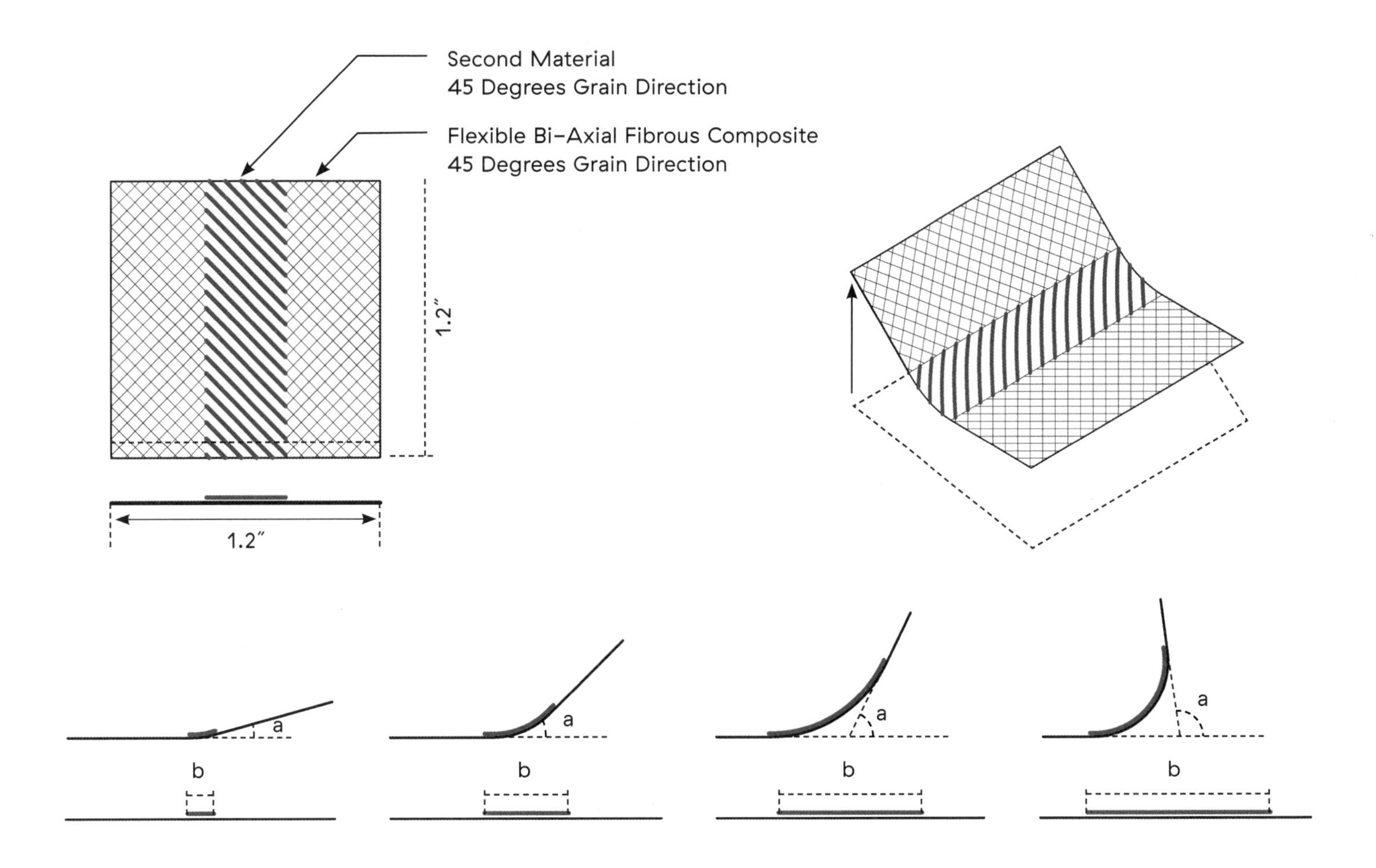

Figure 15.7 (left page, top)
Photographs showing different folding angles during the transformation of heat-activated programmable carbon fiber.

Figure 15.8 (left page, bottom)
Photograph showing a light- and temperature-active carbon fiber screen that was printed and CNC-cut into a tessellated pattern.

Figure 15.9 (top)
Diagrams showing the transformation angle of a flat square of programmable carbon fiber when a different amount of active polymer is printed on its surface.

material is exposed to such energies, a precise-shape transformation can be induced. Thus, programmable materials offer an accessible, sustainable, and low-cost solution to developing transformable smart materials that can be easily applied in both industrial and design practices.

In this paper, the application of the general principles for programming materials were demonstrated through the development of three programmable materials: programmable wood, programmable textiles, and programmable carbon fiber. These are only three of many possible materials than can be programmed using the outlined methods. Apart from wood, carbon fiber, and textiles, the Self-Assembly Lab is currently researching the programmability of various polymers, foams, and rubbers, seeking to broaden the range of programmable materials. By outlining the common principles for programming materials, we aim to provide a general framework and methodology that will make it possible to program almost any material.

Programmable materials have found a wide range of applications across different industries. The variety of applications hints at the broad applicability: to the sole of a shoe whether for traction or orthotics, to the fabric of clothing for breathability, to a table for flat shipping or ease of storage, to indoor panels of office spaces that help control visibility and acoustics, to transformable architectural structures that adapt to the environment, and to building facades that physically adjust shading and ventilation.

Programmable materials create a future where computation won't be exclusive to machines but a universal material quality. Programmable materials highlight the programmability already embedded in our surrounding environment and show that programmability of materials will be as accessible and universal as our computers and machines today.

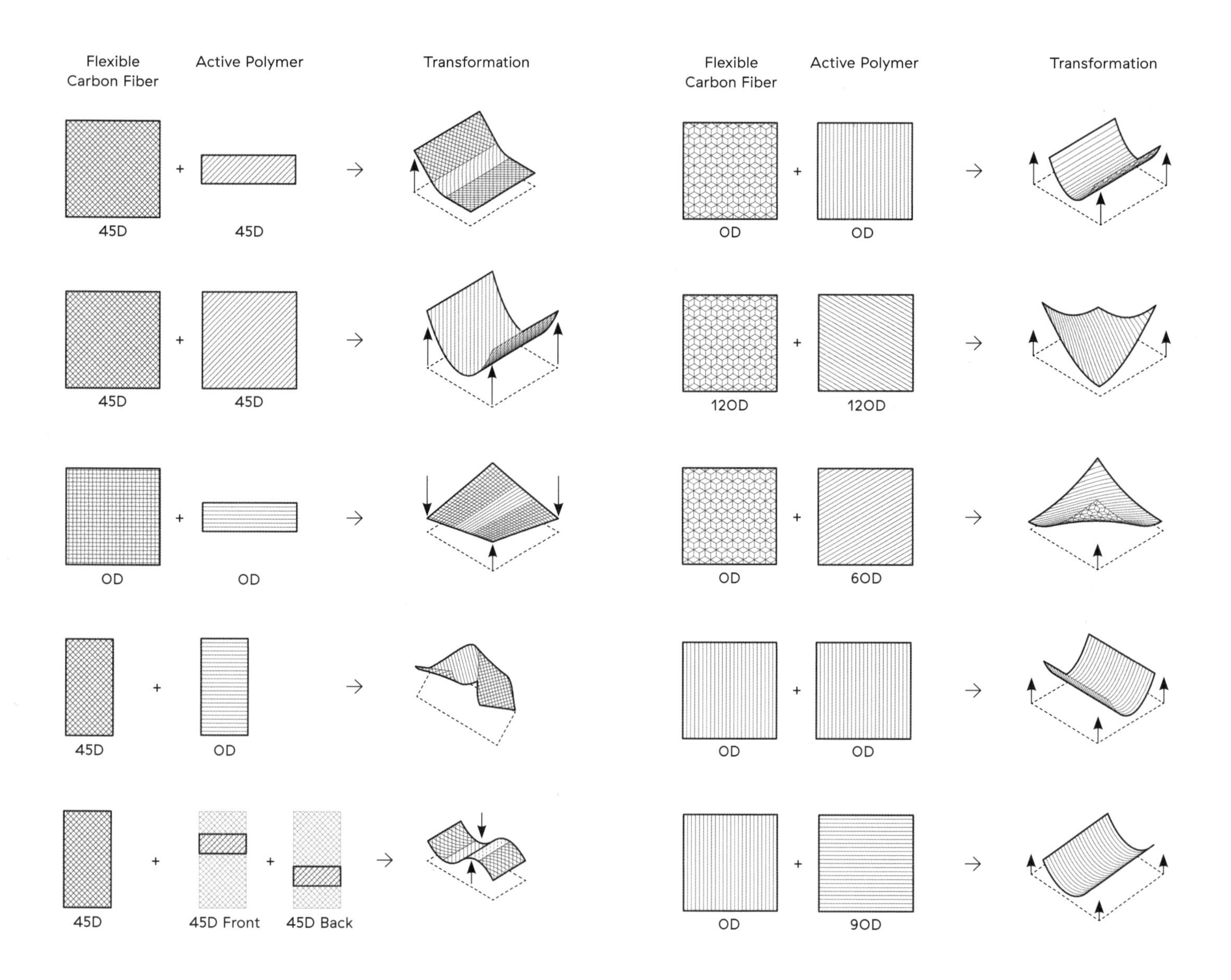
Flexible
Carbon Fiber
Active Polymer
Transformation
45D
45D
45D
45D
OD
OD
45D
OD
45D
45D Front
45D Back
Flexible
Carbon Fiber
Active Polymer
Transformation
OD
OD
120D
120D
OD
60D
OD
OD
OD
90D

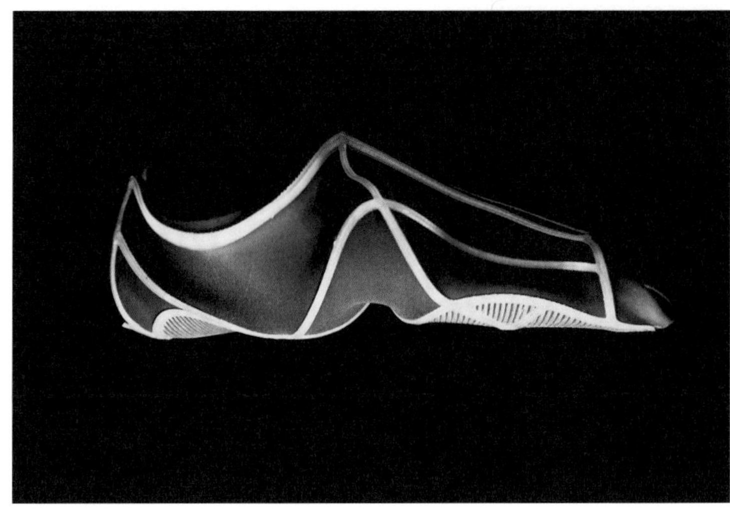

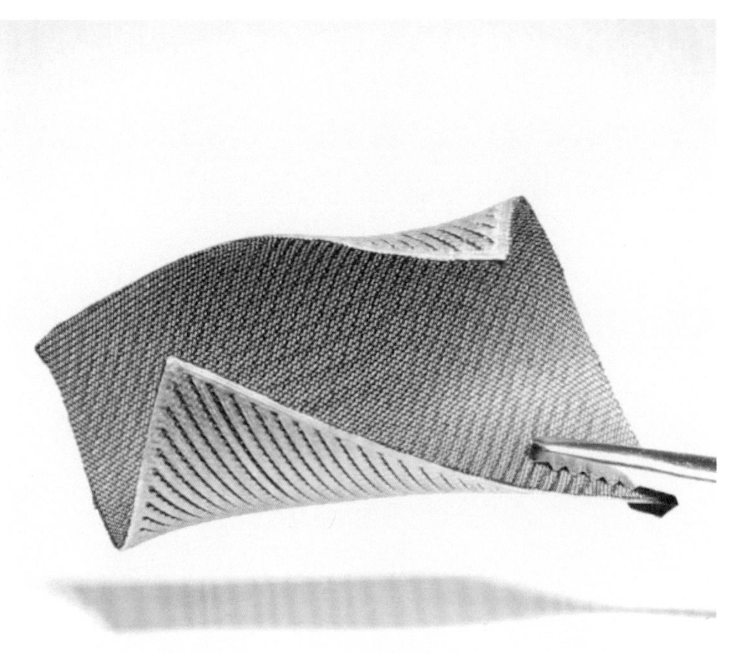

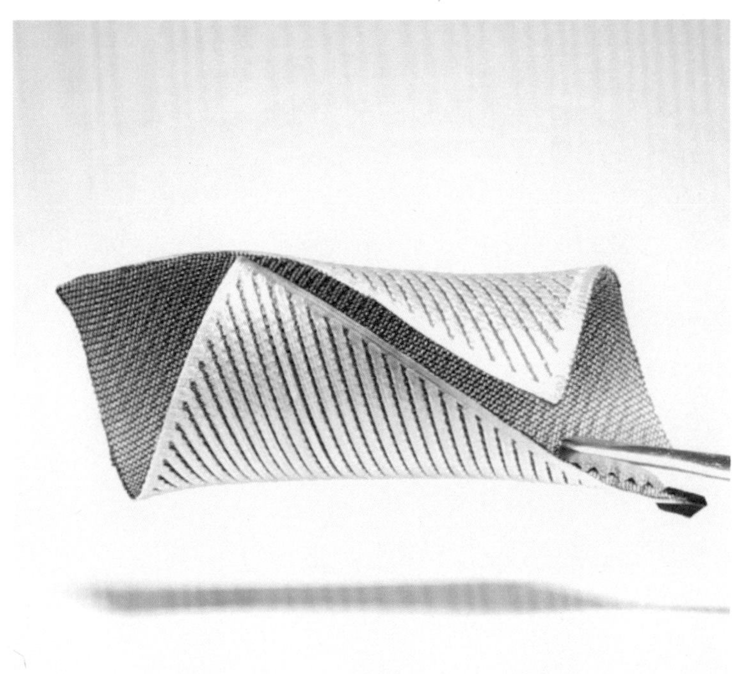

Figure 15.10 (left page)
Diagrams showing various orientations of carbon fiber and printer polymer layers resulting in different angles and transformations when the composites are exposed to heat.

Figure 15.11 (top left)
Prestressed active textile that self-transforms from a flat sheet to the desired three-dimensional shape.

Figure 15.12 (top right)
The Active Shoe project explores the possibility of producing the upper part of a shoe by printing a specific pattern on stretched fabric. The 2D pattern emerges after cutting the 2D pattern and then transforms into a 3D form. The combination of stretchable fabric and plastic offers both flexibility and stability.

Figure 15.13 (bottom)
A programmable textile made from a printed active polymer on a stretched textile base layer. The left image shows the nonheated state where the textile is open; the image on the right shows the heated state where the textile closes.

ACKNOWLEDGMENTS

We would like to thank all of our MIT researchers and outside collaborators who have contributed in the development of the programmable materials research: Christophe Guberan, Carlo Clopath, Erik Demaine, Carbitex LLC, Autodesk Inc., Schendy Kernizan, Bjorn Sparrman, Carrie McKnelly, Christopher Martin, Nynika Jhaveri, Filipe Campos, and David Costanza. We would also like to thank Achim Menges, David Correa, and Steffen Reichert from the Institute for Computational Design at the University of Stuttgart, for their collaboration on the programmable wood research. Finally, we would like to thank MIT's Department of Architecture and MIT's International Design Center for their continued support.

NOTES

1 Skylar Tibbits, "4D Printing: Multi-Material Shape Change," *Architectural Design* 84, no. 1 (2014): 116–121.

2 Neil A. Gershenfeld, *Fab: The Coming Revolution on Your Desktop—from Personal Computers to Personal Fabrication* (New York: Basic Books, 2007).

3 Daniela Rus and Michael T. Tolley, "Design, Fabrication and Control of Soft Robots," *Nature* 521, no. 7553 (2015): 467–475.

4 Sahab Babaee, Jongmin Shim, James C. Weaver, Elizabeth R. Chen, Nikita Patel, and Katia Bertoldi, "Metamaterials: 3D Soft Metamaterials with Negative Poisson's Ratio," *Advanced Materials* 25, no. 36 (2013): 5116.

5 M. R. Jones, N. C. Seeman, and C. A. Mirkin, "Programmable Materials and the Nature of the DNA Bond," *Science* 347, no. 6224 (2015): 1260901.

6 Dan Raviv, Wei Zhao, Carrie Mcknelly, Athina Papadopoulou, Achuta Kadambi, Boxin Shi, Shai Hirsch, Daniel Dikovsky, Michael Zyracki, Carlos Olguin, Ramesh Raskar, and Skylar Tibbits, "Active Printed Materials for Complex Self-Evolving Deformations," *Scientific Reports* 4 (2014): 7422.

7 David Correa, Athina Papadopoulou, Christophe Guberan, Nynika Jhaveri, Steffen Reichert, Achim Menges, and Skylar Tibbits, "3D-Printed Wood: Programming Hygroscopic Material Transformations," *3D Printing and Additive Manufacturing* 2, no. 3 (2015): 106–116.

8 Michael H. Ramagea, Henry Burridgeb, Marta Busse-Wicherc, George Feredaya, Thomas Reynoldsa, Darshil U. Shaha, Guanglu Wud, Li Yuc, Patrick Fleminga, Danielle Densley-Tingleye, Julian Allwoode, Paul Dupreec, P. F. Lindenb, and Oren Schermane, "The Wood from the Trees: The Use of Timber in Construction," *Renewable and Sustainable Energy Reviews* 68 (2017): 333–359.

9 William A. Keyser Jr., "Steambending: Heat and Moisture Plasticize Wood," in *Fine Woodworking on Bending Wood* (Newtown, CT: Taunton Press, 1985).

10 Fritz Hans Schweingruber, *Wood Structure and Environment* (Berlin: Springer, 2007).

11 Andrey V. Chumaevskii, Evgeny A. Kolubaev, Sergei Yu. Tarasov, and Alexander A. Eliseev, "Mechanical Properties of Additive Manufactured Complex Matrix Three-Component Carbon Fiber Reinforced Composites," *Key Engineering Materials* 712 (2016): 232–236.

12 Taslim Badeghar, "Lightening Up with Carbon Fiber-Reinforced Plastics," *Adhesives and Sealants Industry* 23, no. 1 (January 2016): 25–27.

13 T. Ellringmann, C. Wilms, M. Warnecke, G. Seide, and T. Gries, "Carbon Fiber Production Costing: A Modular Approach," *Textile Research Journal* 86, no. 2 (2015): 178–190.

14 Haibao Lu, Yongtao Yao, and Long Lin, "Carbon-Based Reinforcement in Shape-Memory Polymer Composite for Electrical Actuation," *Pigment and Resin Technology* 43, no. 1 (2013): 26–34.

15 Rebeccah Pailes-Friedman, *Smart Textiles for Designers: Inventing the Future of Fabrics* (London: Laurence King, 2016).

16 Jinlian Hu, Harper Meng, Guoqiang Li, and Samuel I. Ibekwe, "A Review of Stimuli-Responsive Polymers for Smart Textile Applications," *Smart Materials and Structures* 21, no. 5 (2012): 053001.

17 S. Ahlquist, "Sensory Material Architectures: Concepts and Methodologies for Spatial Tectonics and Tactile Responsivity in Knitted Textile Hybrid Structures," *International Journal of Architectural Computing* 14, no. 1 (2016): 63–82.

16

Combinatorial Design of Floxelated Metamaterials

Corentin Coulais and Martin van Hecke

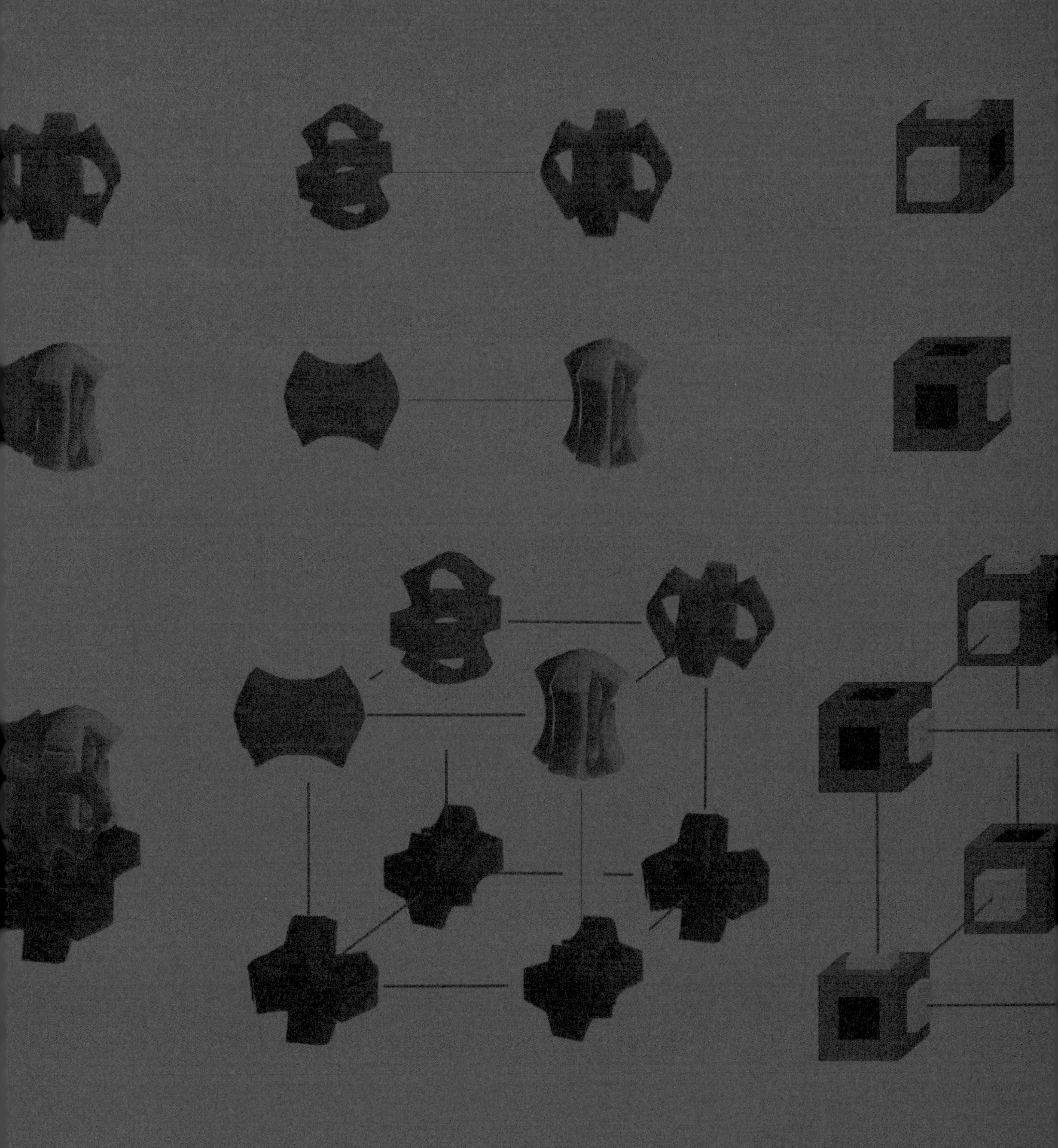

16

Combinatorial Design of Floxelated Metamaterials

Corentin Coulais and Martin van Hecke

The structural complexity of metamaterials is limitless in principle, although in practice most designs comprise periodic architectures that lead to materials with spatially homogeneous features. More advanced tasks, arising in soft robotics, prosthetics, and wearable tech, involve spatially textured functionalities that require aperiodic architectures. A naıve implementation of such structural complexity, however, invariably leads to frustration, which prevents coherent operation and impedes functionality. Here we introduce a combinatorial strategy for the design of aperiodic, yet frustration-free, mechanical metamaterials with spatially textured functionalities. We implement this strategy using cubic building blocks called floxels, a local stacking rule that allows cooperative shape changes by guaranteeing that deformed building blocks fit as in a 3D jigsaw puzzle, and 3D printing. These aperiodic metamaterials exhibit long-range holographic order, with the 2D pixelated surface texture dictating the 3D interior voxel arrangement. Moreover they act as programmable shape shifters, morphing into spatially complex but predictable and designable shapes when uniaxially compressed. Finally, their mechanical response to compression by a textured surface reveals their ability to perform sensing and pattern analysis. Combinatorial design thus opens a new avenue toward mechanical metamaterials with unusual order and machine-like functionalities.

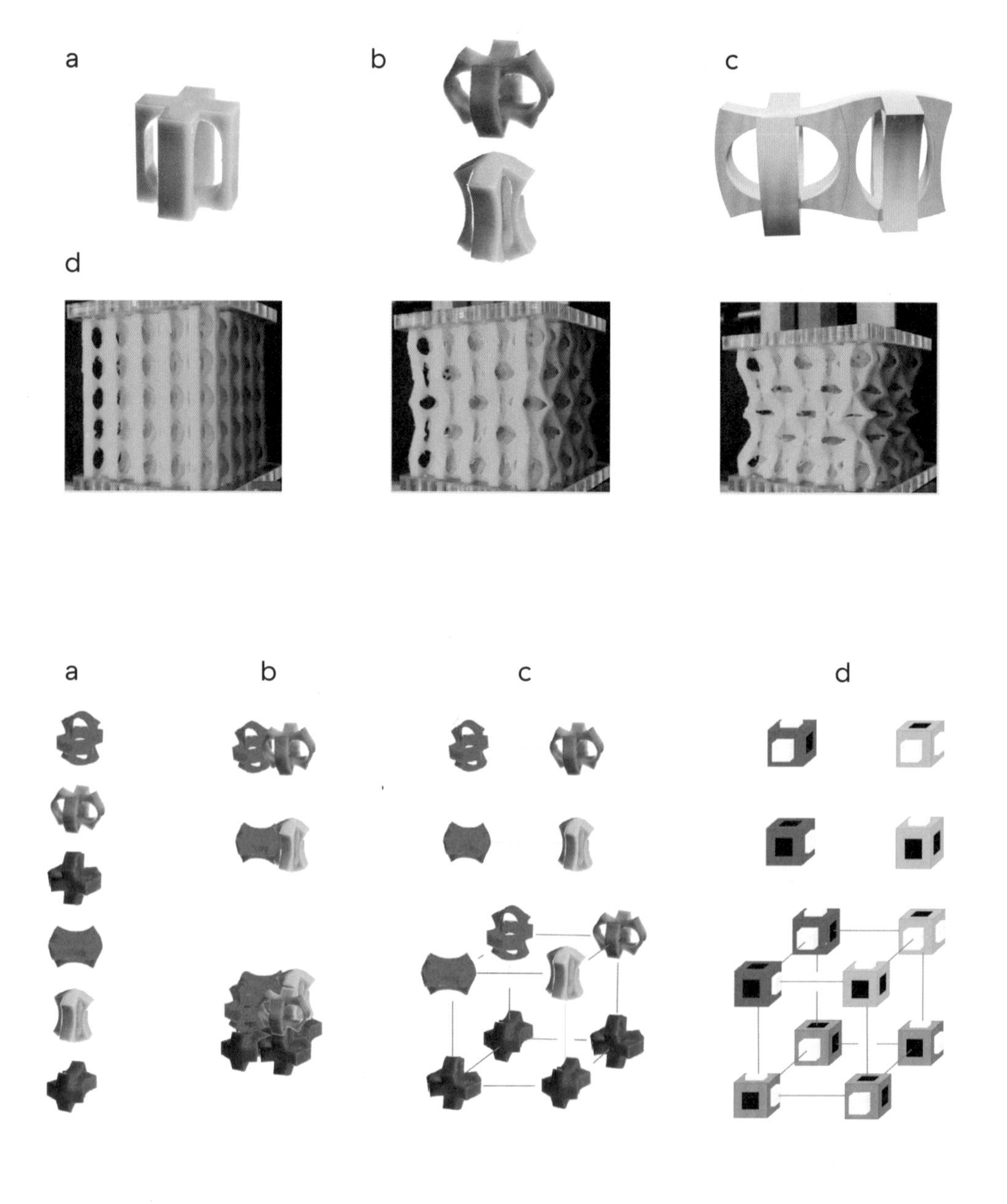

a

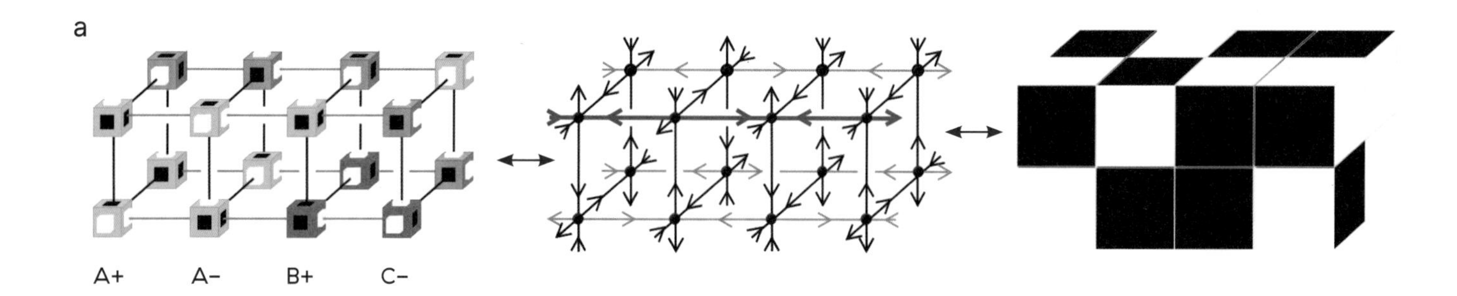

b

c

(i) $A^+ A^- A^+ A^- A^+ A^- A^+ A^-$...

(ii) $A^+ C^- A^+ B^- A^+ C^- A^+ B^-$...

(iii) $A^+ B^- A^+ A^- B^+ A^+ B^- A^+$...

(iv) $A^+ C^- E^+ D^- B^+ E^- A^+ D^-$...

a

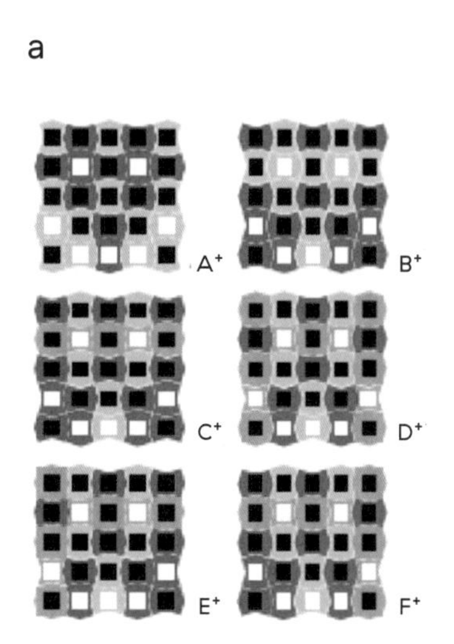

b

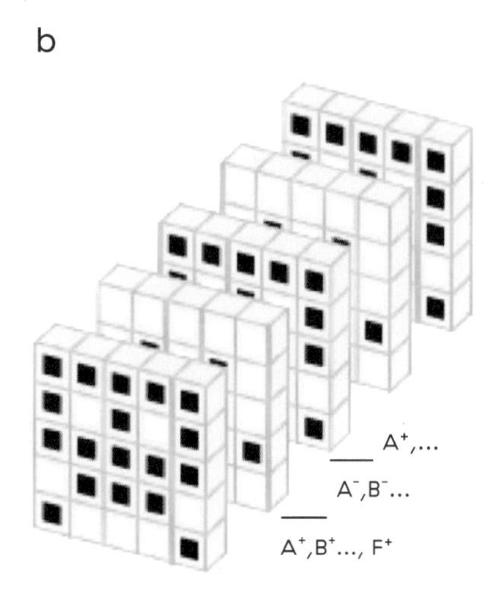

c

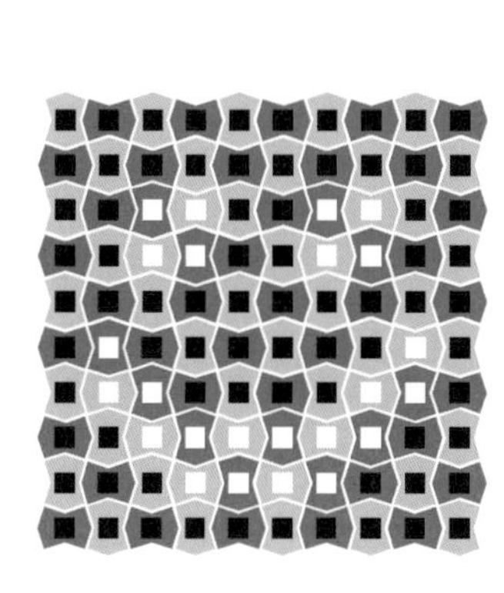

Figure 16.1 (left page, top)
3D mechanical metamaterial. (A) Cubic flexible building block in its undeformed state. (B) Deformed building blocks take the shape of either a flattened (top) or elongated (bottom) brick. (C) A stacking of two deformed blocks, which can flex in unison according to their individual preferred modes of deformation. (D) Silicon rubber metacube (a 3D-printed metacube made of 5 × 5 × 5 building blocks shaped in silicon rubber) under increasing uniaxial compression. The three pictures show three different stages of compression: undeformed (left), moderately compressed (middle), and fully compressed (right). All the building blocks flex in unison as a result of a collective buckling instability throughout the system. Scale bar is 1 cm.

Figure 16.2 (left page, bottom)
Voxelated mechanical metamaterial. (A) Deformed building blocks in their three possible orientations, depicted in red (*x*), green (*y*), and blue (*z*). (B—D) Different representations of 2 × 1 × 1 (top and middle) and 2 × 2 × 2 (bottom) complex stackings of the differently oriented, deformed building blocks demonstrate that these act as flexible voxels in a compact (B), splintered (C), and schematic (D) representation. The schematic representation depicts the bulges by a white beveled pad and the hollows by a black flat padhole.

Figure 16.3 (top)
Holographic structure. (A) Mapping between the RGB voxelated structure (left) to a spin problem (middle) to BW pixelated textures (right). Thanks to the three mirror symmetries of the building blocks, the internal bulging and hollows are alternating along lines across the structure. As a result, the boundary spins or, equivalently, BW pixels (bump or hollow) completely determine the internal spins and RGB voxels. This so-called holographic correspondence allows us to simplify the problem of finding compatible 3D stackings to that of listing compatible 2D slices. (B) Exhaustive library of the 1 × 2 × 2 slices which are compatible with a prescribed BW pixel texture in the *x* direction. The + and — superscripts denote the texture or its negative. (C) Examples of periodic (i—ii), quasi-periodic (iii) (beginning of the Fibonacci sequence), and aperiodic (iv) stackings can be created using the library in panel (B)—the only condition is that positive and negative textures must be stacked alternately.

Figure 16.4 (bottom)
Combinatorial design. (A) Library corresponding to a 5 × 5 smiley texture. The RGB flexible blocks are represented in a 2D flattened "front view" representation for ease of visualization. (B) Stacked 5 × 5 × 1 slices chosen from this library can be used to design a 3D structure with an arbitrarily large number of slices and designable periodicity (or lack thereof).
(C) One of the two elements of the library corresponding to a 10 × 10 smiley texture.

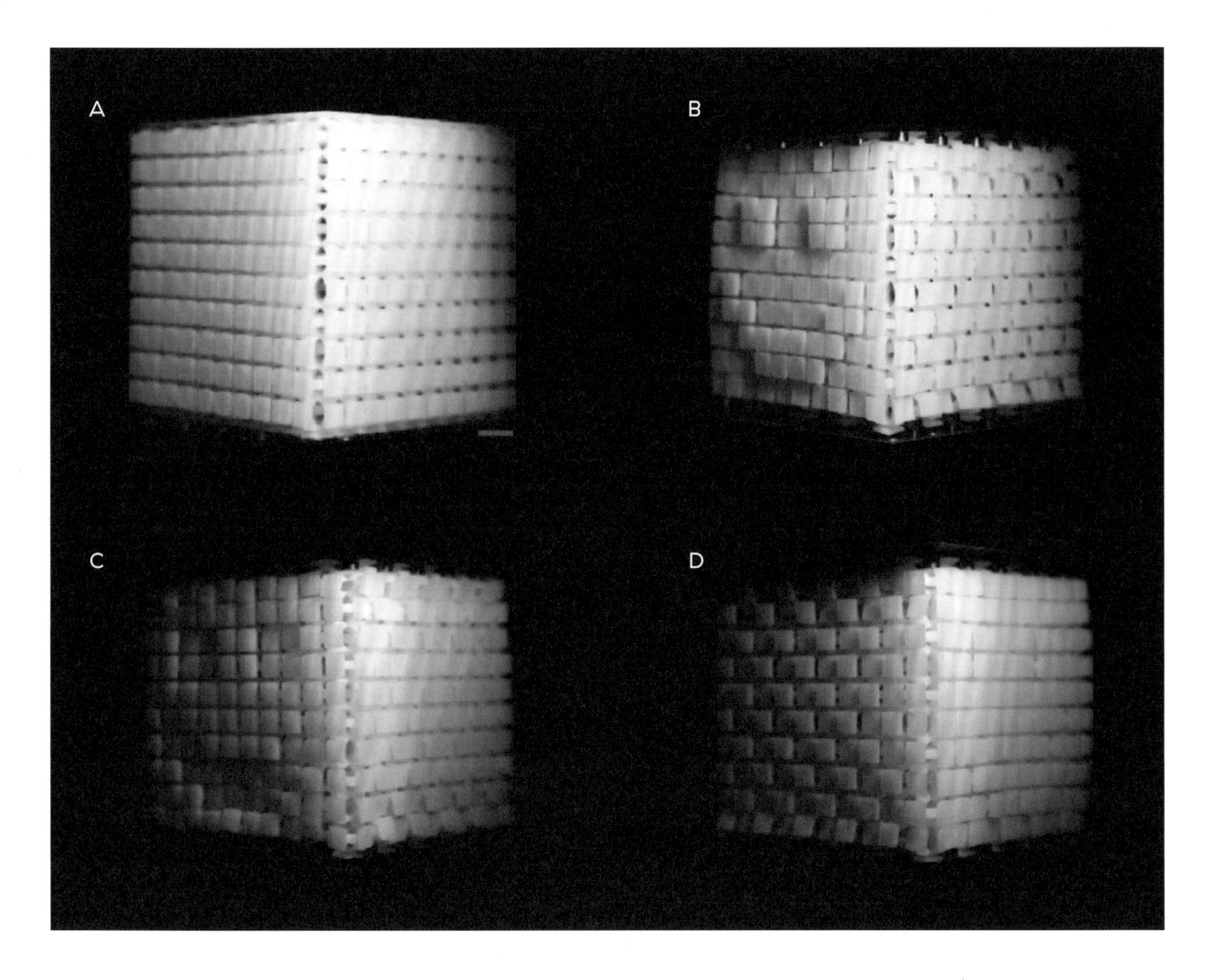

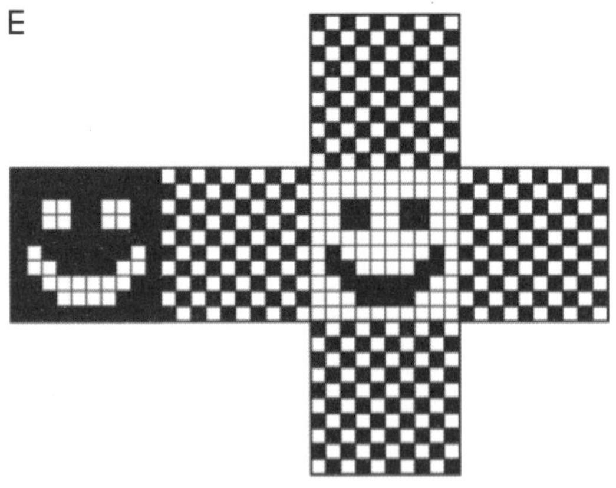

Figure 16.5
3D-printed 10 × 10 × 10 smiley metamaterial. This metacube is made of 10 × 10 × 10 building blocks where the outside blocks have been decorated with square pedestals to clearly visualize surface deformations. The 3D design follows from stacking slices as it has been designed by stacking 10 of the slices shown in figure 16.4(C) (together with their negative, alternatively) and has been 3D-printed commercially out of sintered TPU (Materialize, Leuven, Belgium). (A) Undeformed state. Scale bar is 2 cm. (E) Schematic representation of the programmed, lowest-energy programmed texture of all the faces of the cube. (B—D) Cube under uniaxial compression by textured clamps (with a checkerboard pattern) showing the front face (B), the back face (C), and a side face (D).

NOTES

Reprinted by permission from Macmillan Publishers Ltd. (Nature Publishing Group). C. Coulais, E. Teomy, K. de Reus, Y. Shokef and M. van Hecke, Combinatorial design of textured mechanical metamaterials, Nature 535 (2016): 529–531.

17

Hands–Free Origami: Self–Folding of Polymer Sheets

Ying Liu, Sally Van Gorder, Jan Genzer, and Michael D. Dickey

17

Hands-Free Origami: Self-Folding of Polymer Sheets

Ying Liu, Sally Van Gorder, Jan Genzer, and Michael D. Dickey

Self-folding is an emerging and attractive technique to create three-dimensional (3D) origami structures from two-dimensional (2D) surfaces in a hands-free manner. It is inspired by both origami (the ancient art of paper folding) and the autonomous folding or bending motions present in nature in Venus flytraps, flower blossoms, and protein or DNA folding. Self-folding often employs materials that respond to external stimuli to make 3D shapes with high fidelity and in a hands-free manner. Self-folding has applications such as in actuators and sensors, robotics, biomedical devices, drug delivery, smart packaging, and deployable devices. It is a great yet compelling challenge to create synthetic self-folding systems that can undergo shape change in a controlled response to external stimuli such as light, heat, moisture, mechanical forces, and other sources of energy.[1]

We have developed a simple approach to self-folding. Figure 17.1 highlights a representative cube formed by self-folding. Our process begins with a prestrained planar polymer sheet (also known commercially as "shrink film" or Shrinky-Dinks). If placed in an oven, these sheets would shrink in plane to ~50% of their original size. Instead, we only heat local portions. We achieve this localized heating by first patterning ink on the sheets that can absorb light. The ink can be applied in a number of ways, including using inkjet printing and other planar printing techniques like screen printing. The ink, which is black in figure 17.1, absorbs light selectively relative to the rest of the sheet, which is transparent to light. The absorbed light causes the ink to heat and therefore the polymer beneath the ink to also heat. The resulting heating causes the polymer to shrink, but only in regions beneath the ink. The local shrinking of the polymer sheets behaves like hinges and causes the sheet to fold.

Figure 17.2 describes the process developed using our method of self-folding. The appeal of this technique is that it uses mass-produced polymer sheets and common 2D patterning techniques (i.e., inkjet, screen printing, lithography) to create 3D structures.[2]

From a single planar polymer sheet, many different final 3D structures are possible based on the pattern of the ink. Since the prestrained polymer sheets are transparent, it is possible to use light to heat "hinges" patterned on the back side of the sheet. Thus, bidirectional folding is possible; that is, folding toward the light and away from the light (in origami, these are called valley and mountain folds, respectively).

Figure 17.3 shows various 3D structures created using this method. Key physical parameters (hinge

Figure 17.1
Self-folding of a cube (from left to right) using light as a stimulus. The black ink, which is inkjet-printed on a plastic sheet, absorbs light selectively and gets hot, which induces localized shrinkage of the plastic, and thus folding. The dimension of the cube is 1 cm × 1 cm × 1 cm. The widths are 2 mm and the lengths are 10 mm for all black hinges.

geometry, inked hinge width, and support temperature) affect the rate, angle, and direction of folding. To date, we have modeled the thermal profiles inside the polymer film, and investigated folding dynamics based on thermal shrinkage and the rheological properties of the polymer sheet.[3] We have also modeled the scaling laws of folding, the rate of folding, and the mechanics of folding to develop compliant folding mechanisms.[4]

While the hinges in the 3D structures (figure 17.3) fold simultaneously, it is of great interest to self-fold 3D structures in a sequential manner so that more complicated structures can be achieved. Our method can be extended to different-color inks (instead of just black) that absorb light at prescribed wavelengths. For example, yellow ink will absorb blue light but not red light. In contrast, cyan ink will absorb red light, but not blue. Therefore, some nested structures and layer-by-layer collapsed structures can be created in this way. Figure 17.4 demonstrates structures that can only be formed by sequential self-folding.

We have extended this technique to other types of polymer sheets (e.g., poly(methyl methacrylate)/Plexiglas, polycarbonate, polystyrene), different prestrain directions (biaxial prestrain,[2] uniaxial prestrain[5]), and both thin (hundreds of microns) and thick (~mm) polymer sheets,[6] to induce folding, rolling, and even planar-shape programmable structures. We also explored various stimuli sources including laser light,[7] microwaves,[8] and LEDs[1, 9] to activate self-folding using our strategy, which provides alternative strategies to self-folding polymer films in a hinge-free way or in a sequential pathway. Therefore, the approach has been validated as a controllable, predictable, and general strategy for self-folding (figure 17.5).

The appeal of our approach is its simplicity and versatility for converting surface patterns on prestrained polymer sheets into 3D objects within seconds upon exposure to a stimulus. This method is able to harnesses the multitude of 2D patterning techniques (i.e., inkjet, screen printing, lithography) to pattern the folding hinges. Our approach has commercial relevance because it is compatible with low-tech, high-throughput 2D patterning techniques (e.g., roll-to-roll patterning) as well as 2D patterning of electronics and photonics. There are plenty of opportunities to involve engineers, scientists, and artists in advancing this interdisciplinary approach to self-folding.

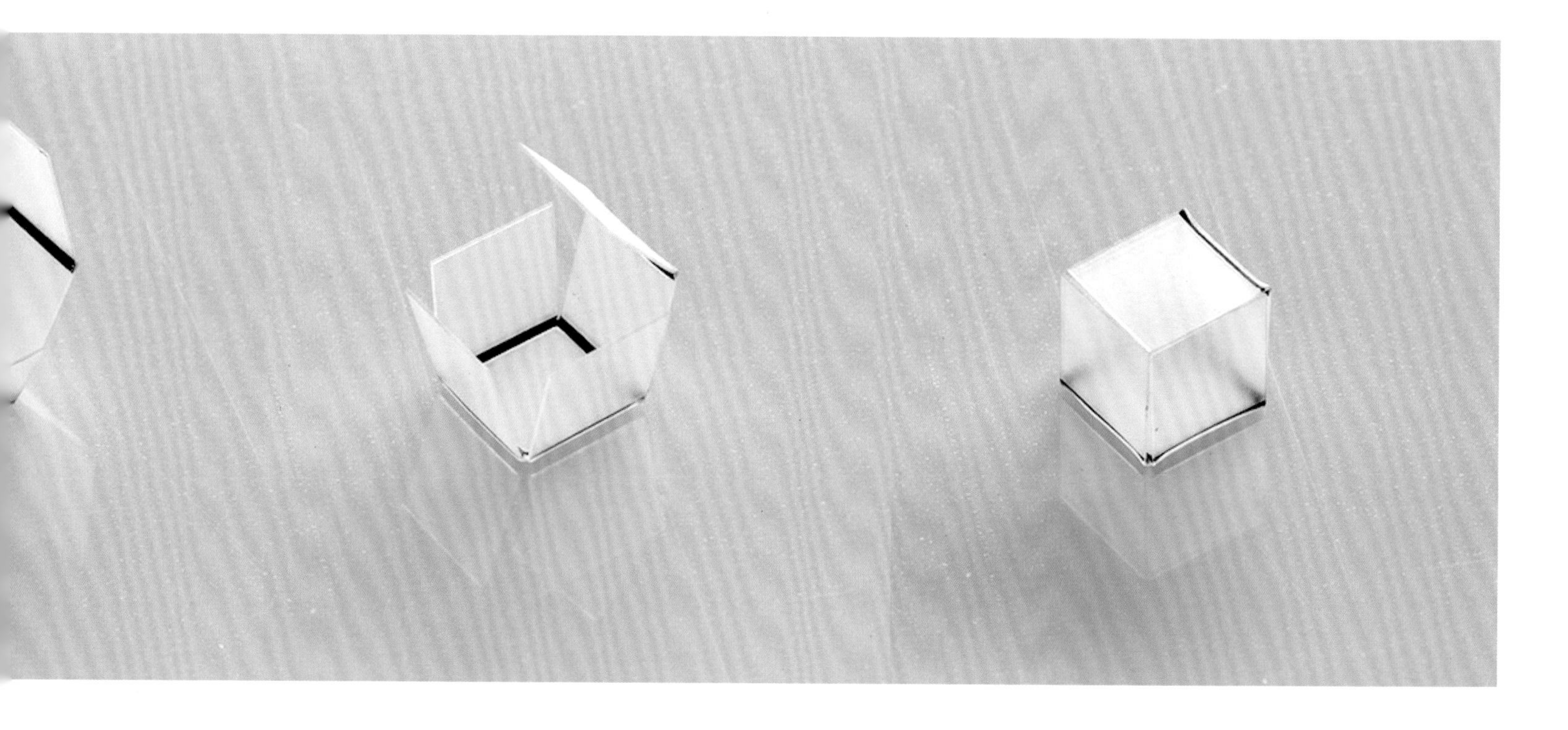

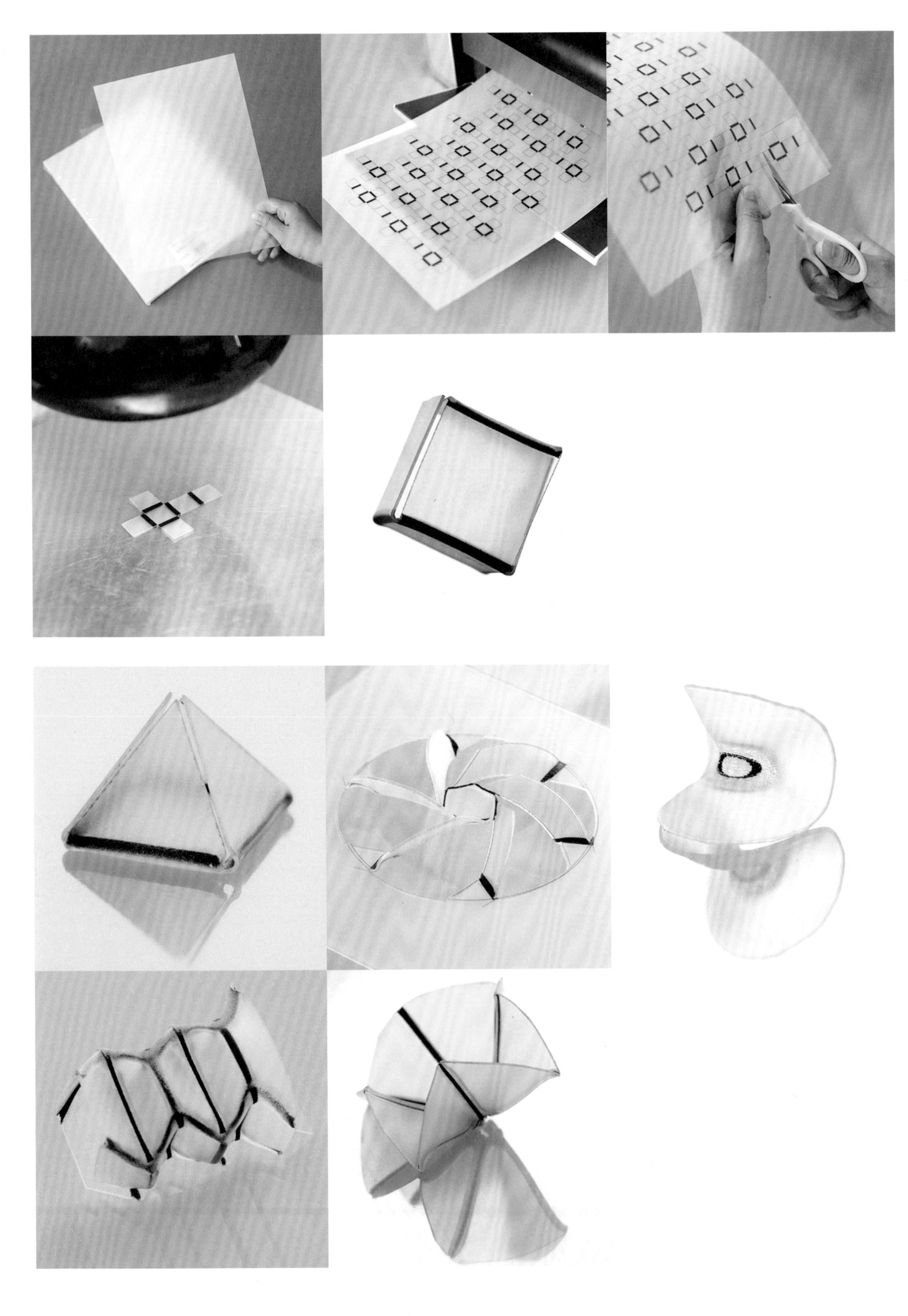

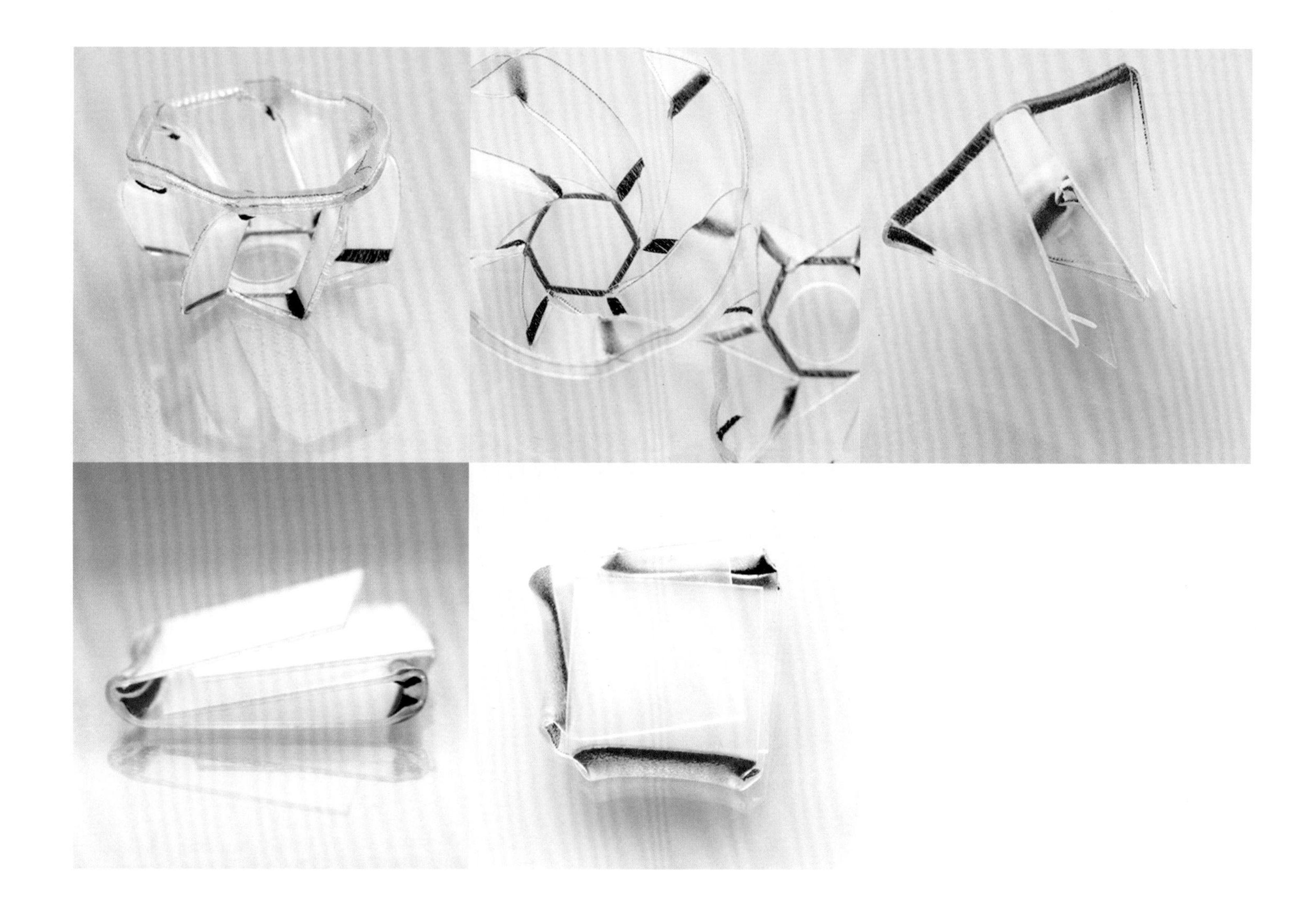

Figure 17.2 (left page, top)
Process of self-folding. From left to right: the prestrained polymer sheets used for self-folding; the polymer sheets are printed with black ink using an inkjet printer; the shapes are cut by scissors; the planar template is placed on a hotplate at 90°C and exposed to an infrared lamp; a 3D structure (a tube as an example) can be folded within seconds. The dimension of the cube here is 1 cm × 1 cm × 1 cm.

Figure 17.3 (left page, bottom)
Collage of 3D structures self-folded using this novel method. The tetrahedral box has each side of 1 cm on the bottom. The widths of all black hinges are 1.5—2 mm.

Figure 17.4 (top)
Complex structures achieved by sequential self-folding. The widths of all color hinges are 1.5—2 mm.

Figure 17.5
Generalization of this self-folding strategy utilizing other materials, various ink absorbers, and different stimuli.

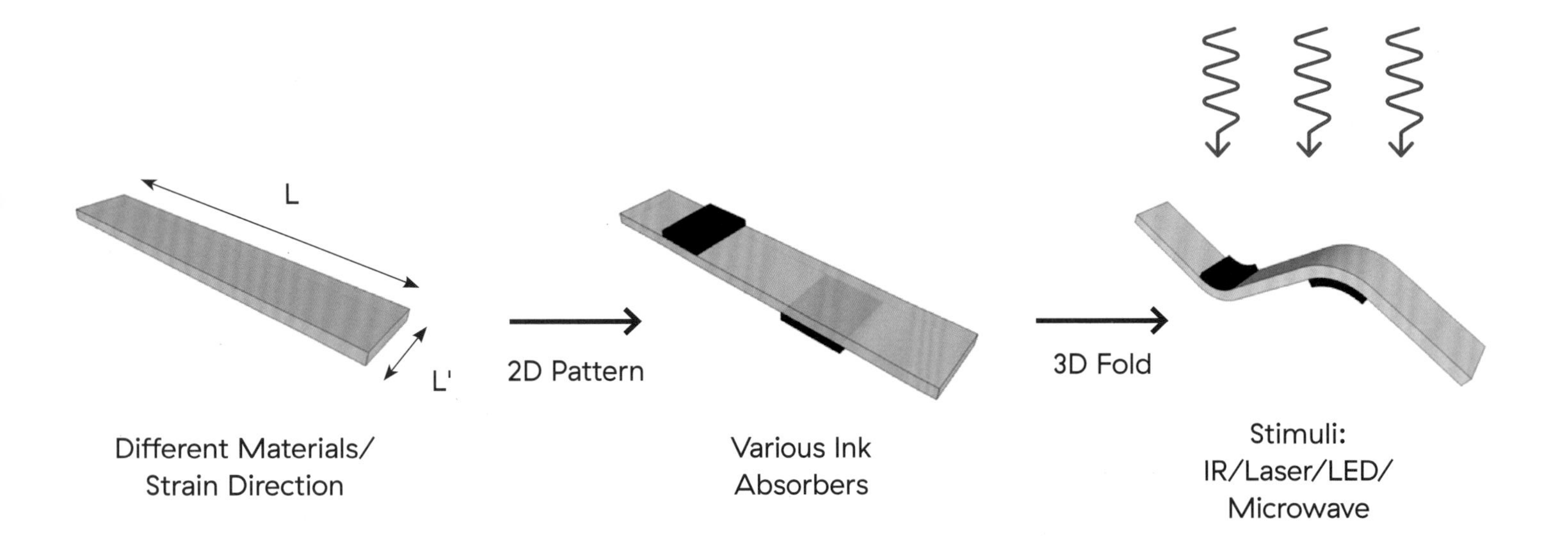

NOTES

1 Ying Liu, Jan Genzer, and Michael D. Dickey, "2D or Not 2D: Shape Programming of Polymer Sheets," *Progress in Polymer Science* (2015).

2 Ying Liu et al., "Self-Folding of Polymer Sheets Using Local Light Absorption," *Soft Matter* 8, no. 6 (2012): 1764–1769.

3 Ying Liu et al., "Simple Geometric Model to Describe Self-Folding of Polymer Sheets," *Physical Review E* 89, no. 4 (April 10, 2014): 42601.

4 Russell W. Mailen et al., "Modelling of Shape Memory Polymer Sheets That Self-Fold in Response to Localized Heating," *Soft Matter* 11, no. 39 (2015): 7827–7834.

5 James R. Allensworth et al., "In-Plane Deformation of Shape Memory Polymer Sheets Programmed Using Only Scissors," *Polymer* 55, no. 23 (2014).

6 Duncan Davis et al., "Self-Folding of Thick Polymer Sheets Using Gradients of Heat," *Journal of Mechanisms and Robotics* 8, no. 3 (March 7, 2016): 31014.

7 Ying Liu et al., "Three-Dimensional Folding of Pre-Strained Polymer Sheets via Absorption of Laser Light," *Journal of Applied Physics* 115, no. 20 (2014).

8 Duncan Davis et al., "Self-Folding of Polymer Sheets Using Microwaves and Graphene Ink," *RSC Advances* 5, no. 108 (October 19, 2015): 89254–89261.

9 Ying Liu et al., "Sequential Folding of Polymer Sheets," in preparation.

18

Growing and Morphing Shapes: Using Swelling and Geometry to Control the Shape of Soft Materials

Douglas P. Holmes

18

Growing and Morphing Shapes: Using Swelling and Geometry to Control the Shape of Soft Materials

Douglas P. Holmes

Slender structures are ubiquitous—they include carbon nanotubes, airplane wings, blood vessels, spider silk, lipid membranes, contact lenses, and human hair. The behavior of these thin objects is fascinating because their geometry can change dramatically in response to small amounts of force—hair may curl and tangle, skin will wrinkle, and soda cans can crumple. If these shapes are active and adaptable, their bending, folding, and twisting will provide advanced engineering opportunities for deployable structures, soft robotic arms, mechanical sensors, and the rapid manufacturing of 4D materials.

At play in many soft and active systems is the coupling between fluid motion and structural deformation. The fluids may provide fuel or contain materials necessary for self-healing or sensing. These fluid-structure interactions occur across many length scales within synthetic and biological systems. At large scales, inertial flows and fluid weight can cause substantial structural deformations, while at small length scales, surface tension may dominate a material's deformation. In addition to the forces that a fluid exerts externally onto a flexible structure, the diffusion of fluid into the material can cause substantial swelling and deformation.

CURLING WITH DROPLETS

When you put a straw into a liquid, the liquid rises via capillary action—surface tension draws the fluid up while gravity pulls it down. The smaller the straw's diameter, the higher the fluid rises. If the walls of that straw are flexible, the fluid rises higher still as surface tension pulling on the walls is strong enough to bend them closer together. This is known as elastocapillarity, and it is what you see when bristles of a paintbrush or wet hairs clump together. Now, if the material is flexible and absorbent, like a sponge, the fluid will swell the walls, causing them to curl apart when wetted. So there are two competing effects—surface tension pulling the flexible objects together, and

Figure 18.1
Capillary curling. Fibers of silicone rubber are lowered into a bath of silicone oil. Coupling capillary action with flexibility pulls the fibers together, while swelling of the beams begins to slowly curl them apart. Finally, the curling fibers peel from the bath, entrapping a droplet of fluid which is ratcheted upward. photo credit: P.T. Brun and D.P. Holmes

swelling curling them apart. In figure 18.1, two flexible and absorbent silicone rubber fibers are dipped into a bath of silicone oil.

Initially, a balance between elasticity and capillarity pulls the fibers together, and then the swelling of the fluid into the material slowly curls the fibers apart. Eventually, the fibers peel from the surface of the bath, and a fluid droplet moves upward. These large deformations are caused by an amount of fluid smaller than the volume of water in a raindrop. The addition of swelling to the study of elastocapillarity may bring new insights to the swelling and drying of many soft, porous engineered materials, such as textiles and paper, as well as swellable biological structures, such as hair, certain types of plants, and other soft tissues. Elastocapillary swelling could also lead to the design of new types of soft actuators involving liquid transport and shape changes, building on recent advances in capillary origami.

GROWING SHEETS INTO SHELLS

For thin structures, geometry is paramount. It's easy to roll a sheet of paper into a cylinder, but impossible to wrap it around a sphere. Wrapping a sheet (which is intrinsically flat) around a sphere (which is intrinsically curved) requires you to stretch it, while bending it around a cylinder does not. Thin structures will deform in a way that avoids changing their intrinsic geometry whenever possible, because it is far easier to bend a thin structure than to stretch one. In engineering design, this constraint is important because it provides a means for large, controllable shape changes with minimal energy input. So flat sheets of any shape will morph into cylinders (figure 18.2), while shells bend into spindles (figure 18.3), mimicking the shape transition of a drying pollen grain. Much of swelling is slow, as the permanent structural morphing is driven by the diffusion of viscous fluid—a process that may

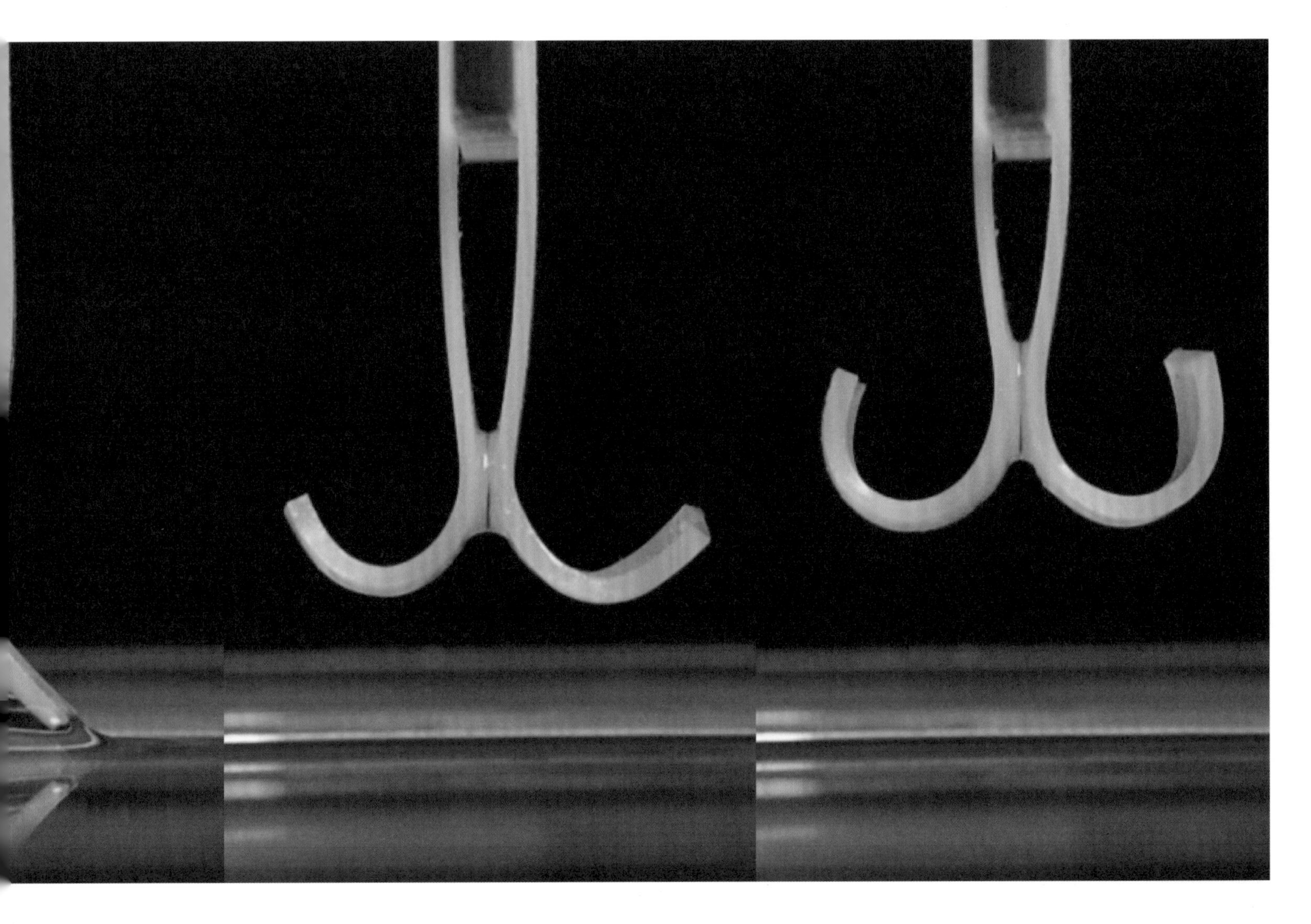

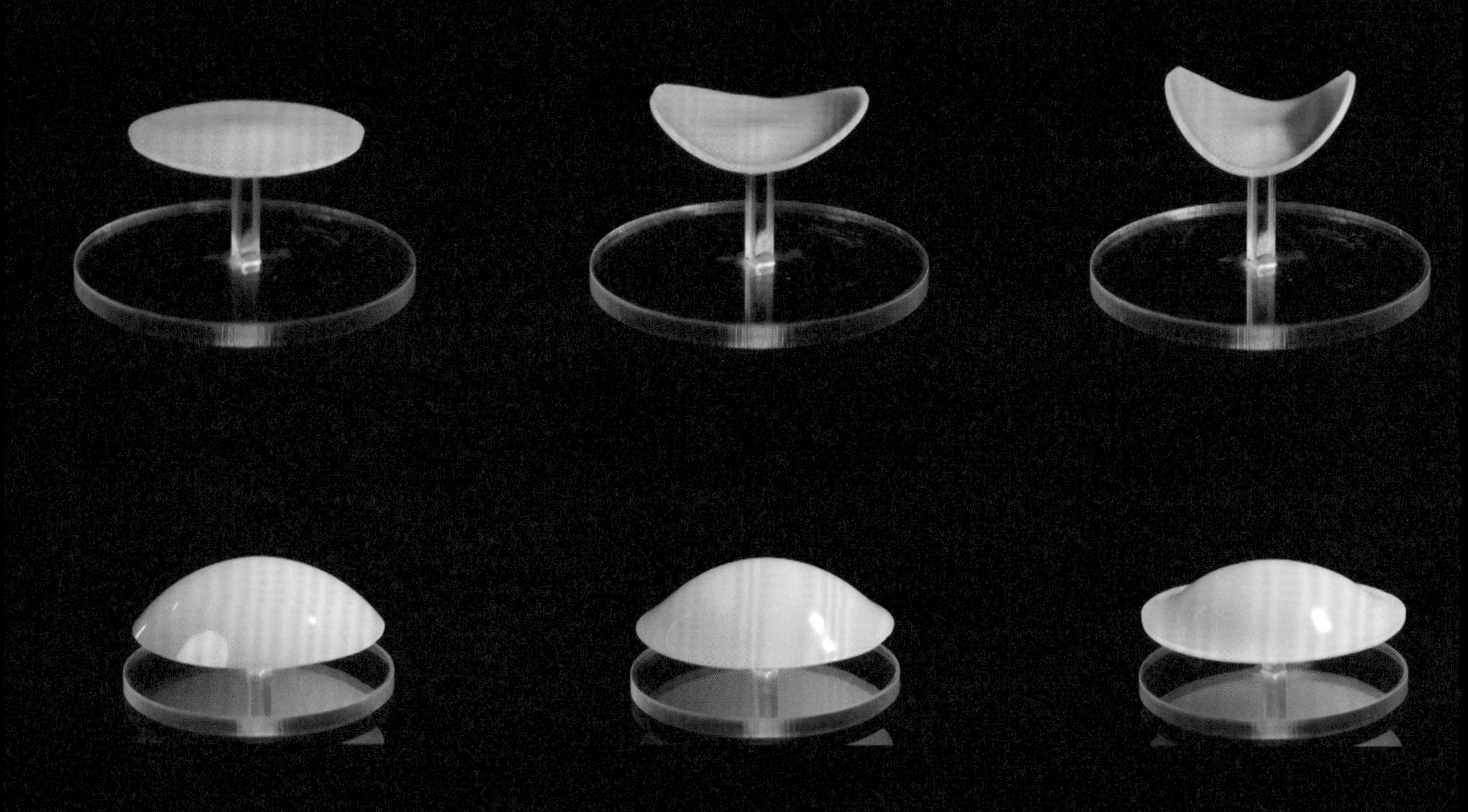

Figure 18.2
Morphing bilayers. Top: A flat, circular plate morphs into a cylinder as its top layer shrinks and its bottom layer swells. Bottom: A spherical cap snaps into a spindle as its outer layer shrinks. Both morphing bilayers adopt deformed shapes that require a minimal amount of stretching. photo credit: M. Pezzulla, M. Steranka

take hours. However, when aided by structural instability, large shape changes can occur in fractions of a second. A shell can dynamically snap between shapes (figure 18.2), similar to an umbrella on a windy day. These transitions are governed by geometry rather than diffusion, and can occur at the speed of a sound wave propagating through the material.

To utilize the swelling of soft materials to dynamically morph sheets and shells, the swelling fluid can be incorporated directly into the material. Since thin structures can significantly deform in response to swelling by small volumes of fluid, we can actively deform structures by moving small amounts of fluid around within a material, thereby causing specific regions to swell or shrink. This process is analogous to the heating of bimorphs—materials that are bound together but expand to different lengths when heated. Using a process known as residual swelling, where residual amounts of fluid are left within an elastomer and free to swell nearby regions, we can permanently grow bilayer sheets into shells.

In some cases, stretching is unavoidable. If residual swelling occurs in the plane of the object, rather than through its thickness, bending alone will not suffice. These shapes will change their intrinsic curvature, morphing from flat sheets into saddles (figure 18.4). Here all the fluid is moving from the center of the disk to its edges, and since the disk's circumference is increasing in length more than its radius, it buckles with wavy edges. This structural morphing is driven by the same mechanism of differential growth that governs the wavy edges in growing leaves and flowers, such as a blooming lily.

Behind all of these shape transitions is the interplay between fluid and structure; swelling and geometry. Thin structures bend easily and soft materials can morph on command in response to the movement of small amounts of fluid. Perhaps what is most compelling is that the mechanics and mathematics that describe these shape changes are indifferent to the chosen stimulus—we can replace swelling with heat, or voltage, or magnetism, and the structures will morph accordingly.

Figure 18.3 (left)
Pollen grains. Before and after images of a spherical shell (top) rolling into a spindle and (bottom) pinching closed due to residual swelling. photo credit: A.Bade, M. Pezzulla

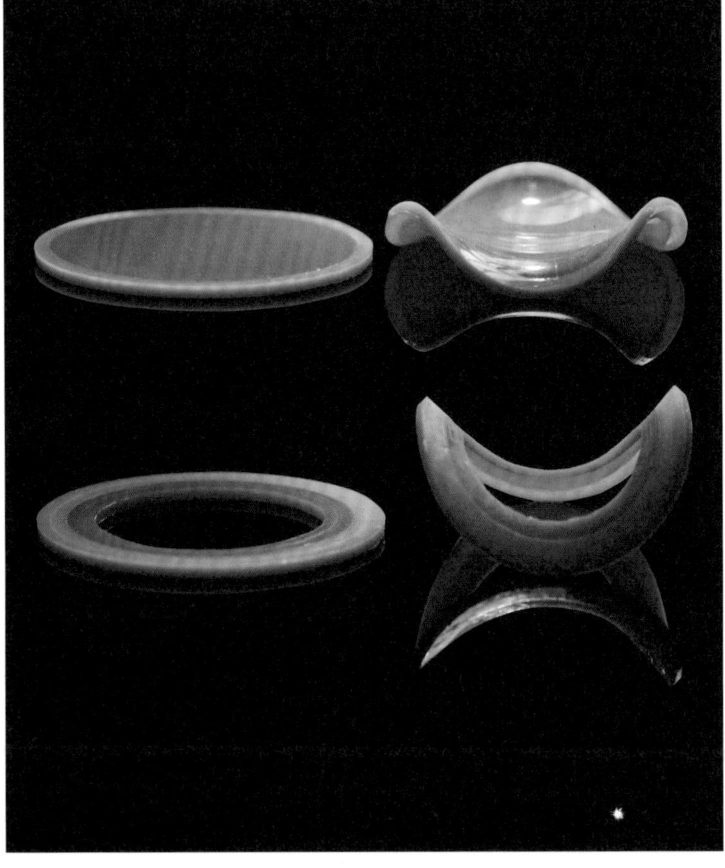

Figure 18.4 (right)
Soft saddles. Top: A flat sheet undergoes differential growth and buckles into "monkey saddle." Bottom: A ring experiencing residual swelling morphs into a saddle that while curved in space is locally flat on the surface. photo credit: M. Pezzulla

NOTES

Figure 18.1

1 D.P. Holmes, P.-T. Brun, A. Pandey, and S. Protière, "Rising beyond elastocapillarity," *Soft Matter*, **12**, 4886–4890, (2016).

Figure 18.2

2 M. Pezzulla, N. Stoop, X. Jiang, and D.P. Holmes, "Curvature-Driven Morphing of Non-Euclidean Shells," *Proceedings of the Royal Society A*, **473** (2201), (2017).

3 M. Pezzulla, G.P. Smith, P. Nardinocchi, and D.P. Holmes, "Geometry and Mechanics of Thin Growing Bilayers," *Soft Matter*, **12**, 4435–4442, (2016).

4 M. Pezzulla, S.A. Shillig, P. Nardinocchi, and D.P. Holmes, "Morphing of Geometric Composites via Residual Swelling," *Soft Matter*, **11**, 5812–5820, (2015).

19

Coffee Bags and Instinctive Active Materials: The Path to Ubiquity

Greg Blonder

19
Coffee Bags and Instinctive Active Materials: The Path to Ubiquity

Greg Blonder

"Instinctive" active materials respond to their environment in predictable ways, sensing their surroundings and modifying their geometries to achieve a class of desired, if rigid, goals. Neither smart nor dumb, such materials could find applications in the agricultural, energy, and construction fields, providing they are reliable and very low-cost. We've developed a range of instinctive bimorph films (IBFs) based on modifying existing food packaging films and production lines. They are fast, cheap, and versatile, and well suited for commercial applications.

Open a container of Cheetos, and you might easily dismiss the bag in a powdered orange feeding frenzy. Yet the plastic package is a laminated feat of engineering. An outer layer of aluminized mylar affords a tough mechanical, oxygen, and light barrier, along with a smooth surface for printing. Inner layers provide additional barrier/structural capabilities, and are typically capped by a heat-fusible film to seal the folded and filled bag.

Every year, around 1 terameter^2 of plastic films are produced worldwide, predominantly in single layers. With an average price of ten cents a square meter, they are ubiquitous to the point of becoming a disposal and recycling challenge.

Plastic food packaging films are laminated, printed, folded, and sealed at astonishing speeds on a continuous reel-to-reel web. However, most food packaging films are an asymmetric stack of layers that will curl or twist in response to stresses embedded during the reel-to-reel lamination process, or from unbalanced thermal expansion. While curling is a deleterious annoyance for the plastic bag industry, these thermally responsive twists and bends provide the mechanism behind their "instinctive" behavior in our application.

As seen in figure 19.1, a basic IBF structure consists of a high-expansion plastic film coextruded with a low-expansion layer—a simple two-layer bimorph. The low-expansion layer can be very thin, providing it is stiff and inextensible.

The green film expands or contacts vigorously with temperature, the red layer is nearly stable. At one unique temperature, the layflat temperature (LFT), the film lies flat; above or below the LFT it curls.

The bimorph bending equation is well established. For thin films with low stress, the deflection of a cantilevered beam tip d is:

$$d \simeq \frac{\Delta\alpha\Delta T}{\delta} L^2$$

while the radius of curvature r is:

$$r \simeq \frac{\delta}{2\Delta\alpha\Delta T}$$

L is the beam length, $\Delta\alpha$ is the difference in thermal expansion coefficient of the two plastic layers, ΔT the temperature difference relative to the layflat temperature LFT, and δ is the thickness of the beam.

As a concrete example, consider a polyethylene film laminated to a thin sheet of PET, with a $\Delta\alpha$ of 150 ppm/°C, a thickness of 4 mils (0.01cm), and a length > 10 cm. When heated 20°C above its LFT, this strip will curl into a ~3 cm diameter cylinder. Due to the low mass of the film, response times are of order a second.

A more advanced structure, as seen in figure 19.2, might rely on an adhesive layer to bond the low- and high-expansion films together, along with an integrated thin-film heater layer to electrically activate the structure (the heater can be buried inside the film, or on the surface). Sections of the film can be partially die- or laser-cut. Or the unprocessed lamina may be in the form of a symmetric stack (high-low-high expansion) with the high-expansion film selectively removed from one side so the film can bend in either direction. Finally, stiffener bars or plates can force the film to bend along a specific axis relative to the built-in curl.

The layflat temperature, along with any intrinsic curl stresses, can be adjusted via some combination of roller tension, curing temperatures, feed speeds, and postprocessing annealing. Generally, the film will curl in

Figure 19.1 (top)
Effect of temperature on bimorph curvature. The curvature, which is a linear function of temperature, changes sign above or below the layflat temperature.

Figure 19.2 (middle)
Cross-section of an instinctive bimorph film structure. A bilayer's simple curling behavior can be modified by locally stiffening certain regions, by removing one layer of the film, or by slicing into distinct regions.

Figure 19.3 (bottom)
Flower Surprise Friends, a playset sold by Tucker Toys International, with an IBF "flower" that "blooms" in sunlight. A small flower die-cut from IBF enfolds a plastic character. When placed in the sun, the petals warm, the flower "blooms," and the character is revealed.

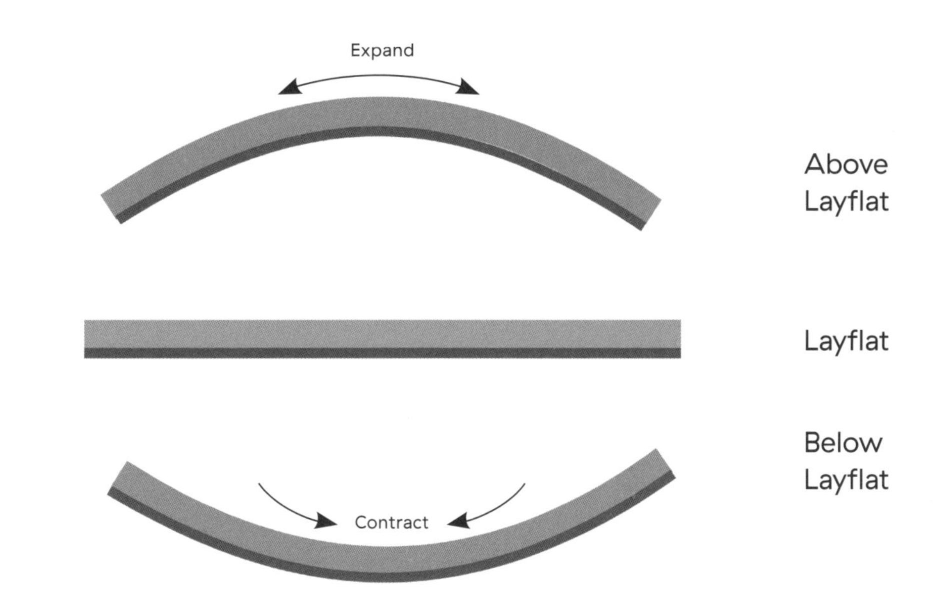

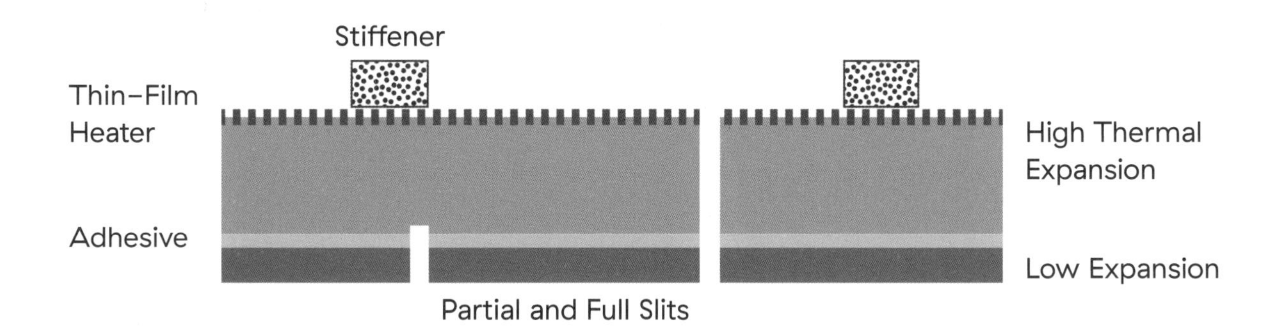

Figure 19.4
Diploria. An art installation with IBF films surrounding a thermometer. A reference to coral reef bleaching and global warming.

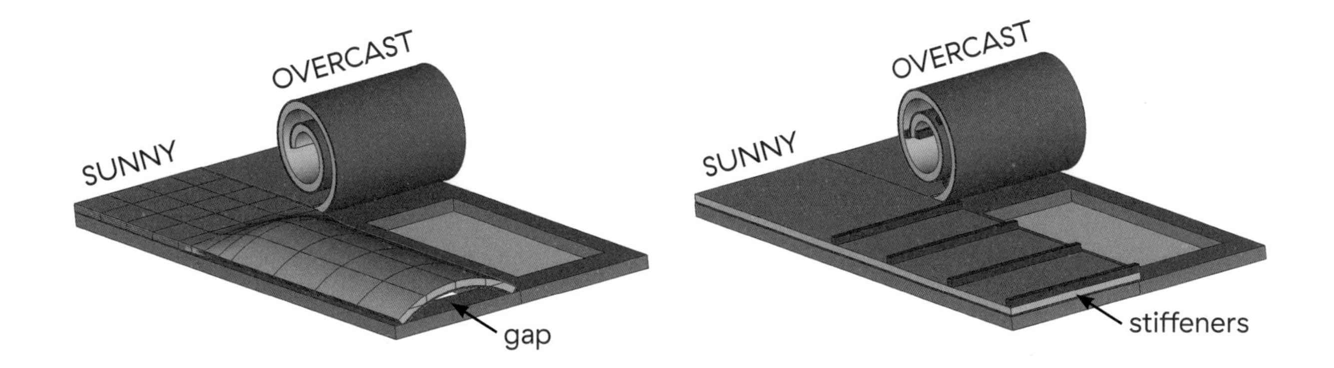

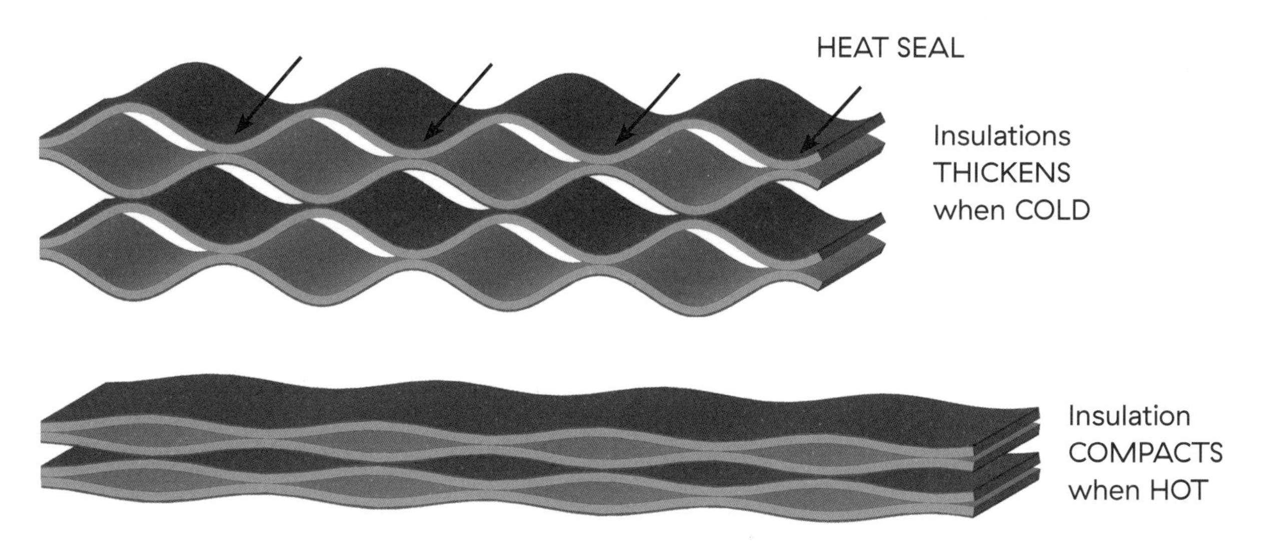

Figure 19.5 (top left)
Solar insolation controlled by instinctive bimorph dampers. When heated in sunlight, the IBF films uncurl, blocking sunlight from entering the room and thus limiting the room's temperature rise.

Figure 19.6 (top right)
Stiffened bimorph damper version of figure 19.5. The bars more tightly control the film's curl direction.

Figure 19.7 (bottom)
Quilted bimorph wall cavity insulation, responding to hot and cold temperatures. Each IBF layer is periodically bonded to its neighbor, such that the film bows open into an air-trapping, insulating batt when cold. When warm, the instinctive film flattens and allows heat to pass.

a direction parallel to the web motion. We have produced IBF structures with LFTs ranging from 0°C to 250°C, and thicknesses from a few mils to a millimeter. Lifetimes are > 100 million cycles.[1]

Since these films move in response to temperature changes (and in some cases humidity or their chemical environment), a desired function can be activated by careful control over its geometry. Our long-term goal is to leverage high-volume packaging to impact the efficient use of energy; however, first applications were in toys (a sunlight-blooming flower shown in figure 19.3) and kinetic art (figure 19.4).

Two examples of future energy applications follow.

SOLAR LOADING

Windows are often a source of significant solar loading during daylight hours. An instinctive active material would sense the heat produced by high solar loads and reflect excess solar energy during the day, while allowing light to enter at all other times.

For example, a large die-cut array of slots (figure 19.5, the gray film) is overlaid with a semireflective IBF film. Annealing and curing temperatures during film lamination adjust the LFT to 90°F. This screen might be placed within a double-glazed window pane. Exposed to bright sunlight, the film heats and flattens as shown, almost blocking the slots.

Under less intense solar heating, the film curls in the perpendicular direction (due to the intrinsic built-in stress from lamination), uncovering the slot and allowing light to enter. All at a cost that is acceptable to the construction industry.

In the improved design shown in figure 19.6, small stiffener bars are laminated to the light valve. These stiffeners prevent the flap from curling into a tight cylinder in sunny conditions, but allow the flap to roll up on overcast days. Longer flaps are made possible with this approach, increasing the modulation range of the flaps to near 80%. A similar structure instinctively controls ventilation.

INSULATION APPLICATION

Conventional wall cavity insulation affords a fixed R value, independent of weather conditions. However, by alternating, stacking, and laminating IBF films into a quiltlike structure (figure 19.7), we can produce an air-filled insulation batt. This batt changes thickness depending on temperature, and thus instinctively adapts its effective R value to the ambient conditions. For example, in a northern climate, the quilt would insulate an exterior wall on a cool day, and on a hot day would flatten and let heat escape, thus moderating temperature swings without the need for active control.

IMPLICATIONS

Laminated plastic films permeate our industrial society, enabling sophisticated chemical and physical protection from the environment. However, when incorporated into more complex mechanical structures, where the strips are free to respond to the temperature and humidity of their surroundings, they are transformed into instinctive bimorph films. These IBFs transcend their intended passive roles and are capable of adapting the structure's function to a changing environment, with potentially dramatic improvements in performance.

NOTES

1 See patents US 6,966,812; US 8,991,026; US 20050284588 A1; US 6966812 B1.

20

Hydro–Fold

Christophe Guberan

20
Hydro–Fold

Christophe Guberan

This research aims to explore the properties of paper and how its structure can change with liquid. In effect, paper is sensitive to moisture and distorts when it gets wet. This is usually perceived as something one should avoid, but Hydro–Fold focuses on this phenomenon.

Extensive experimentation with a wide variety of paper types led to the use of tracing paper for the project, which proved to react the most to humidity. Indeed, air trapped in layers of cellulose is removed thanks to a chemical treatment, and water thus does not penetrate the paper but rather remains on the fiber's surface.

A water printing device was developed in–house. In effect, one needs to apply a precise amount of water at specific locations, a task that was too tedious using conventional tools such as sprays, foam paint roller, or brushes. By exchanging the contents of ink cartridges for a combination of ink and water, different patterns such as grids and line–based shapes can be precisely printed, and parameters such as line thickness and transparency can be adjusted to produce various bending effects. Moisture is allowed to evaporate, thereby causing paper fibers to contract. As it dries, paper distorts and folds around the most printed areas, transforming itself from a two–dimensional sheet to a three–dimensional structure in which lines become edges and surfaces become stable volumes.

With no external physical intervention by the designer, paper sheets seem to come alive before crystallizing into configurations that reveal the magic of the process. As Juhani Pallasmaa puts it, "vision reveals what touch already knows."

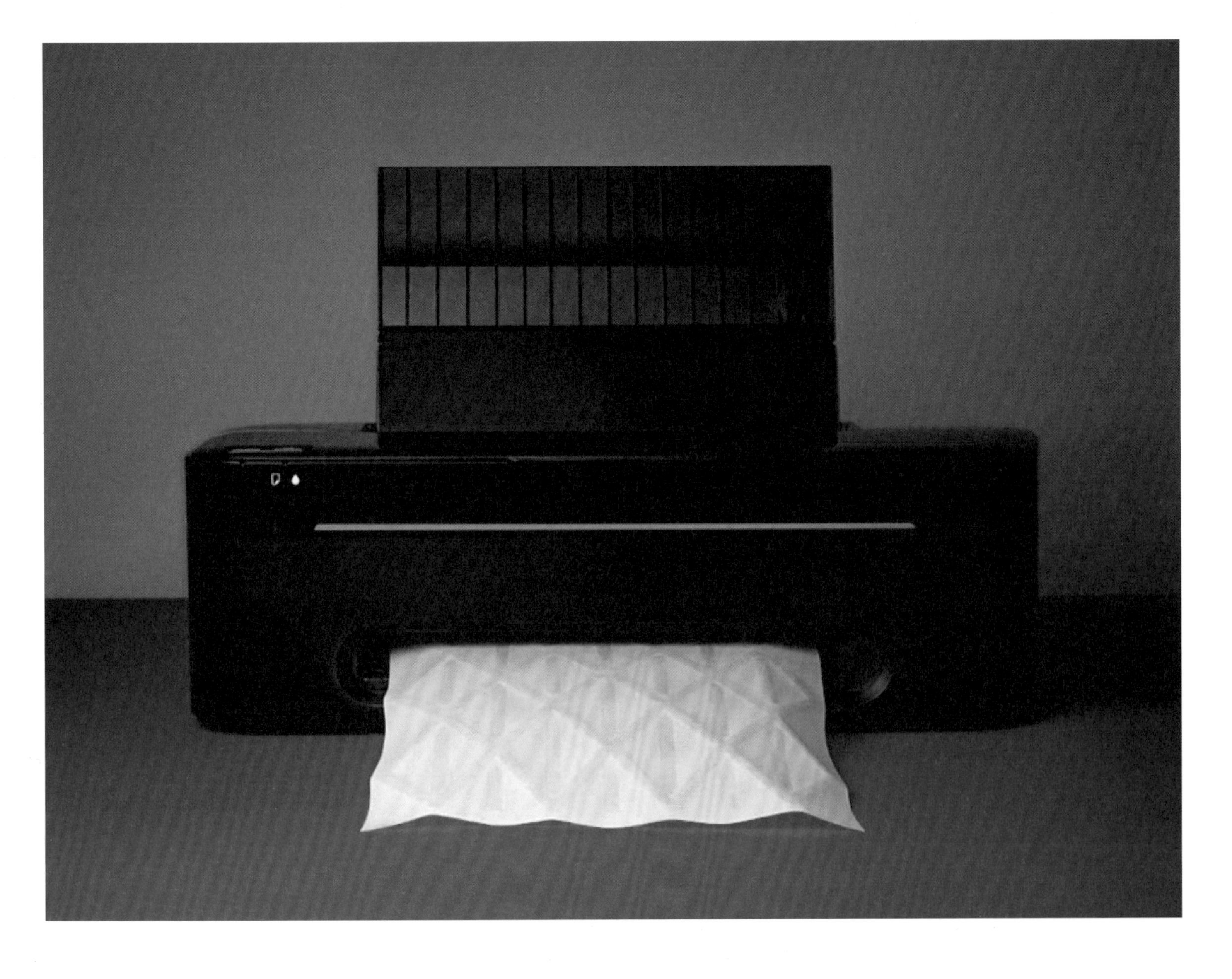

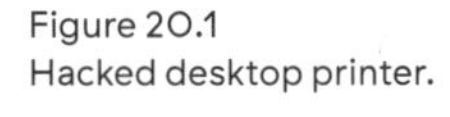

Figure 20.1
Hacked desktop printer.

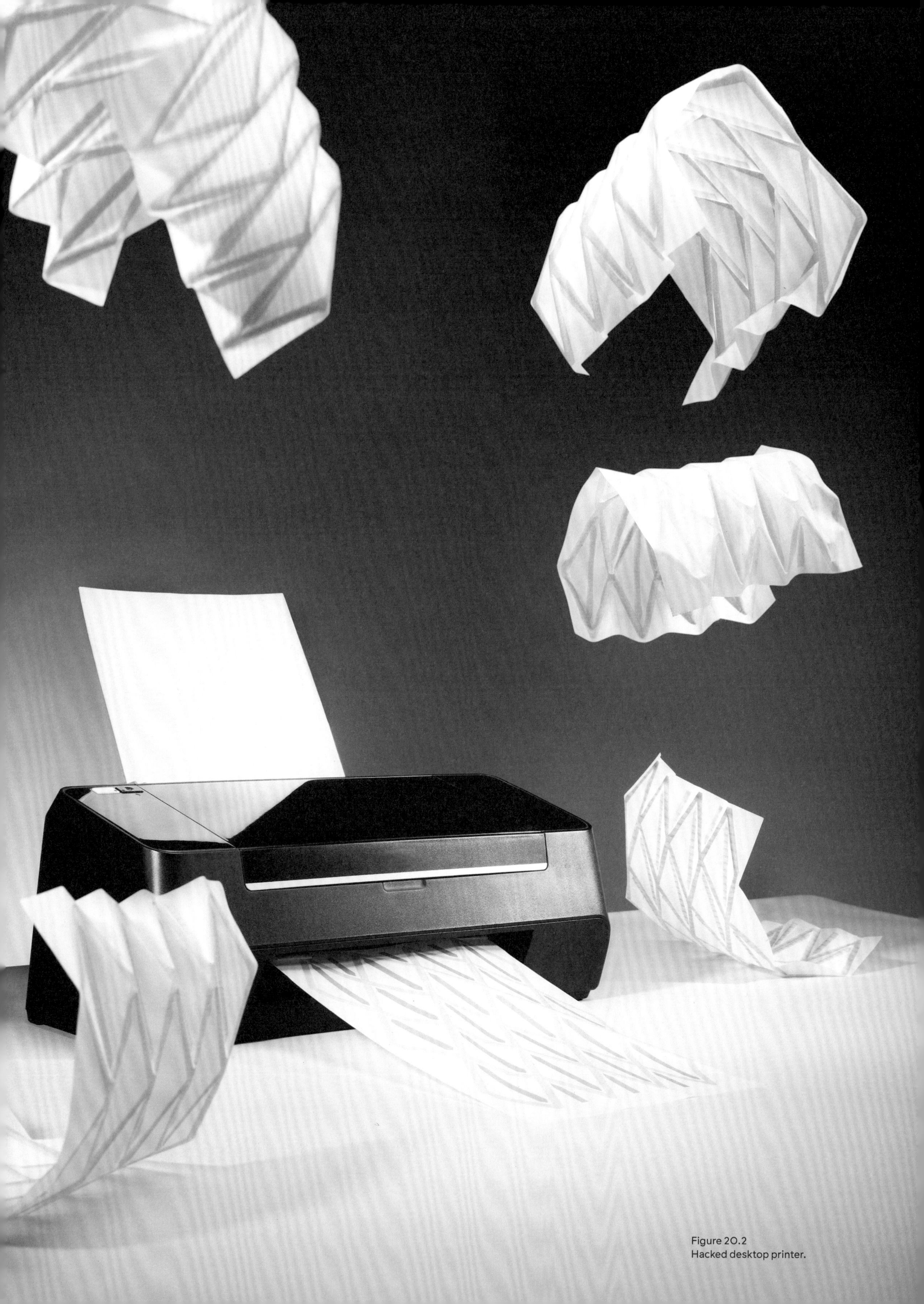

Figure 20.2
Hacked desktop printer.

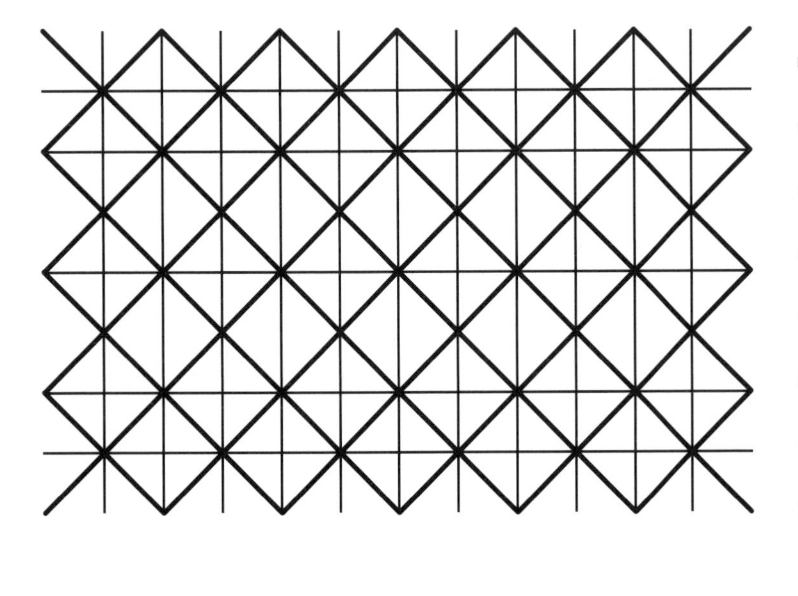

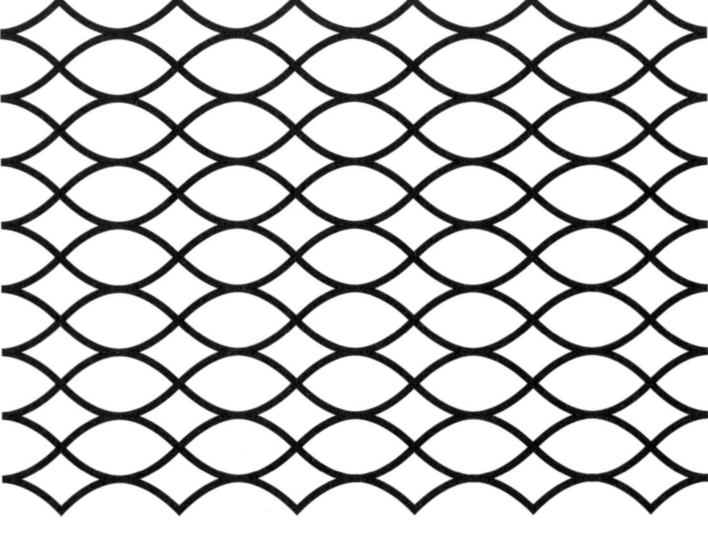

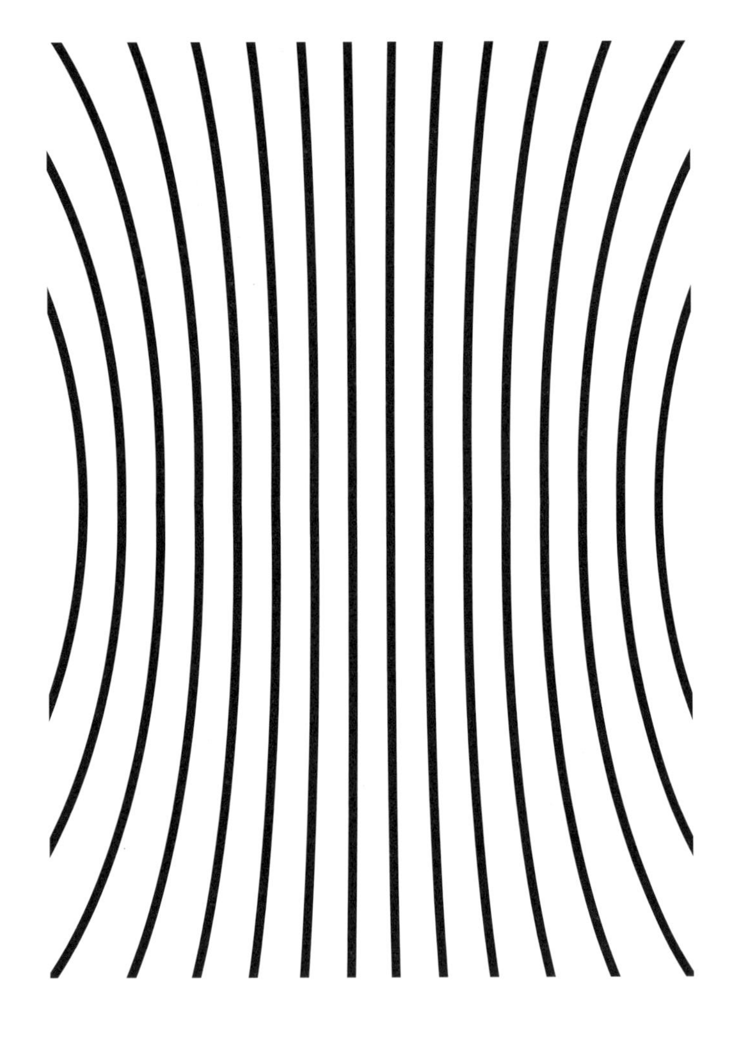

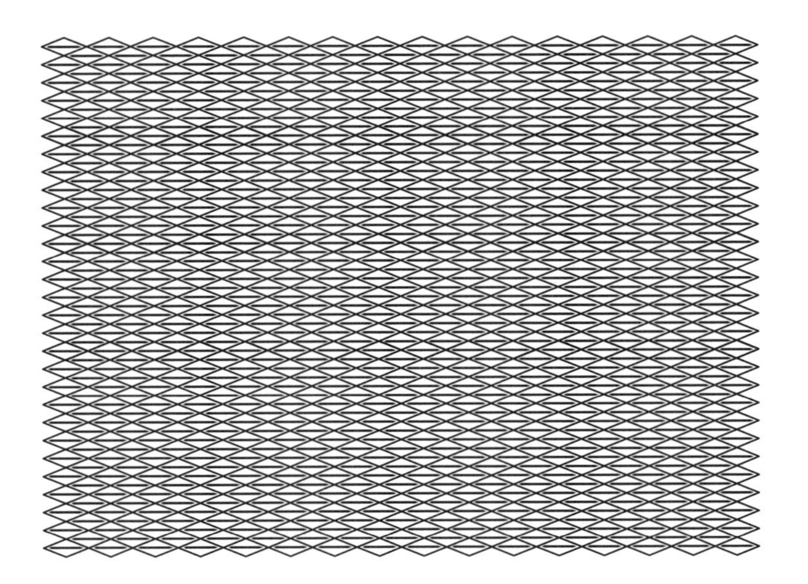

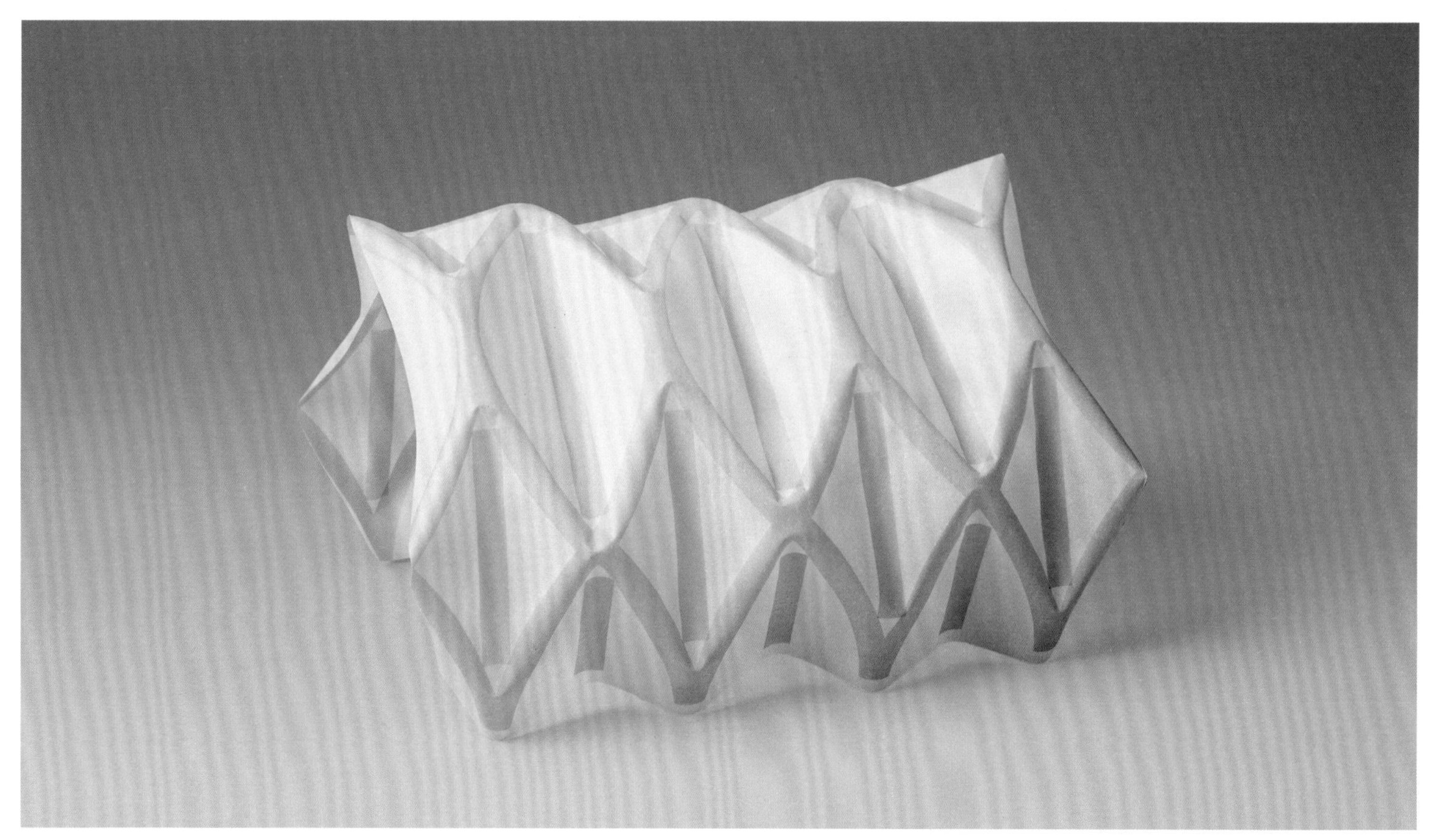

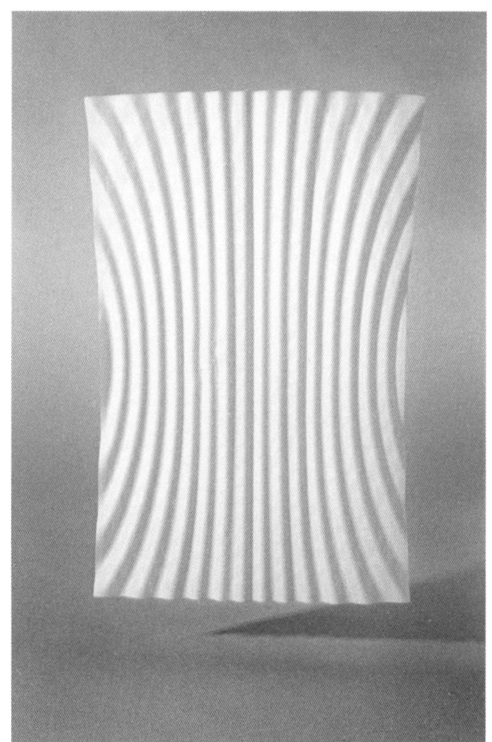

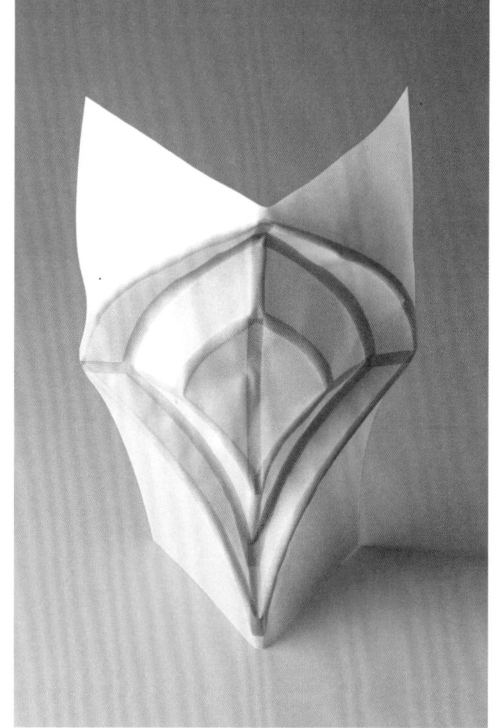

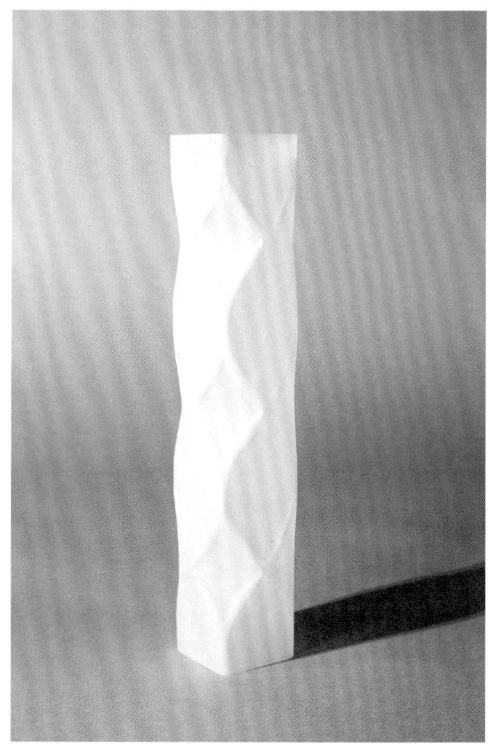

Figure 20.3 (left page, top left)
Drawing, straight lines, verso.

Figure 20.4 (left page, bottom left)
Drawing, curves, recto/verso.

Figure 20.5 (left page, top right)
Drawing, pattern curves, recto/verso.

Figure 20.6 (left page, bottom right)
Drawing, micro pattern test, recto/verso.

Figure 20.7 (top)
Dry-printed paper, pattern curves, recto/verso.

Figure 20.8 (bottom left)
Dry-printed paper, curves, recto/verso.

Figure 20.9 (bottom middle)
Dry-printed paper, curves, recto/verso.

Figure 20.10 (bottom right)
Dry-printed paper, pattern curves, recto.

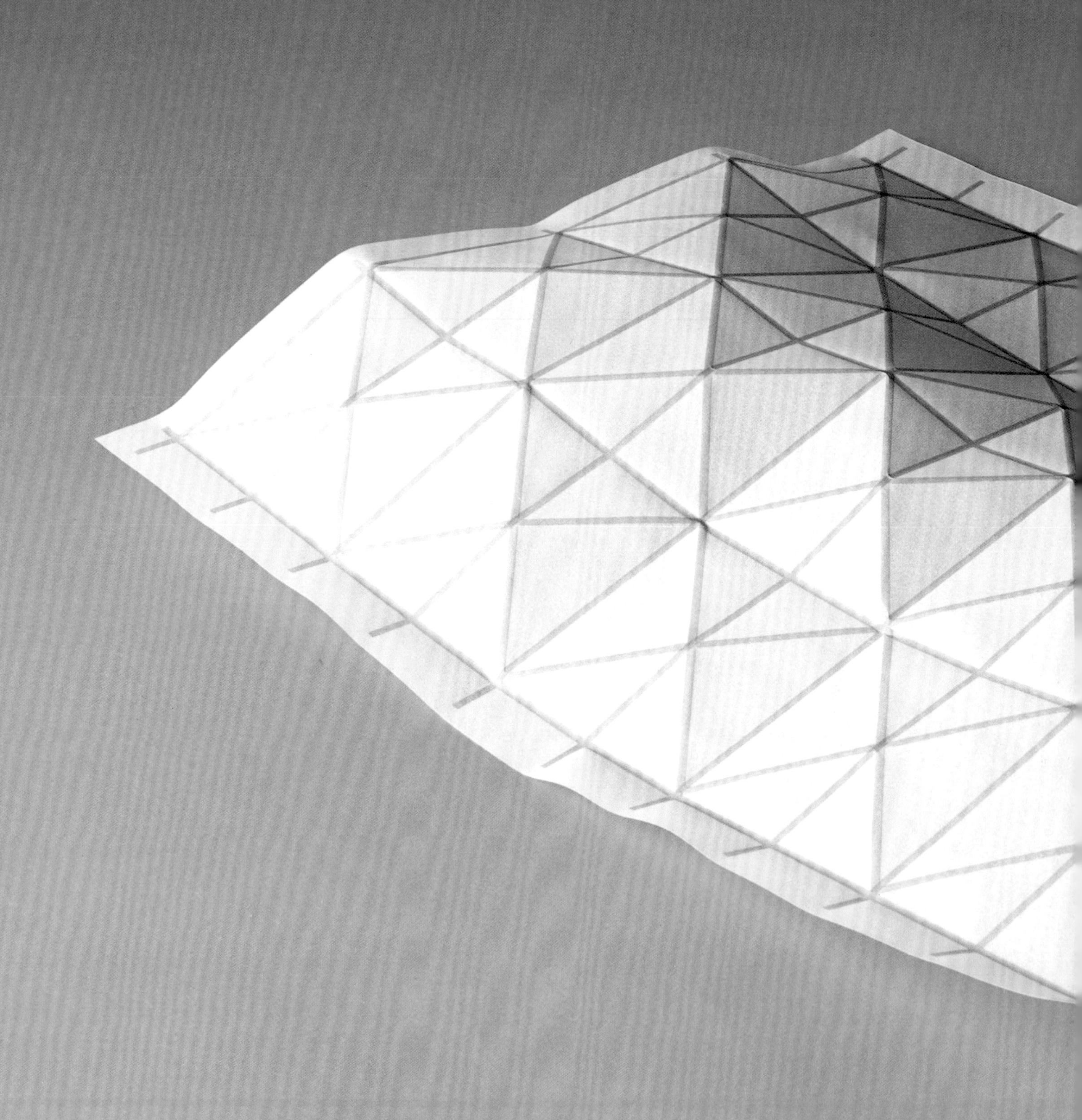

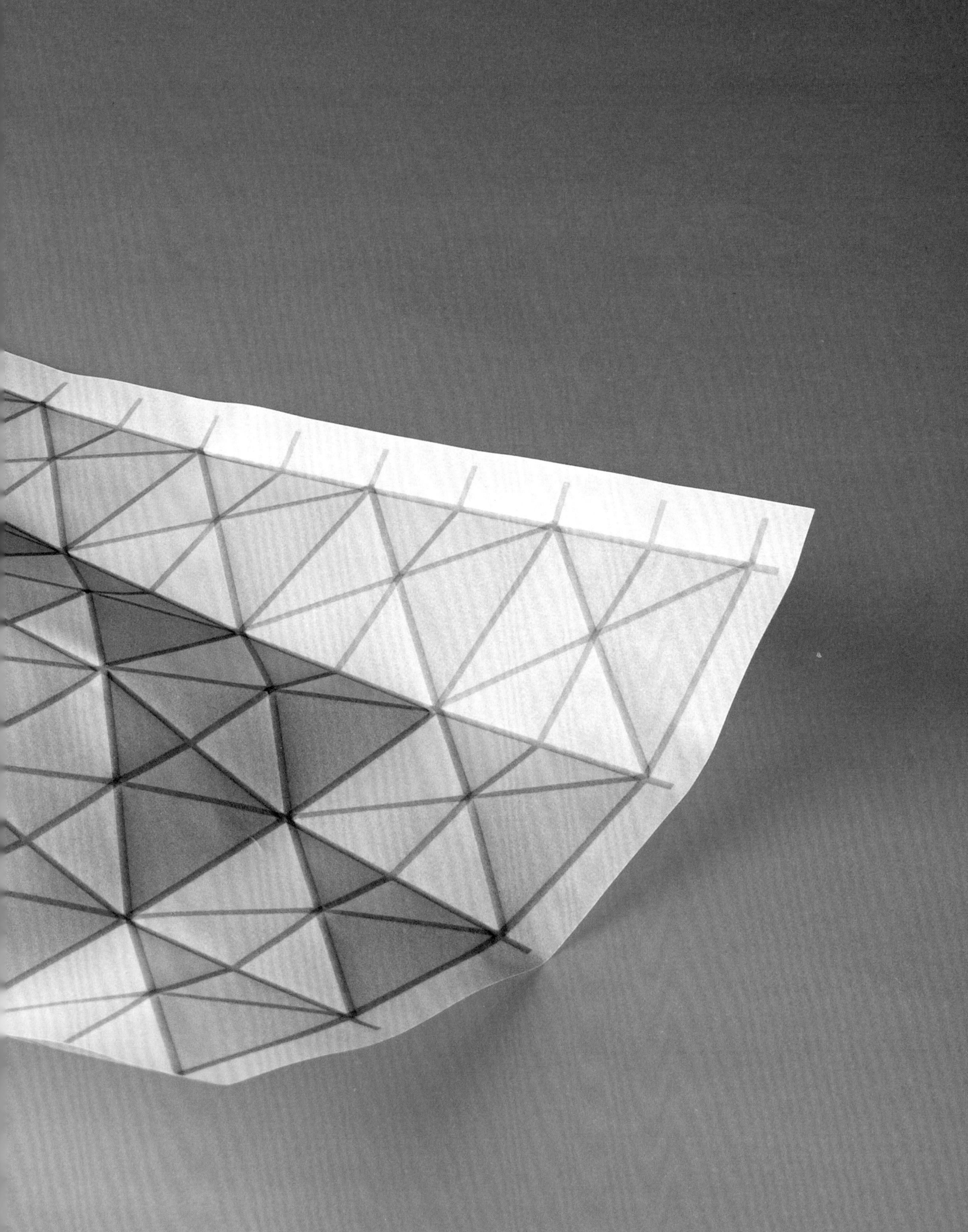

Figure 20.11
Dry-printed paper, straight lines, verso.

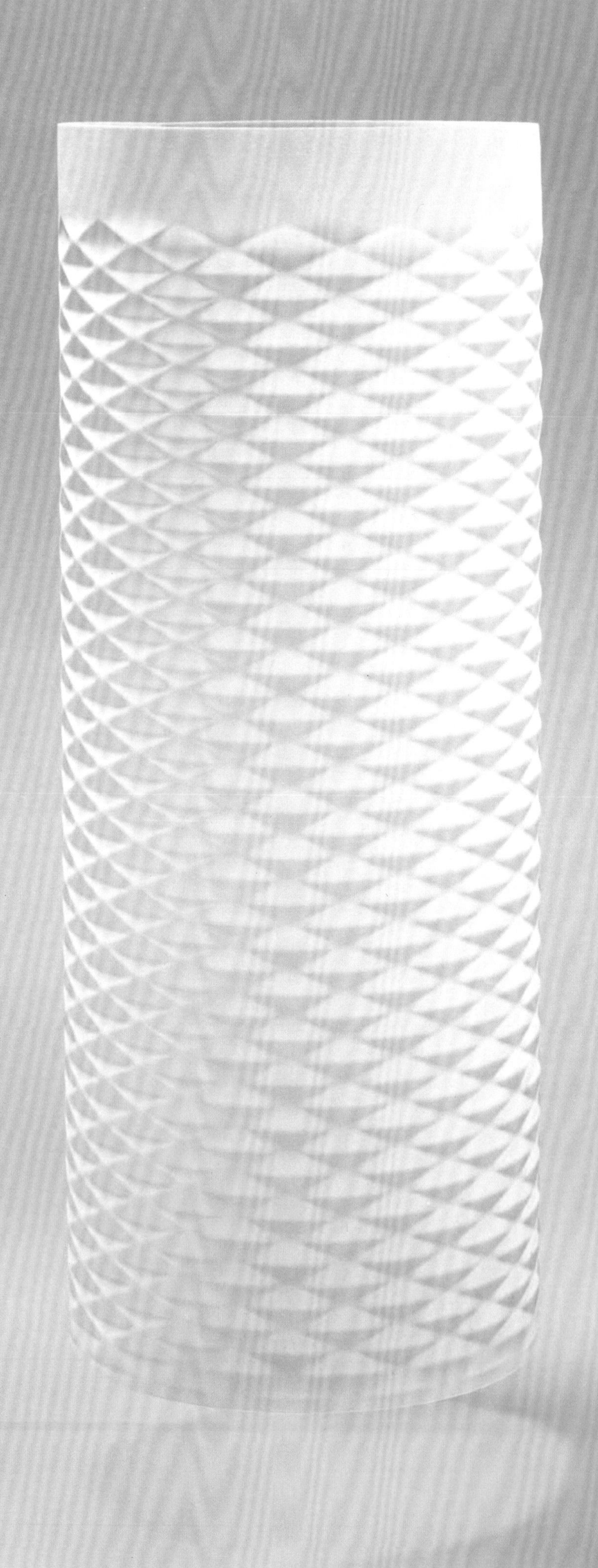

Figure 20.12
Dry-printed paper, micro pattern test, recto/verso.

21

Heat-Active Auxetic Materials

Athina Papadopoulou,
Hannah Lienhard,
Jared Laucks, and
Skylar Tibbits

21
Heat-Active Auxetic Materials

Athina Papadopoulou, Hannah Lienhard, Jared Laucks, and Skylar Tibbits

When a material is being stretched, it usually becomes thinner in the axis perpendicular to the direction of the pulling force. Contrary to common materials, however, auxetic materials when pulled expand in all directions, and when compressed they shrink in all directions. For every material there is a specific ratio of strain in the perpendicular axis to strain in the longitudinal axis, which is called the Poisson ratio. The term "auxetics" refers to materials that exhibit a negative Poisson ratio, as opposed to the more typical positive Poisson ratio.[1]

Heat-active auxetic materials, developed at MIT's Self-Assembly Lab, are materials that exhibit auxetic behavior when exposed to heat. Compared to traditional auxetic materials, heat-active auxetic materials demonstrate autonomous performance, environmental response, easy customization, and greater possibilities for the design and fabrication of material properties.

Although it has been known for over a century that auxetic material are theoretically possible, the first report on the development of auxetic materials was made in 1987 by Lakes, who demonstrated a foam with auxetic behavior.[2] Auxetic materials have several advantages over conventional materials, such as improved mechanical properties, customizable expansion/contraction properties, crash protection, and other interesting mechanical properties. In recent decades, auxetic materials have been the focus of several significant studies in new material behavior.[3]

Heat-active auxetic materials take this research further by proposing self-transforming materials that uniformly (globally or locally) shrink when exposed to heat, and then expand back to their initial state once cooled. Thus, in heat-active materials, the activation force that causes auxetic behavior is the ambient temperature, and not a mechanical force externally imposed on the material, as is the case in conventional auxetic materials. Heat-induced auxetic behavior has many advantages over common mechanically induced auxetic behavior: the material transformation is autonomous and not linked to an external device/human or tethered control; it is responsive to the environment; it is adaptive to a user's needs, and it allows for a high degree of control over complex patterns.

Heat-active auxetic materials transform on demand, varying their surface area, density, and stiffness through the expansion and shrinkage of their apertures' size. These properties enable products to be designed with unique capabilities across different scales and industrial domains: from smart cushioning in footwear and orthotic applications to breathable cushions and temperature-responsive clothing, body-adaptive furniture, load-responsive dynamic structures, smart packaging, and crash protection materials.

Heat-active auxetic materials are made of struts composed of programmable bilayer composites connected in specific geometric patterns. The programmable bilayer composites are made of materials with different thermal expansion properties. A wide range of materials can be

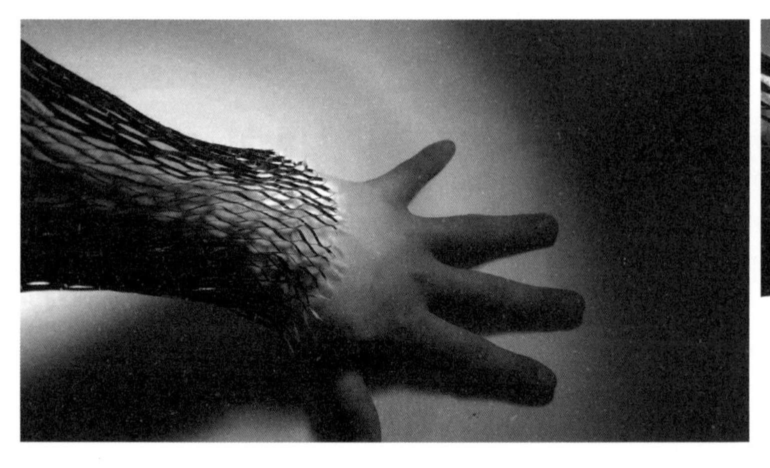

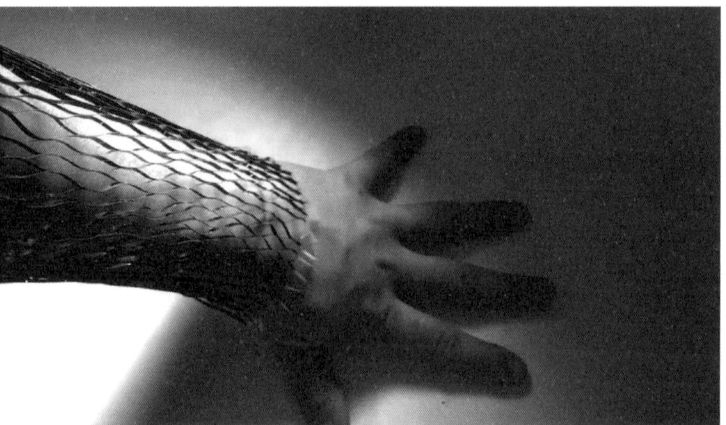

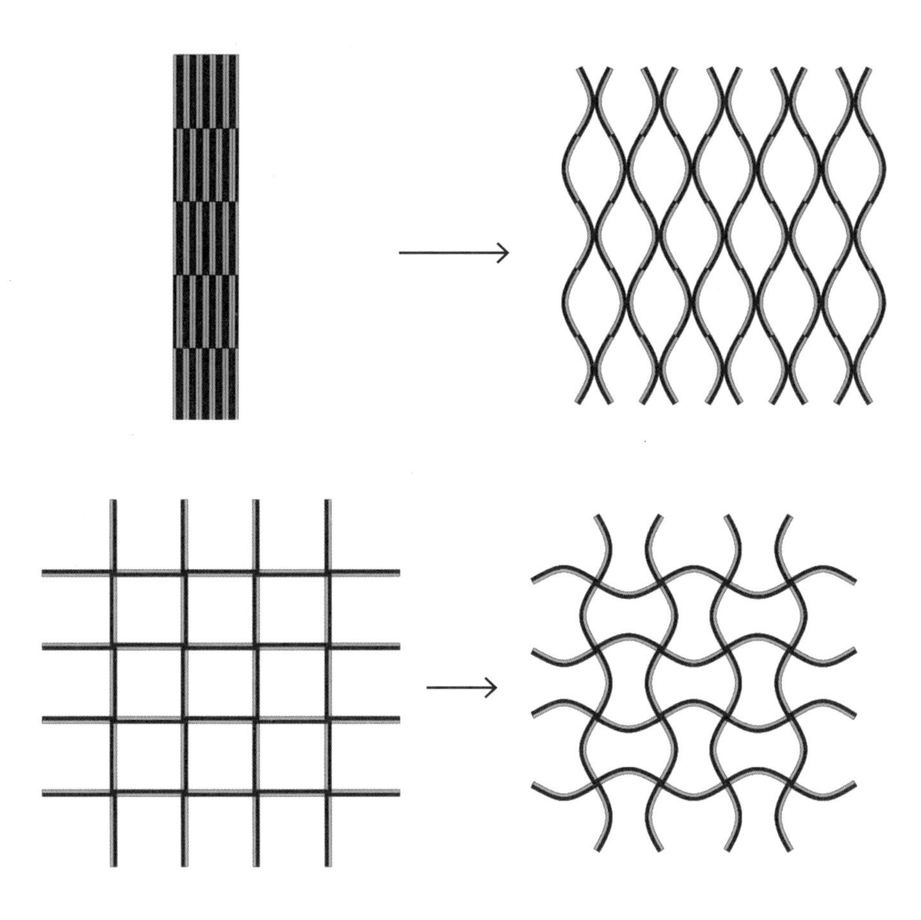

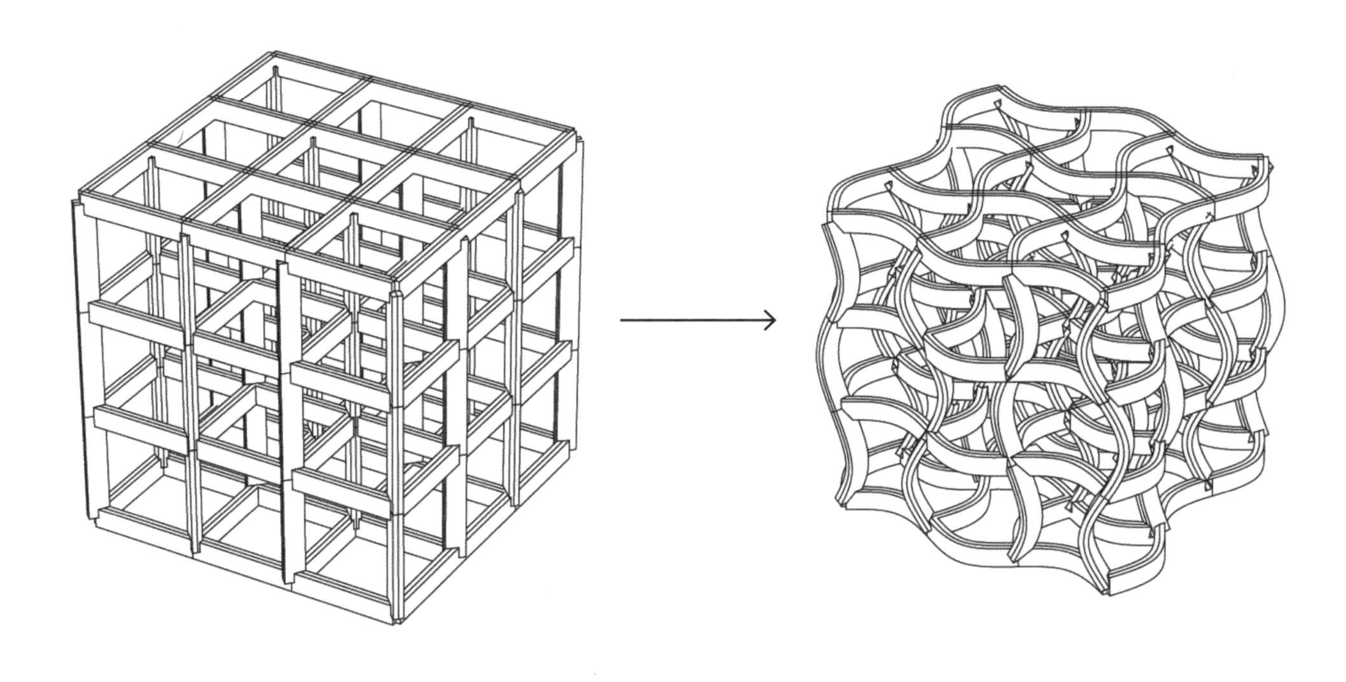

Figure 21.0 (left page)
Photographs demonstrating a temperature adaptive sleeve made of programmable bilayer struts. The sleeve's apertures open up when exposed to high temperatures to allow for breathability.

Figure 21.1 (top)
Diagrams showing 1D heat-active transformation of a planar grid configuration formed by bilayer struts (left) and 2D heat-active auxetic behavior of a planar grid configuration formed by bilayer struts (right), using a bilayer strut as the basic module.

Figure 21.2 (bottom)
Diagrams showing 3D heat-active auxetic behavior of a cubic lattice configuration formed by bilayer struts.

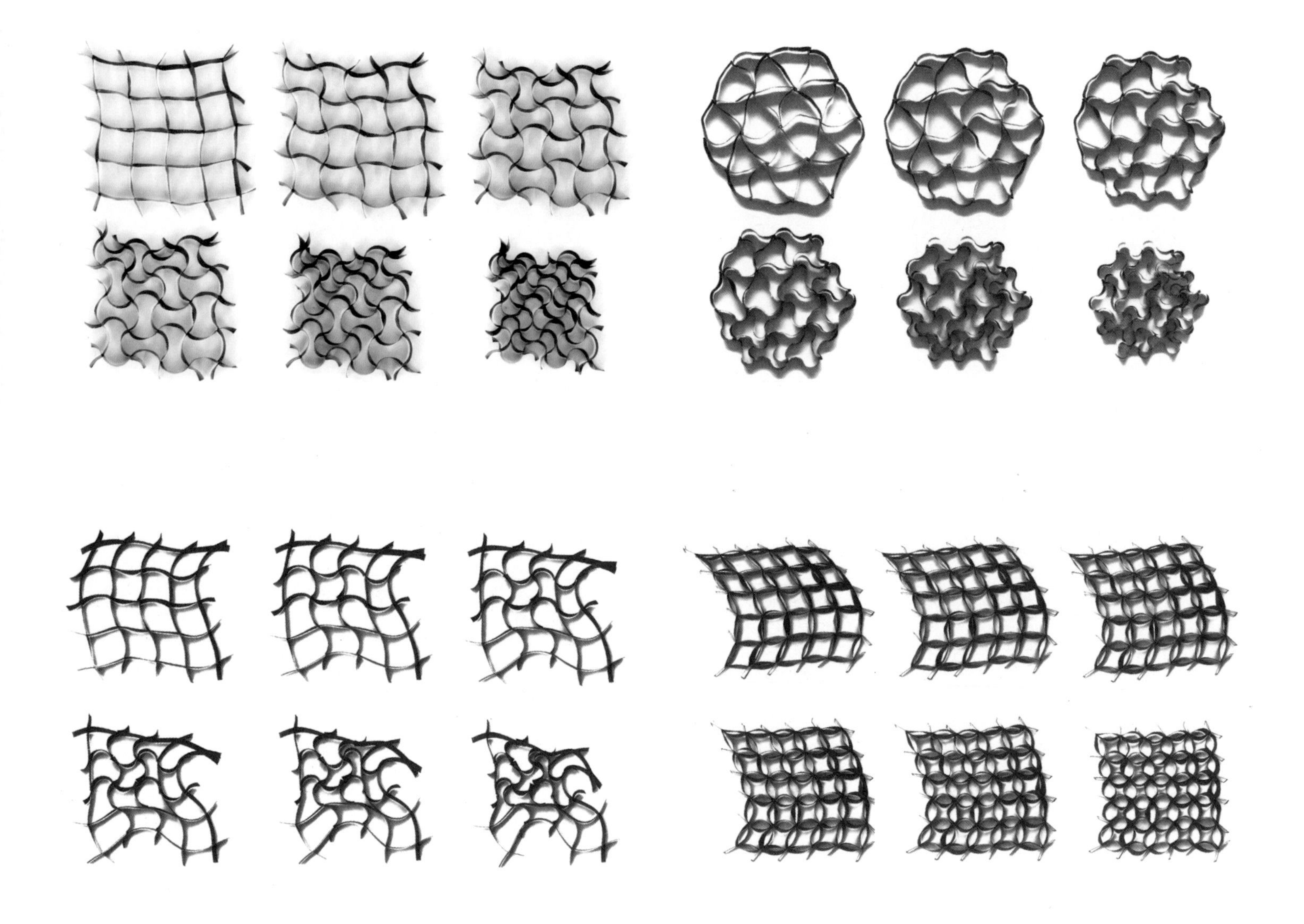

Figure 21.3 (top left)
Photographs of a planar rectangular grid made of programmable bilayer struts, depicting gradual shrinking due to auxetic material behavior when exposed to ambient heat.

Figure 21.4 (top right)
Photographs of a planar hexagonal wheel–form grid made of programmable bilayer struts, depicting gradual shrinking due to auxetic material behavior when exposed to ambient heat.

Figure 21.5 (bottom left)
Photographs of a planar rectangular grid made of programmable bilayer struts, depicting gradual shape change due to local auxetic behavior along one of its rows and columns when exposed to ambient heat.

Figure 21.6 (bottom right)
Photographs of a planar grid made of a double layer of programmable bilayer struts, depicting a gradual pattern change due to auxetic behavior when exposed to ambient heat.

used for the bilayer composites, from polymers to textiles, carbon fiber, or even printed wood. In this case study we have used polymer sheets with significantly different coefficients of thermal expansion. The struts are composed of two polymers. One polymer has a high coefficient of thermal expansion and dramatically expands when exposed to heat; the other polymer has a low coefficient of thermal expansion. The differential thermal expansion of the two polymers results in the transformation of the strut from its natural state to a curved shape.

The struts of the bilayer composites can be fabricated directly through multilayer 3D printing, or the two materials can be fabricated separately and then combined into a composite through lamination or manual means such as stitching. Different methods for heat activation can be used: through contact heating such as, electrical activation using nichrome wire, or noncontact heating such as heat radiation or infrared heat lamps. Heating temperatures can range from 30° to 200° Celsius, depending on the activation method and the bilayer composite used.

To achieve auxetic behavior, the struts of the programmable bilayer composites need to be designed in very specific geometric patterns. Previous research in auxetic materials has provided a palette of possible patterns, from simple square grids and honeycomb patterns to more complex shapes.[4] Beyond this list, new geometric patterns and variations can also be designed to exhibit auxetic material behavior. The patterns of the auxetic materials presented have been designed

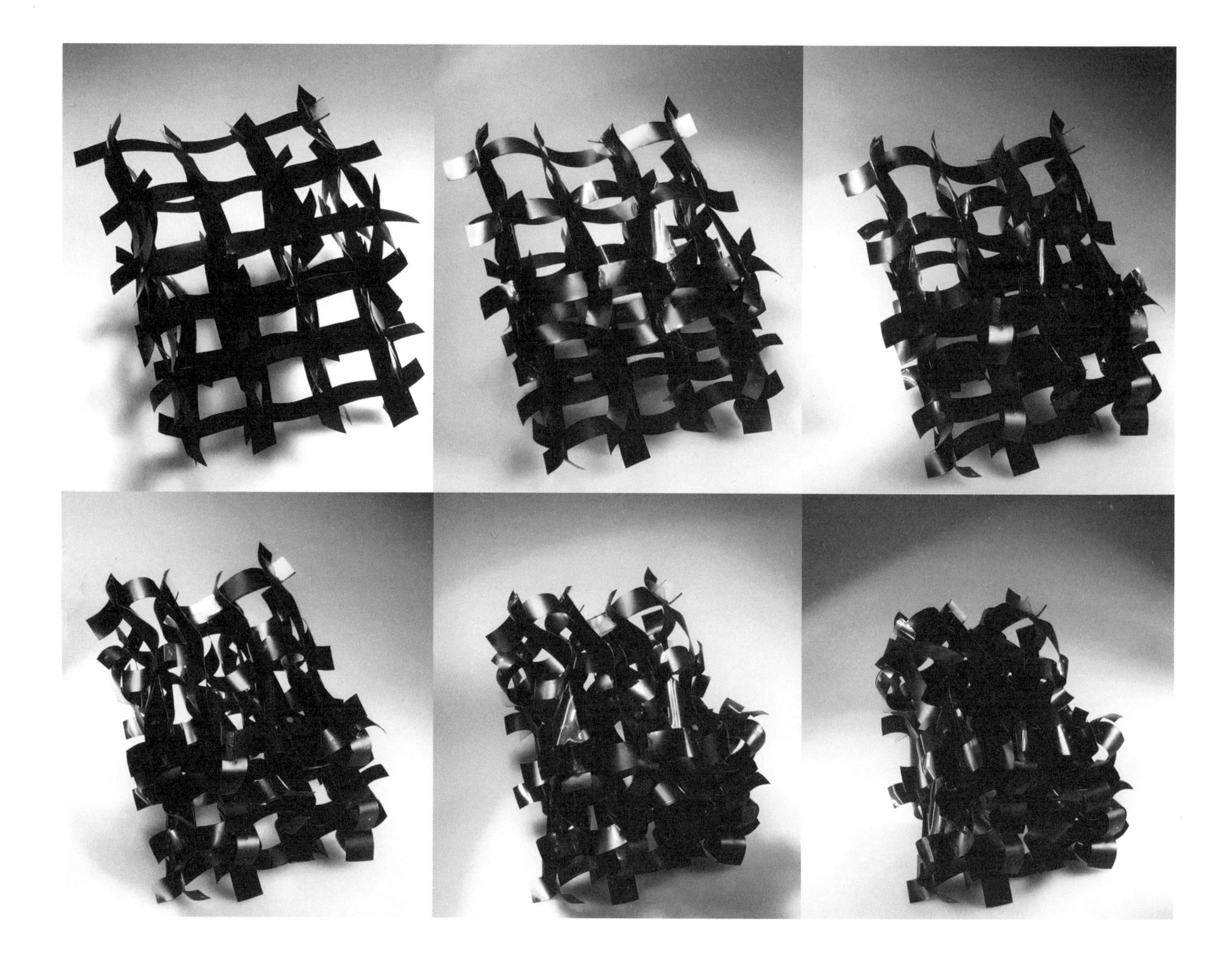

Figure 21.7
Photographs of a three-dimensional lattice made of programmable bilayer struts, depicting gradual shrinking due to auxetic material behavior when exposed to the ambient heat of an infrared heat lamp.

as variations upon a basic strut module consisting of the bilayer composite. The two materials in the bilayer composite have equal lengths on each side of the strut, and alternate in a periodic manner. When the strut is being exposed to heat, it transforms from flat to a sinusoid shape.

One way to produce a heat-activated material that expands/shrinks in one direction is to connect multiple strut modules along their length. When exposed to heat, the connected struts expand in the direction perpendicular to their length, creating a planar grid of sinusoidal shapes (figure 21.1). Although one-dimensional transformations cannot be classified as auxetic behavior, they are nevertheless useful when considering the design of the auxetic system or considering applications for linear expansion/contraction.

A heat-active auxetic material that shrinks/expands in two dimensions can be made of multiple struts connected in various orientations: for example, an auxetic grid can be created if the struts are formed into an orthogonal grid, alternating the positions of the two materials in the bilayer composite (figures 21.1 and 21.3). Another way to form a heat-active auxetic material in two dimensions is to compose a hexagonal grid of the basic strut modules and then repeat the module radially (figure 21.4). For three dimensions, using the same bilayer strut as the basic module, a three-dimensional grid can be created with a vertical and horizontal orientation of two-dimensional planar grids connected with bilayer struts forming a cubic lattice (figures 21.2 and 21.7).

When exposed to ambient

heat, the two-dimensional and three-dimensional grids shrink uniformly in all dimensions, and once they cool down, they expand uniformly, returning to their initial shape. Local and global auxetic behavior can be used for different applications, allowing for adaptability and flexibility in the material behavior (figure 21.5 and 21.6). For example, a curtain made of heat-active auxetic materials could be used for shading of large spaces, adapting to the local environmental conditions with variable apertures. Also, breathable clothing could adapt to the environment or body temperature through expanding or contracting apertures, both locally and globally.

Heat-active auxetic materials offer new opportunities for adaptive material solutions, leading to a wide range of applications in product, fashion, furniture, and architecture design. Future research can expand the palette of heat-active auxetic materials and activation methods to further address functional and design challenges.

ACKNOWLEDGMENTS

We would like to thank Schendy Kernizan and all the Self-Assembly Lab researchers who contributed in the development of programmable materials research. We would also like to thank MIT's Department of Architecture and MIT's International Design Center for their continued support.

Many thanks to Greg Blonder and Doug Holmes for the inspiring conversations and pioneering work in the field.

NOTES

1 Ken E. Evans, "Auxetic Polymers: A New Range of Materials," *Endeavour* 15, no. 4 (1991): 170—174.

2 R. Lakes, "Foam Structures with a Negative Poisson's Ratio," *Science* 235, no. 4792 (1987): 1038—1040.

3 Sahab Babaee, Jongmin Shim, James C. Weaver, Elizabeth R. Chen, Nikita Patel, and Katia Bertoldi, "Metamaterials: 3D Soft Metamaterials with Negative Poisson's Ratio," *Advanced Materials* 25, no. 36 (2013): 5116.

4 Carolin Körner and Yvonne Liebold-Ribeiro, "A Systematic Approach to Identify Cellular Auxetic Materials," *Smart Materials and Structures* 24, no. 2 (2014): 025013.

22

BIOCOUTURE

Suzanne Lee

Suzanne Lee

The concept for BIOCOUTURE™ originated during the research for Suzanne Lee's book *Fashioning the Future: Tomorrow's Wardrobe*. A serendipitous conversation in 2003 with Dr. David Hepworth, a biologist and materials scientist, presented a new vision of future fashion, one that emerges fully formed from a vat of liquid.

Rather than exploit plants or petrochemicals to provide the raw material for fabric, BIOCOUTURE investigates the use of microbes to grow a textile biomaterial. Certain bacteria will spin microfibrils of pure cellulose during fermentation that form a dense layer that can be harvested and dried.

To produce a flexible cellulose mat, a mixed culture of bacterial cellulose, yeasts, and other microorganisms was added to a sugary green tea solution. The bacteria feed on the sugar and spin fine threads of cellulose. As these start to stick together they form a skin on the surface of the liquids. After two to three weeks, when it is approximately 1.5 cm thick, the cellulose skin is removed from the growth bath. It can then be either wet-molded onto a 3D form, like a dress shape, or dried flat to be cut and sewn into a garment.

Large sheets of microbial cellulose were readily dyed and printed with far fewer inputs of dye and water than for other fibers. Experiments included work with composite structures, lattices, 3D formation in solution, etc.

The material is nearest in feel to vegetable leather, and, like your vegetable peelings, it can be safely composted at end of life. Left untreated, the material is superabsorbent, but it can be made water-resistant with hydrophobic finishing. Its hydrophilicity was not the barrier to bringing the material to market; the real challenge proved to be finding a commercially viable way of producing the material at scale at a competitive cost.

BIOCOUTURE garments were exhibited around the world over several years, along with lectures and workshops. Lee's work has been covered by international media and books and has initiated many similar research projects globally.

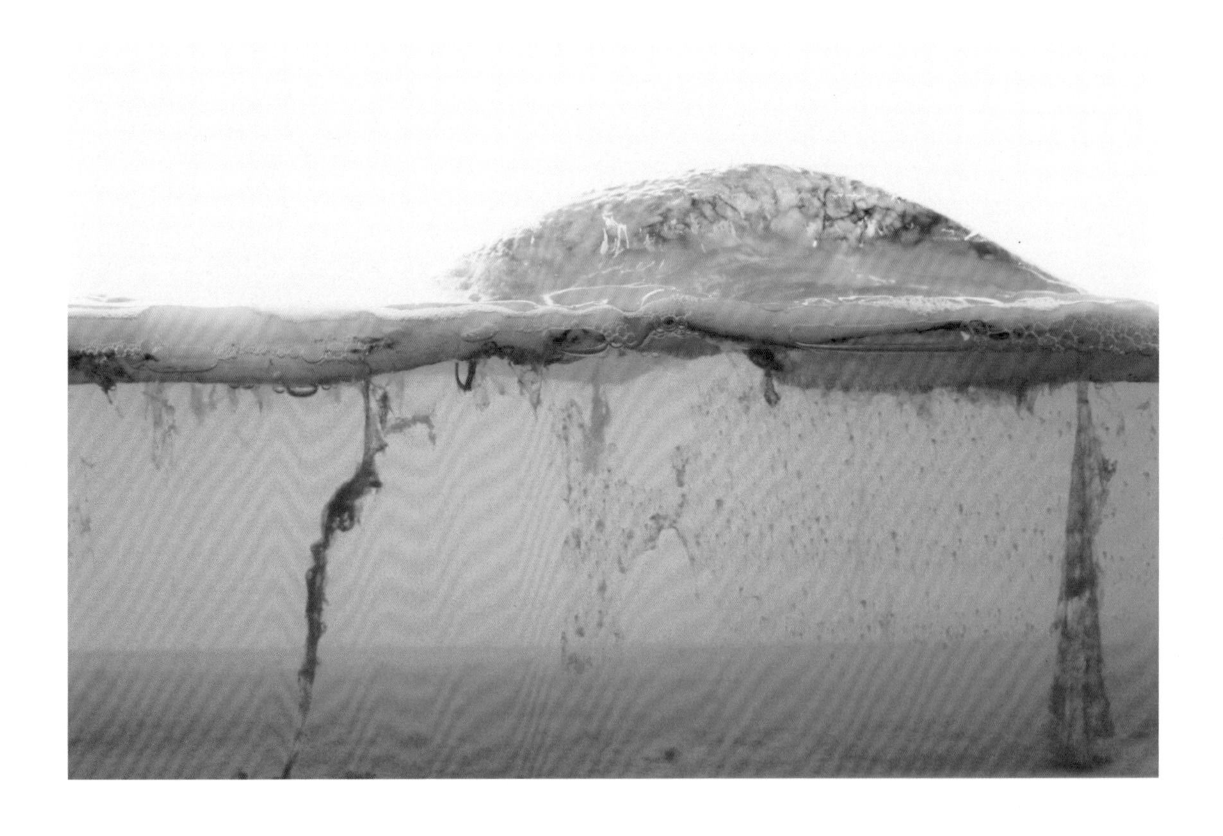

Figure 22.1 (left page)
Biomaterial.

Figure 22.2 (top left)
BioBiker jacket—microbial cellulose, iron oxidation.

Figure 22.3 (top middle)
BioDenim jacket—microbial cellulose, indigo dye.

Figure 22.4 (top right)
BioBomber jacket—microbial cellulose, engineered print by Philip Delamore, laser-cut stencil, fruit and vegetable pulps providing staining.

Figure 22.5 (bottom)
BioShoe, with Liz Ciokajlo—microbial cellulose, traditional brogueing, beetroot dye.

Figure 22.6
EcoKimono—microbial cellulose, kakishibu persimmon tannage, laser-cut stencil, iron oxidation print.

23
From Material Expressivity to Material Interaction

Behnaz Farahi

Figure 23.1
Caress of the Gaze by
Behnaz Farahi, 2015.
photo by Elena Kulikova

From Material Expressivity to Material Interaction

Behnaz Farahi

This research shows how active matter can contribute to nonverbal communication in interaction design through material expressivity.

What is material expressivity? Nature is full of examples: from the way in which a single atom interacts with a ray of light, as studied in the science of spectroscopy, to the manner in which a mimosa tree closes its leaves to protect itself, or a male bowerbird gathers blue-colored items to attract a mate. Manuel DeLanda, who coined the term "material expressivity," sees forms of expressivity in both nonhuman and human processes. He refers to them as "fingerprints" that can be used at a molecular level to identify and determine the properties of a given material through the process of spectroscopy. This approach, first developed in the nineteenth century, uses various methods to extract fingerprints from materials. It relies, as DeLanda notes, mainly on "the capacities of atoms themselves to produce expressive patterns, through emission, absorption, or other processes."[1] He further elaborates, "These expressive patterns are what scientists call information," and this *information* has no semantic meaning but consists of "linguistically meaningless physical patterns." Material expressivity can therefore be found in both biological and nonbiological materials as an integration of *matter* and *information* rather than simply a combination of *matter* and *form*.

Once we start addressing the expressivity of matter, we can immediately see how these characteristics open up certain potentialities to affect or be affected by other species. In our daily existence we constantly encounter and extract information from the material of our environment, and it is arguably the different material expressivities in our environment that make possible certain "potential" actions.

Advances in material sciences and smart materials have opened up a new perspective for the world of design, by demonstrating new materials that display nonstandard expressivity. Smart materials represent a radical shift from conventional materials in that they behave dynamically in response to external stimuli, such as stress, temperature, moisture, pH, and electric or magnetic fields. Shape memory alloys (SMAs) are an example of smart materials that respond to temperature differentials by changing their shape. These alloys—composed of nickel and titanium—can return to their initial shape after being deformed through exposure to heat. At a molecular level, the SMA goes through a phase transition between austenite (rigid) and martensite (soft) without losing its solidity.

Caress of the Gaze (figure 23.1) offers an example of the effect of material expressivity within interaction design. This is a multimaterial 3D-printed garment equipped with SMA actuators and a facial tracking camera. The garment morphs and changes its shape based on an onlooker's gaze. We can use material behaviors to communicate new potentialities and engage with our social and cultural context, in a manner not dissimilar to how goosebumps on our skin can reveal environmental or emotional information such as temperature, moisture, fear, and excitement.

Inspired by animal/human skin and its complex architecture—the interplay of muscles, hair, feathers, quills, and scales—Caress of the Gaze engages with a series of issues. Firstly, it demonstrates how the very latest and most advanced 3D printing technologies might contribute to the realm of fashion, by exploring the tectonic properties of materials printed using Objet500 Connex 3D printers. This technology allows the fabrication of composite materials with varying flexibilities and densities and can combine materials in several ways, with different material properties deposited in a single print run. Inspired by the flexible behavior of the skin itself, this outfit therefore exhibits different behaviors over various parts of the body, ranging from stiff to soft. Secondly, the project explores the potential of SMA actuation systems. It uses 8 SMA linear actuators located in

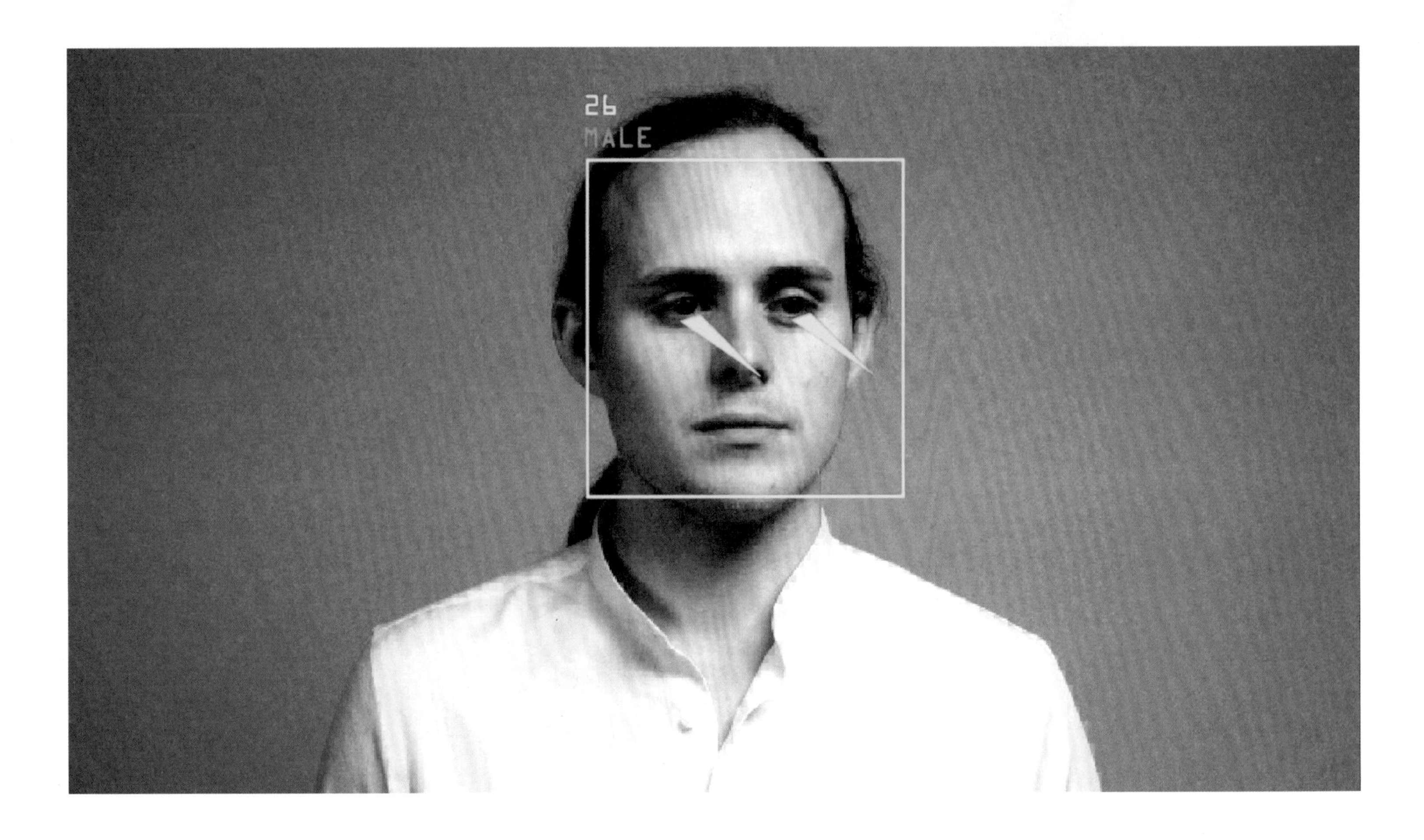

Figure 23.2
Caress of the Gaze equipped with facial tracking camera, detecting age, gender and gaze orientation of the onlookers, photo by: Charlie Nordstrom

a diagonal pattern beneath the surface of the garment. These can be actuated individually to enable transformations such as bulging, swelling, or twisting.[2] Thirdly, it investigates how our clothing can interact with other people as a primary interface activated using facial gaze tracking technologies. By mapping the yaw and pitch values of the onlooker's eye movements[3]—as captured by the facial tracking camera—and relating them to the nodes of the SMA, controlled by a Teensy microcontroller, the garment is able to move in response to the onlooker's gaze.

This project therefore explores the potential of an actuation system, assembled as a form of muscle system using an SMA that informs the motion of 3D-printed quills, which in turn engender certain social interactions. In this case, material expressivity can be detected from micro-scale changes in the SMA through to macro-scale behavior in the garment. This behavior increases the wearers' awareness of their social context by allowing them to feel the garment moving on their bodies according to the onlookers' gaze. It allows them to communicate to onlookers that their actions have been perceived, thereby allowing them to ward off any further disturbing actions. In so doing, this project addresses the emerging field of shape-changing structures and interactive systems that bridge the worlds of fashion, art, technology, and design. It does so in the belief that by implementing morphological principles and behavioral responses inspired by natural systems, we might be able to rethink the relationship between our bodies and the surrounding environment.

Particular attention to material processes and behaviors can lead us to develop materials that can express themselves in an entirely new fashion. Moreover, advances in material sciences and bio/nano computing have opened new doors to nonbiological material expressivity. We can look at these materials as dynamic, shape-changing entities that can foster certain types of interaction. This approach to developing new materials can serve to augment the degree of freedom of expressivity beyond the limitations of traditional media.

Figure 23.3
Caress of the Gaze equipped with facial tracking camera, detecting age, gender, and gaze orientation of the onlookers.
photo by: Charlie Nordstrom

NOTES

1 Manuel DeLanda, "Material Expressivity," *Domus*, no. 893 (June 2006): 122—123.

2 The heat generated by passing an electrical current through an SMA can activate the material if it meets the activation temperature. Activation temperature of the SMA depends on intrinsic properties of the material. In other words, they demonstrate various behaviors in relation to their internal properties.

3 Yaw value refers to the horizontal angle between the onlooker's gaze and the center of camera lens. This value can be measured in degrees. Pitch value refers to the vertical angle between the onlooker's gaze and the center of the camera lens.

REFERENCES

Addington, D. Michelle, and Daniel Schodek. *Smart Materials and Technologies: For the Architecture and Design Profession*. Amsterdam: Elsevier; Boston: Architectural Press, 2005.

DeLanda, Manuel. "Material Expressivity." *Domus*, no. 893 (June 2006): 122—123.

24
Wanderers

Neri Oxman and
The Mediated
Matter Group

24
Wanderers

Neri Oxman and The Mediated Matter Group

Traveling to destinations beyond planet Earth involves voyages to hostile landscapes and deadly environments. Crushing gravity, amonious air, prolonged darkness, and temperatures that would boil glass or freeze carbon dioxide all but eliminate the likelihood of human visitation. The Wanderers project explores the possibility of voyaging to worlds beyond by visiting the worlds within. 3D-printed wearable capillaries designed for interplanetary pilgrims are infused with synthetically engineered microorganisms to make the hostile habitable and the deadly alive. Each design is a codex of the animate and inanimate, with an origin and a destination: the origin being engineered organisms, which multiply to create the wearable within 3D printed skins; and the destination being a unique planet in the solar system. The setting for this exploration is the solar system, where, with the exception of planet Earth, no life can exist. The series represents the classical elements understood by the ancients to sustain life (earth, water, air, and fire), and

Figure 24.1 (left page) Mushtari (from the Wanderers collection), by Neri Oxman and Mediated Matter, 2014.

Figure 24.2 (top, left) Zuhal (from the Wanderers collection), by Neri Oxman and Mediated Matter, 2014.

Figure 24.3 (top, right) Qamar (from the Wanderers collection), by Neri Oxman and Mediated Matter, 2014.

offers their biological counterpart in the form of microorganisms engineered to produce life-sustaining elements. The wearables are designed to interact with a specific environment characteristic of their destination and generate sufficient quantities of biomass, water, air, and light to sustain life: some photosynthesize, converting daylight into energy, others biomineralize to strengthen and augment human bone, and some fluoresce to light the way in pitch darkness. Each wearable is designed for a specific extreme environment where it transforms elements that are found in the atmosphere to one of the classical elements supporting life: oxygen for breathing, photons for seeing, biomass for eating, biofuels for moving, and calcium for building. Design research at the core of this collection lies at the intersection of multimaterial 3D printing and synthetic biology.

The medieval Arabs are known for their fascination with astronomy. They took a keen interest in the study of celestial bodies; motivated to better comprehend divine creation, they also appreciated the knowledge of the constellations as guidance in their journeys. In honor of these early contributions to the science of astronomy, the Wanderers in this collection are named in Arabic after their respective destination planets: Mushtari (a wearable for Jupiter), Zuhal (a wearable for Saturn), Otaared (a wearable for Mercury), and Qumar (a wearable for the Moon). The word "planet" comes from the Greek term *planetes* (*πλανήτης*) meaning "wanderer."

Research carried out for Wanderers lies at the intersection of design, additive manufacturing, and synthetic biology, bringing together digital growth and biological growth. The designs for Wanderers were digitally grown through a computational process capable of producing a wide variety of structures. Inspired by natural growth, the process creates shapes that adapt to an environment. Starting with a "seed form," the process simulates growth by continuously expanding and refining its shape.

This project is the first of its kind

Figure 24.4 (top)
Qamar (from the Wanderers collection), by Neri Oxman and Mediated Matter, 2014.

Figure 24.5 (right)
Zuhal (from the Wanderers collection), by Neri Oxman and Mediated Matter, 2014.

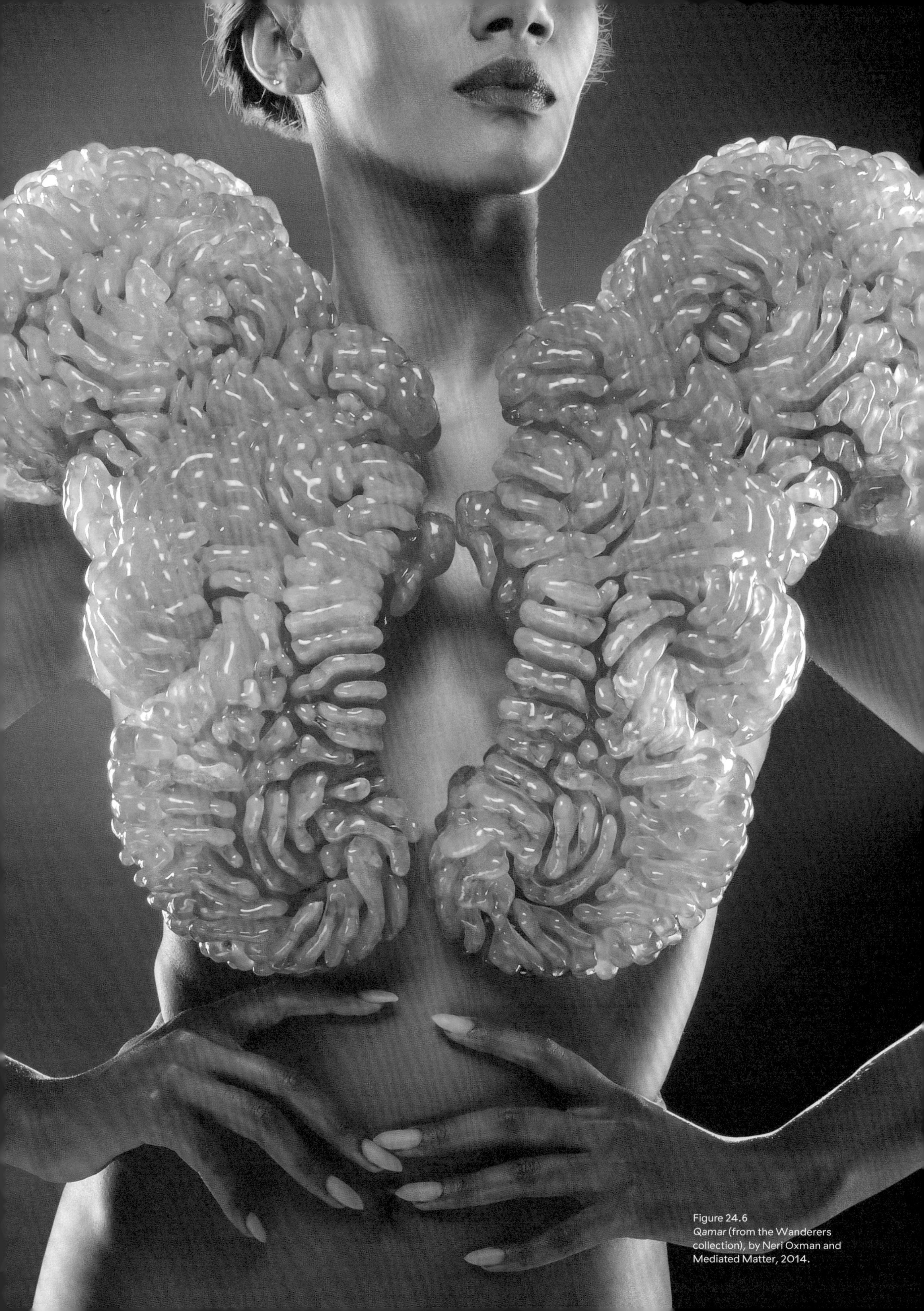

Figure 24.6
Qamar (from the Wanderers collection), by Neri Oxman and Mediated Matter, 2014.

Figure 24.7
Otaared (from the Wanderers collection), by Neri Oxman and Mediated Matter, 2014.

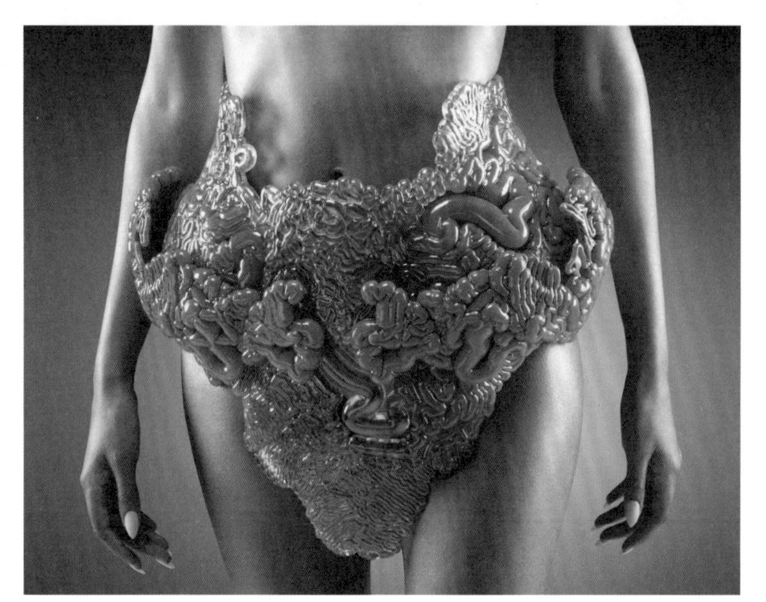

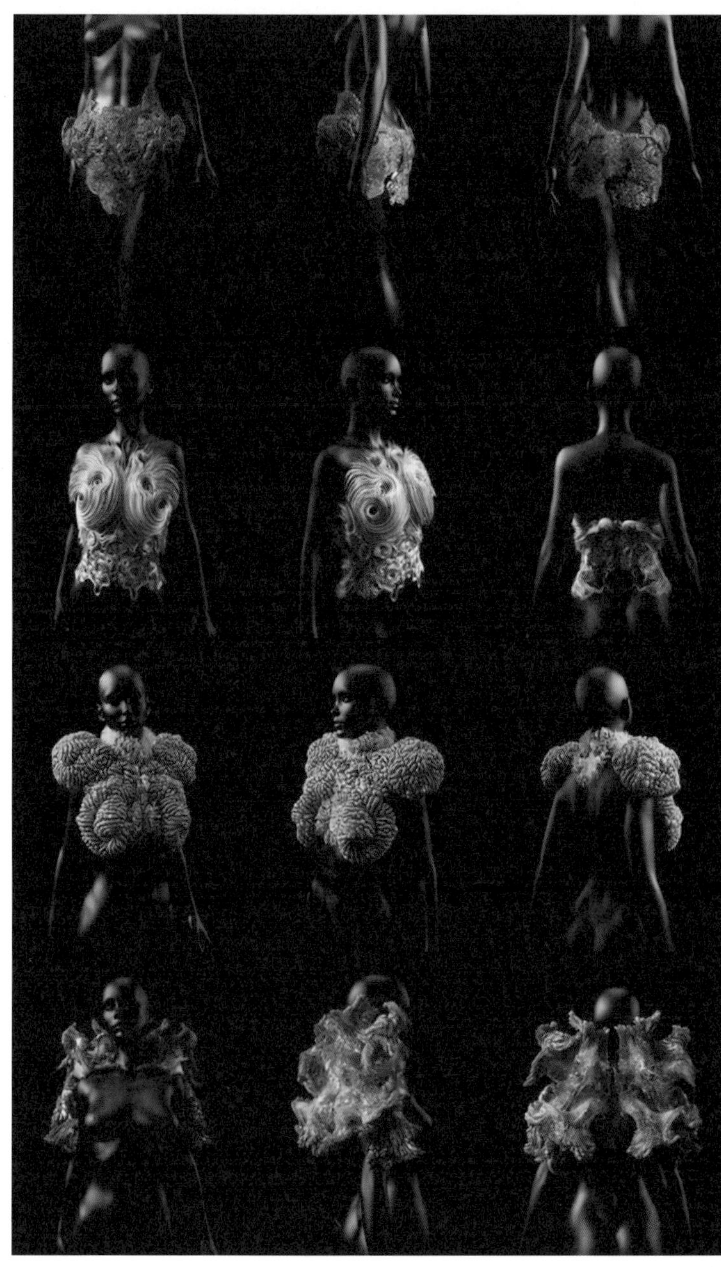

Figure 24.8 (left, top)
Otaared (from the Wanderers collection), by Neri Oxman and Mediated Matter, 2014.

Figure 24.9 (left, bottom)
Mushtari (from the Wanderers collection), by Neri Oxman and Mediated Matter, 2014.

Figure 24.10 (right)
All Wanderers, general views, by Neri Oxman and Mediated Matter, 2014.

to achieve volumetric translucency gradients at extremely high resolution (a few microns) using a process known as bitmap printing. In this process, material composition is given on voxel resolution and used to fabricate a design object with locally varying properties such as color, rigidity, or opacity. Voxel resolution is set by the Stratasys printer's native resolution, making the need for path planning obsolete. Controlling geometry and material property variation at the resolution of the printer provides greater control over structure–property relationships.

The Wanderers series was unveiled as part of Stratasys' collection "The Sixth Element: Exploring the Natural Beauty of 3D Printing," organized by Naomi Kaempfer, on display at EuroMold, Frankfurt, Germany, November 25—28, 2014.

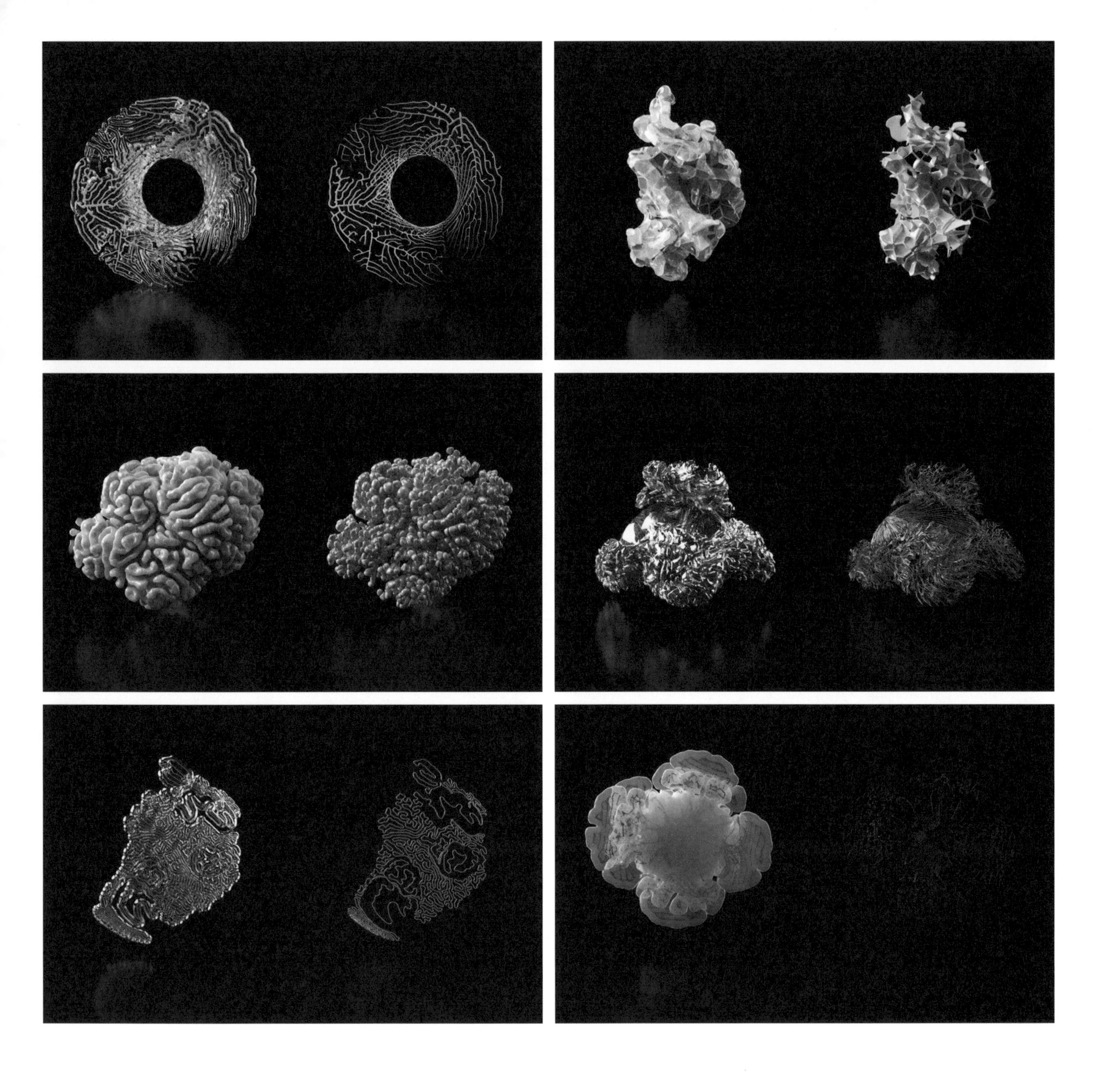

Figure 24.11
Detail Wanderers, view of internal vasculature, by Neri Oxman and Mediated Matter, 2014.

25
Kinematics

Nervous System—
Jessica Rosenkrantz and
Jesse Louis-Rosenberg

25
Kinematics

Nervous System—
Jessica Rosenkrantz and
Jesse Louis-Rosenberg

The Kinematics project fuses fashion, simulation, and 3D printing to examine how digital fabrication and computation can impact the way we create clothing. The project embraces a new approach to manufacturing that tightly integrates design, simulation, and digital fabrication to create complex, customized products. Intricate garments are generated from body measurements or scans, directly customized by users through a powerful and intuitive design app, simulated to understand their movement and find efficient folded forms for fabrication, and finally 3D-printed as a single piece that requires no assembly.

KINEMATICS DESIGN SYSTEM

The Kinematics design system pairs a constructional logic of hinged panels with a simulation strategy of folding and compression to produce customized designs that can be fabricated efficiently by 3D printing. The system is an example of the developing field of 4D printing, where 3D printing is used to create objects that transform in shape. It also explores the idea of programmable matter: the textiles created by Kinematics vary through space to create specific effects, differing in drape, porosity, rigidity, and pattern based on customer body shape and preference. Several components work together to advance our core concept:

(1) Kinematics structure—a hinge mechanism that allows a complex structure to be 3D-printed as one part, creating movable designs with hundreds (or thousands) of interconnected pieces, but which require no assembly,

(2) Kinematics Fold—a folding simulation that can compress the structure into a smaller and more efficient configuration for fabrication by 3D printing,

(3) Kinematics Cloth—web-based software that allows people to create their own designs online for wearables that can be inexpensively 3D-printed.

KINEMATICS DRESS

Kinematics is a system for 4D printing that creates complex, foldable forms composed of articulated modules. It produces designs composed of tens to thousands of unique components that interlock to construct dynamic mechanical structures. Though made of many distinct pieces, these designs require no assembly. Instead, the hinge mechanisms are 3D-printed in place and work straight out of the machine. This enables the production of intricately patterned wearables that conform flexibly to the body.

Bodies are three-dimensional, but clothing is traditionally made from flat material that is cut and painstakingly pieced together. In contrast, Kinematics garments are created in 3D, directly from body scans, and require absolutely no assembly. We employ a smart folding strategy to compress Kinematics garments into a smaller form for efficient fabrication. By folding the garments prior to printing them, we can make complex structures larger than a 3D printer can, which unfold into their intended shape.

The Kinematics Dress is an example of garments created by the system. It's a custom-fit gown with an intricately patterned structure made up of 2,279 unique triangular panels interconnected by 3,316 hinges, all 3D-printed as a single piece in nylon. While each component is rigid, in aggregate they behave as a continuous fabric, allowing the dress to fluidly flow in response to body movement. Unlike traditional fabric, this textile is not uniform; it varies in rigidity, drape, flex, porosity, and pattern through space. The entire piece is customizable, from fit and style to flexibility and pattern, with our Kinematics Cloth design app.

Figures 25.1 (right)
Kinematics Dress—3D-printed as a single piece at Shapeways Factory in New York, with 2,279 unique triangular panels and 3,316 hinges. Material: nylon; print technology: selective laser sintering.

Figures 25.2 (bottom)
Kinematics Dress Detail—3D-printed as a single piece at Shapeways Factory in New York, with 2,279 unique triangular panels and 3,316 hinges. Material: nylon; print technology: selective laser sintering.

Figures 25.3
Kinematics Dress in Motion—3D-printed as a single piece at Shapeways Factory in New York, with 2,279 unique triangular panels and 3,316 hinges. Material: nylon; print technology: selective laser sintering.

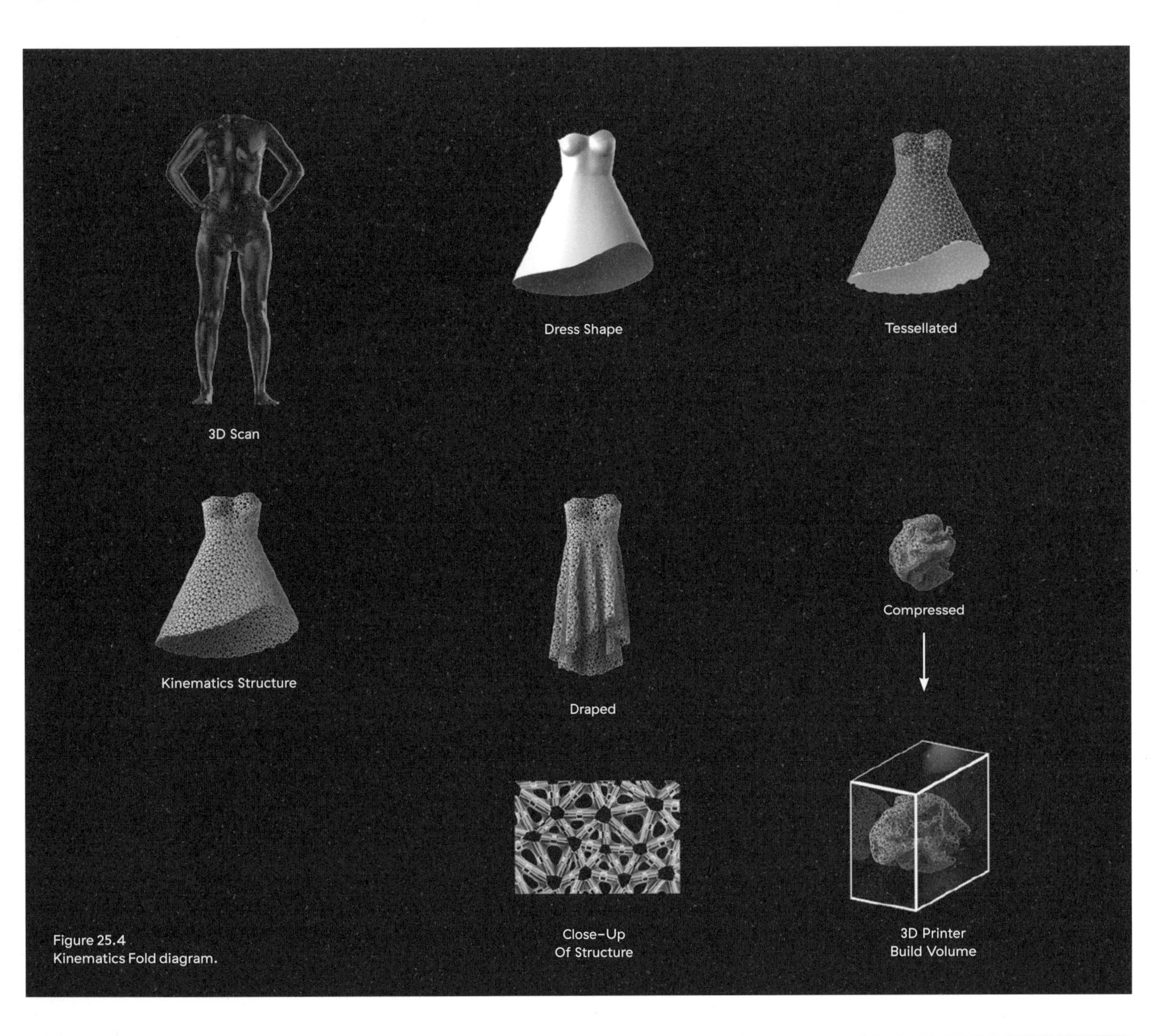

Figure 25.4
Kinematics Fold diagram.

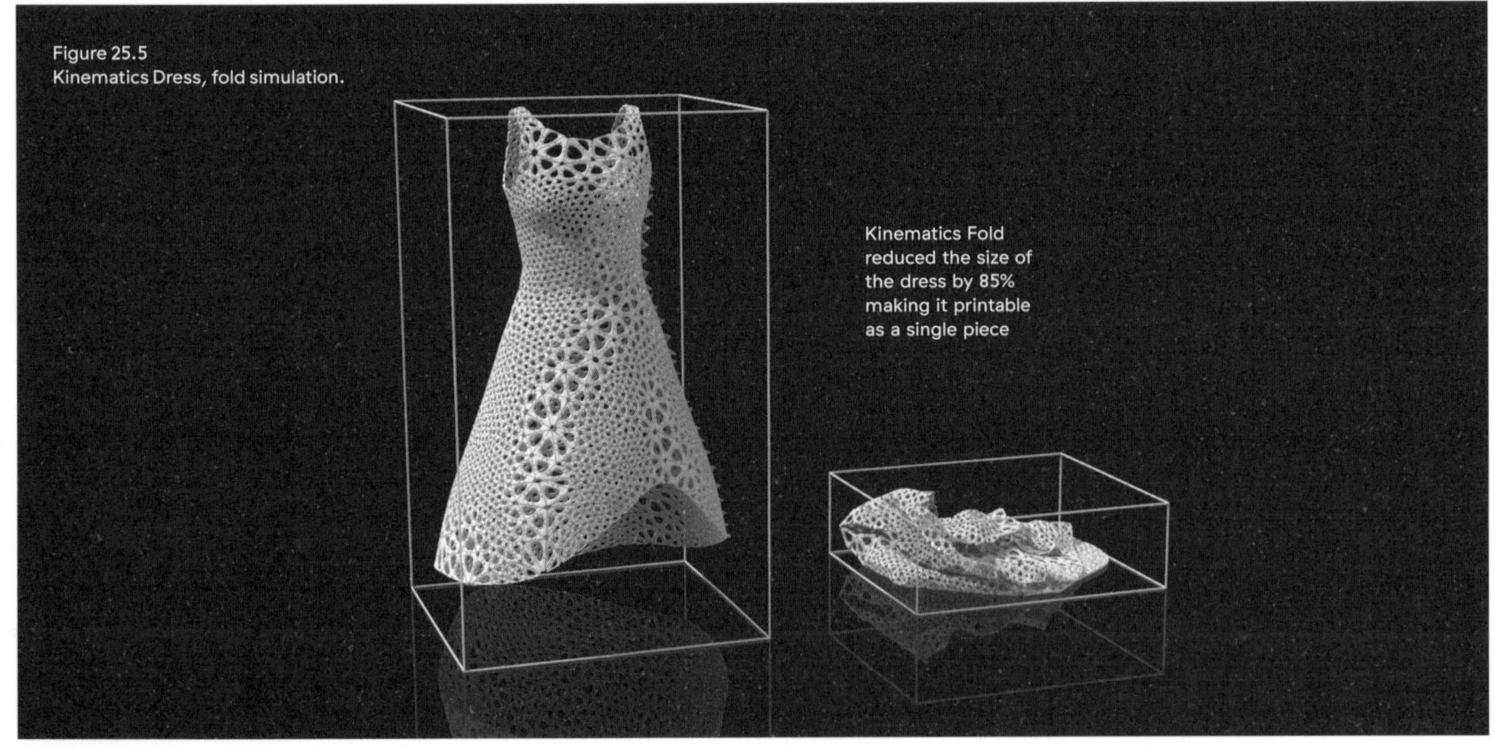

Figure 25.5
Kinematics Dress, fold simulation.

Figure 25.6
Kinematics Dress variations.

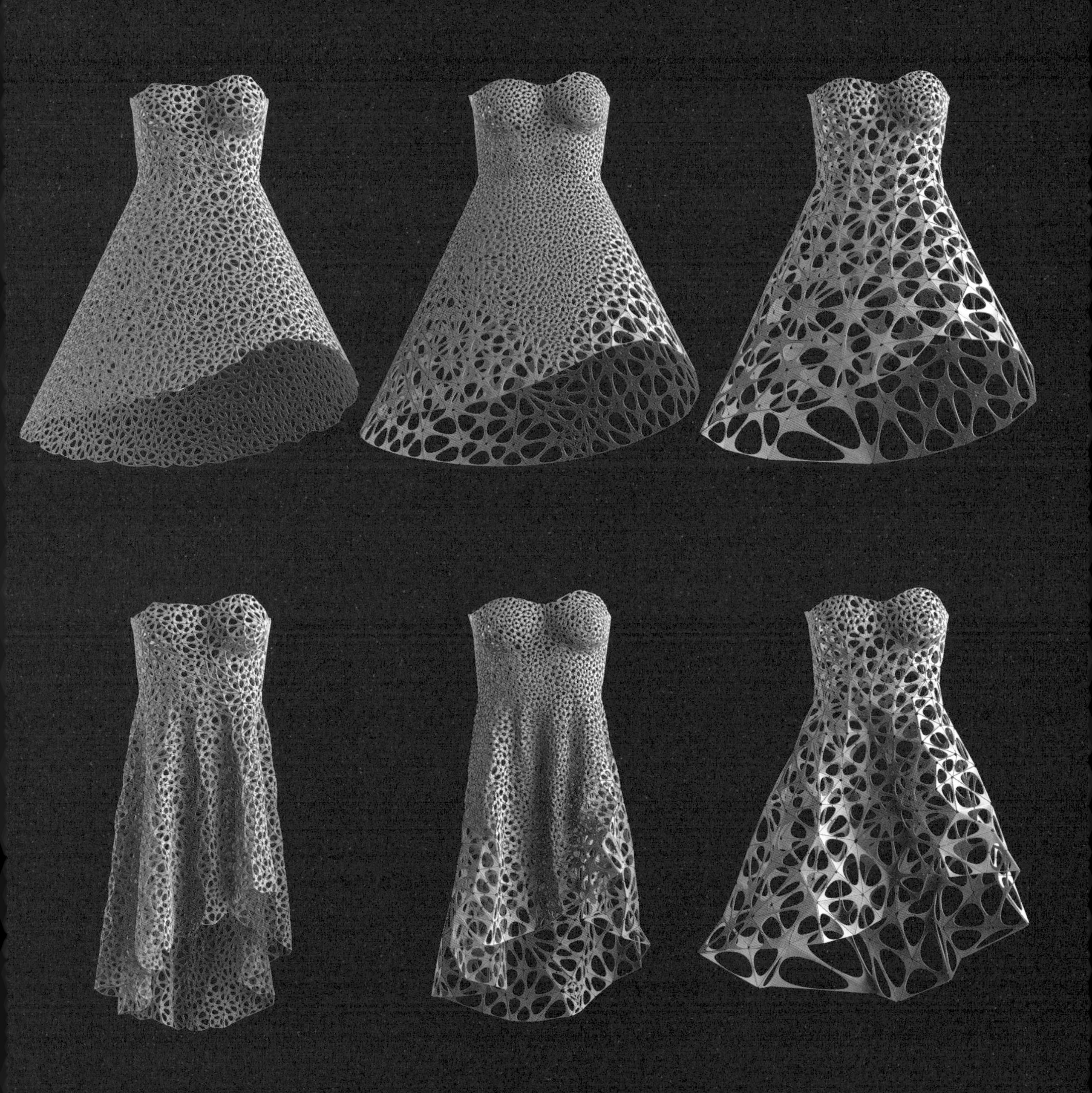

Figure 25.7
Kinematics hinge system.

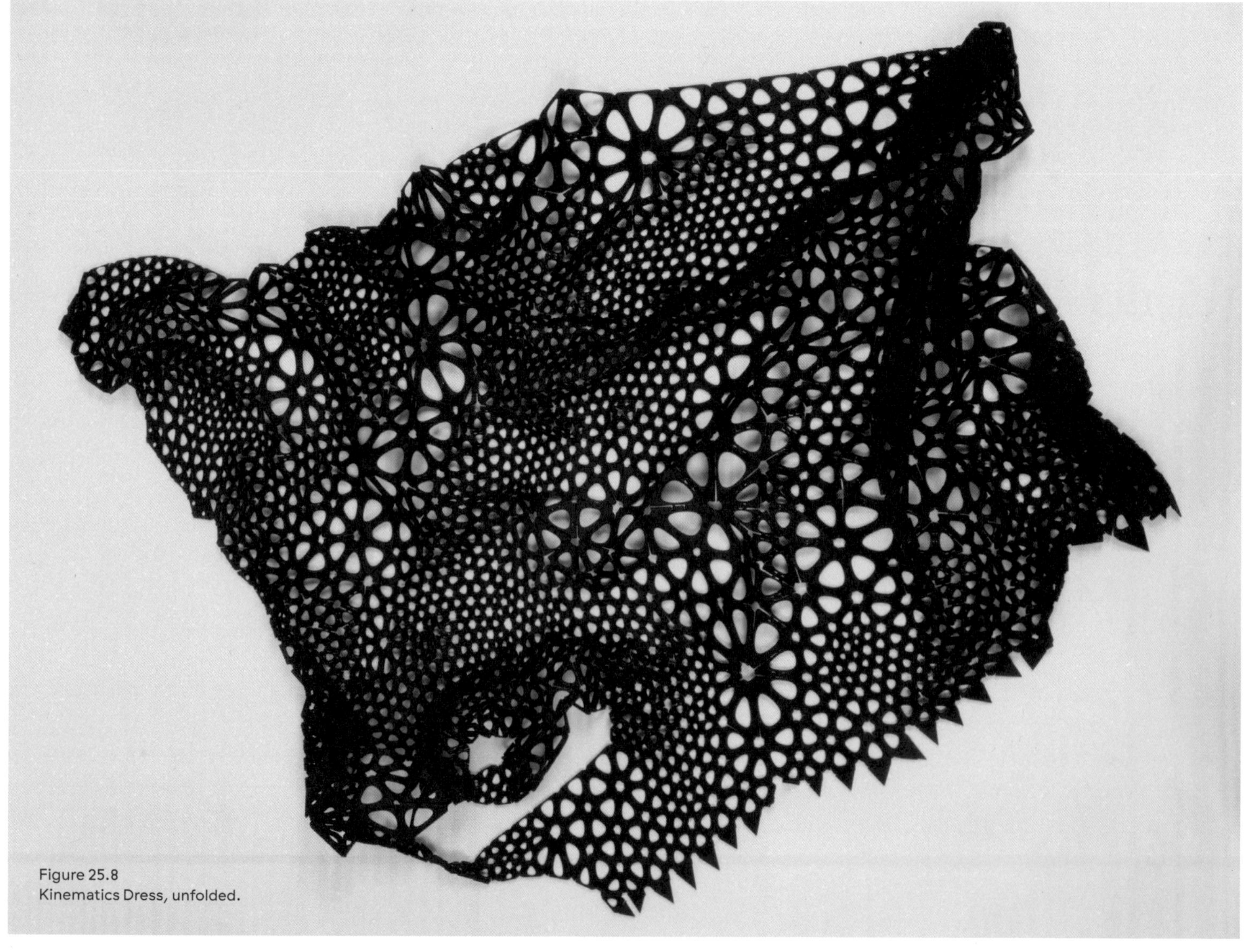
Figure 25.8
Kinematics Dress, unfolded.

26

Knitted Heat-Active Textiles: Pixelated Reveal and the Radiant Daisy

Felecia Davis and Delia Dumitrescu

26
Knitted Heat-Active Textiles: Pixelated Reveal and the Radiant Daisy

Felecia Davis and
Delia Dumitrescu

KNITTED HEAT-ACTIVE TEXTILES

The Pixelated Reveal and the Radiant Daisy are two knitted responsive tension structures that demonstrate two different expressive responses. These tubes were part of the 2012 exhibition "Patterning by Heat: Responsive Textile Structures" at the Keller Gallery, MIT. (See figure 26.1.) Each structure showed one of two material responses that were activated by an electrical current triggered by a proximity sensor. This current irreversibly changed the pattern and surface appearance of the material. The core questions guiding this design work were: How can we design lightweight textiles for use in architecture that can change structural properties in response to their environment? For example, how might we make textiles that open up if the temperature surrounding the textile becomes hot or if one wants more transparency to see the view? Similarly, we asked, How can we make a textile that closes the view, cuts down light coming through its surface, and thickens itself, which slows down energy transfer through its surface? We have been asking *what* these materials can be *when*.[1, 2]

The first typology of material developed was pixelated, designed with yarn that melts at high temperature; accordingly, the fabric *opens or breaks* when it receives current. The opening allows designers flexibility to experiment with see-through effects on the fabric or to "write" upon the fabric, making apertures, collecting foreground and background through the qualities of the material. The second material has been designed with yarn that *shrinks or closes* into solid lines in the fabric when it receives current. The shrinking reveals a more opaque patterning in the textile, *closing off* parts of that textile, transforming the material and the quality of space framed by that material. (See figure 26.2.)

PIXELATED REVEAL: OPENING EXPRESSION

The Pixelated Reveal tube was made on a circular knitting machine, a machine that produces tubes of constant diameter. (See figure 26.3.) The pixel pattern was made by including Grillon VLT yarn in five courses of jersey knit stitches which, when heated to 60°C or 140°F by current running through a continuous copper-coated conductive yarn, opened in small pixel-like rips. The left side of figure 26.4 shows fabric that is closed; the right side shows fabric that has been opened by running current through it. The snaps you see are where we connected a positive current on one side with a negative current connected on the other side. The fabric was designed so that other designers could make their own patterns and interactive scenarios by selecting the pixels they wanted to open on the fabric.

The design work proceeded at two scales at all times: the scale of the overall tube and the scale of the individual stitch geometry. At the scale of the overall tube we made an unfolded drawing of our tube so that we could draw a pattern on it. We called this a choreographic drawing. This drawing helped us to see the time of the opening response in a larger context. The drawing was made after testing samples of the material to see how long it took to open up the small and longer lines of fabric. (See figure 26.5.) On the circular knitter, the tube diameter and the tube's appearance could be altered by changing the yarn type as well as the stitch geometries that made for looser or tighter knits. At the scale of the stitch we diagrammed the structure and geometry of the stitch. What master knitters call "the binding" was the important drawing or information. This binding was given to the knitting technician who was working with us on an industrial knitting machine. The stitch geometry, along with the combinations of yarns, gives the fabric its character. (See figure 26.6.)

The interaction concept for the tube was that, when a person walked past the proximity sensor connected to our Arduino microcontroller, it would send a signal through our microcontroller to send positive and negative power

Figure 26.1
"Patterning by Heat: Responsive Textile Structures," exhibition, Keller Gallery, MIT, November 5—14, 2012.

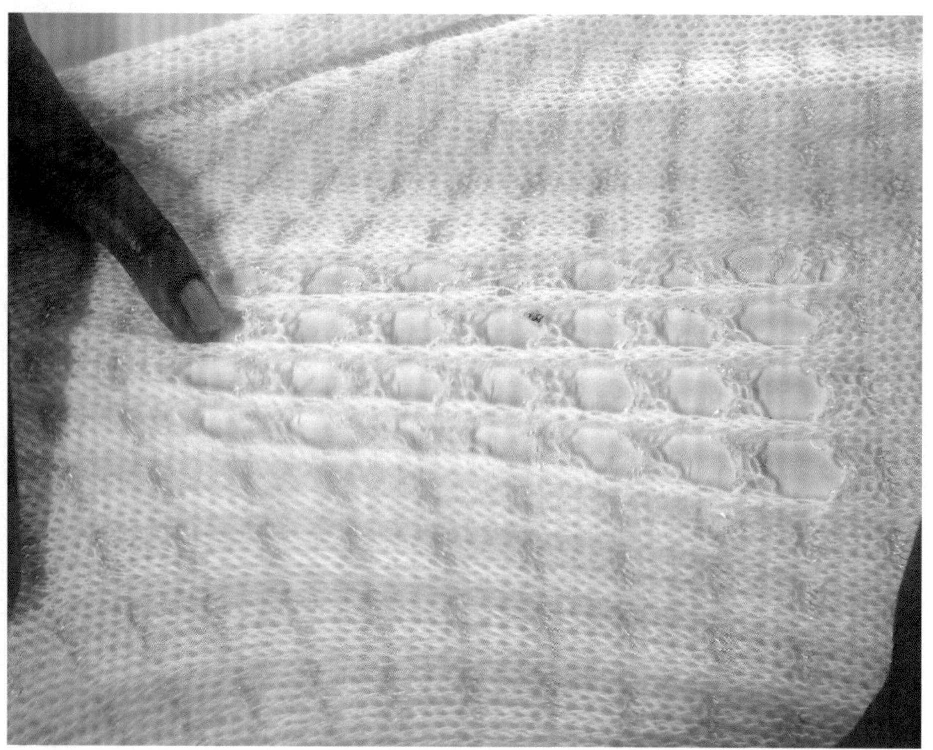

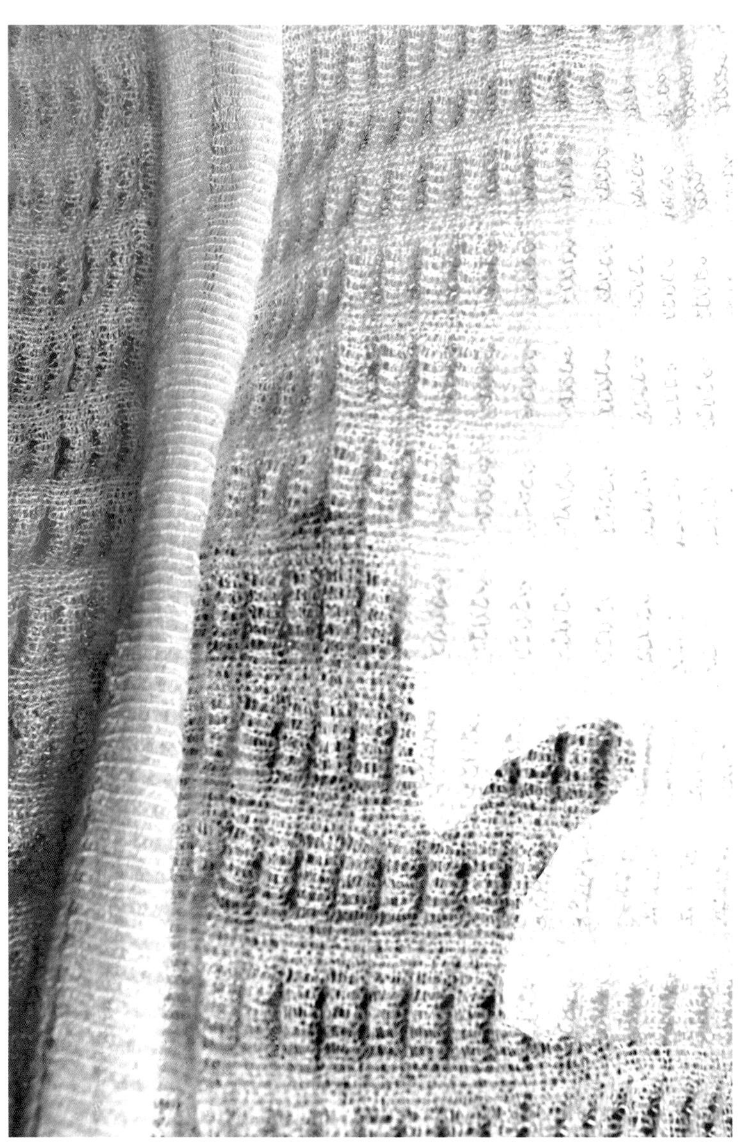

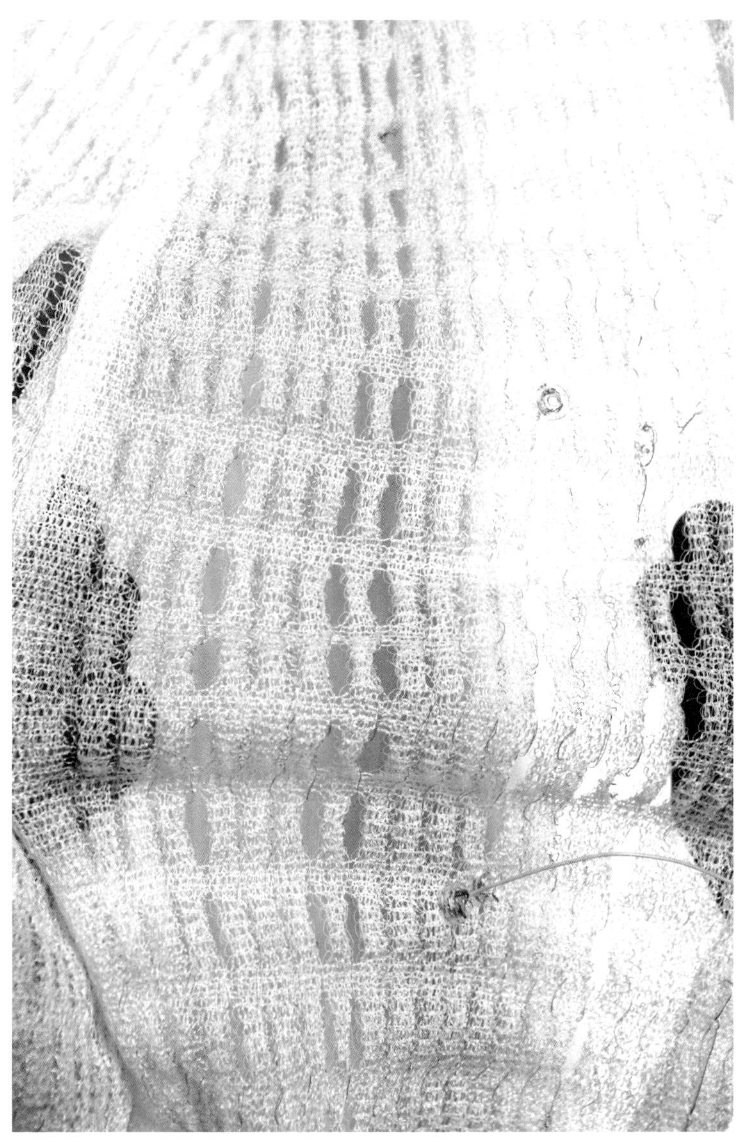

Figure 26.2 (left page, top)
Opening yarn or Grillon VLT (left); closing yarn or Trevira CS Pemotex (right).

Figure 26.3 (left page, bottom)
Circular knitting machine (left) and yarn feed (right).

Figure 26.4 (top)
Pixelated Reveal textile: closed (left); opened (right).

to each end of a connected pixel strip. Gradually, over time, the entire tube would expand in length because it was slowly ripping open, registering the presence of people in the gallery space. (See figure 26.7.)

THE RADIANT DAISY: CLOSING EXPRESSION

The Radiant Daisy textile tube was also made on a circular knitting machine, although some of the other tubes on exhibit used a flat, double-needle knitting bed that could change the tube diameter as it knitted. (See figures 26.8 and 26.9.) The structure for the Radiant Daisy fabric is formed by five courses of Trevira CS Pemotex yarn using a tuck pattern and a stainless steel yarn knitted as single jersey every sixth course. The Pemotex shrinks 40% at 90°C, which is the maximum heat put on it.

The Radiant Daisy Tube material starts out as a transparent volume that, when activated by heat or electrical current, closes the cells defined along horizontal bands in the structure. When patterned, the textile can be used to create closed areas in a surface. The programming for the Radiant Daisy was written so that the microcontroller received a signal from the proximity sensor and then sent current to make one petal of the daisy opaque. Each time a person was sensed near the tube, the next petal would become opaque, until the entire daisy pattern was revealed through opacity. See figure 26.10 for the choreographic drawing of the Radiant Daisy.

The Radiant Daisy knit structure had much higher resistance than that of the Pixelated Reveal material. The pattern was sewn as a parallel circuit, because it was a higher-resistance material. After tension was applied to

Figure 26.5 (top)
Pixelated Reveal
choreographic drawing.

Figure 26.6 (bottom)
Binding drawing (lower left). Above are axonometric drawings of the knitted yarn structure and types of yarns.

Figure 26.7 ((right page, right)
Pixelated Reveal, bottom of the tube, opened.

Figure 26.8 (right page, left, top)
Radiant Daisy tube.

Figure 26.9 ((right page, left, bottom)
Close-up of Radiant Daisy textile in tension.

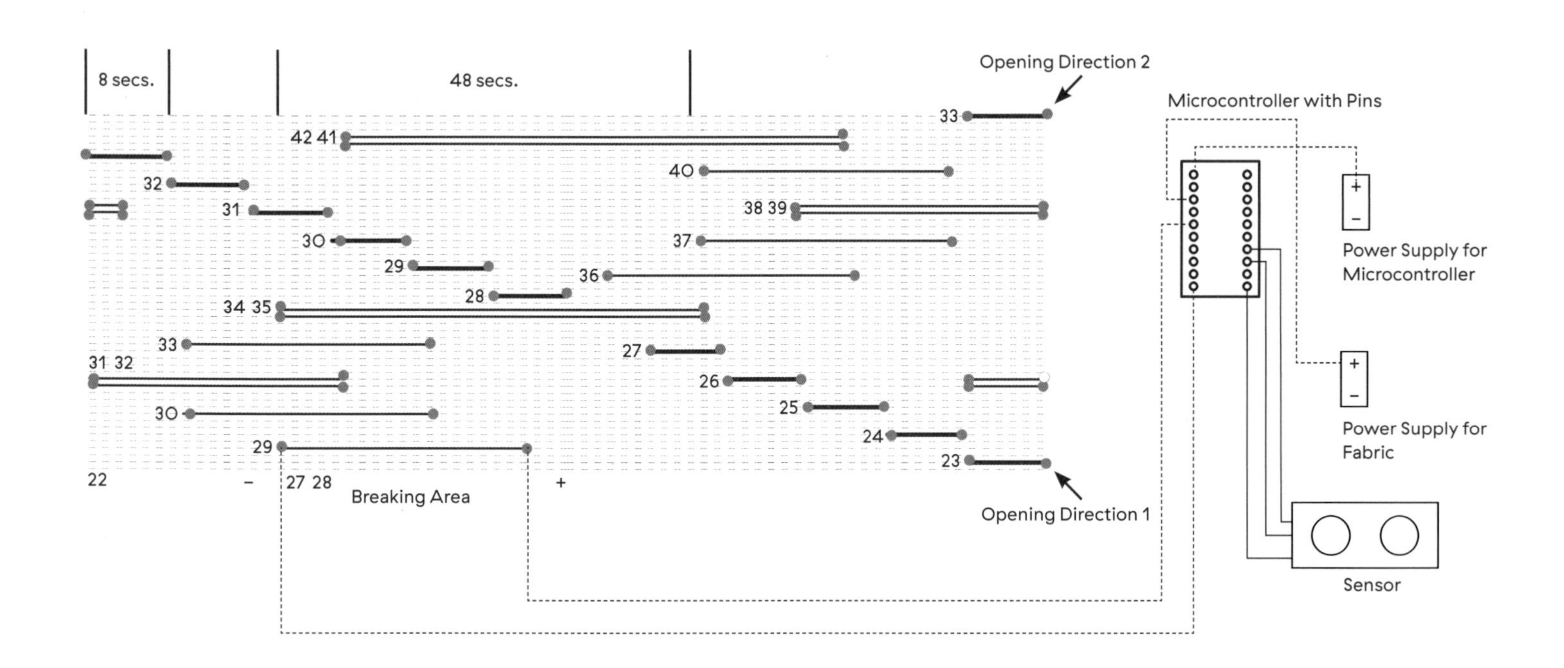

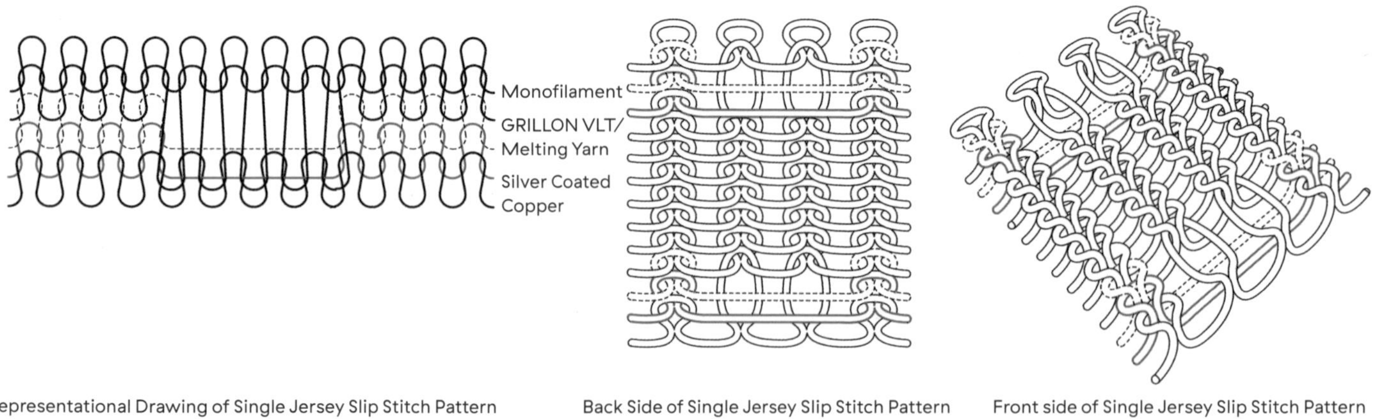

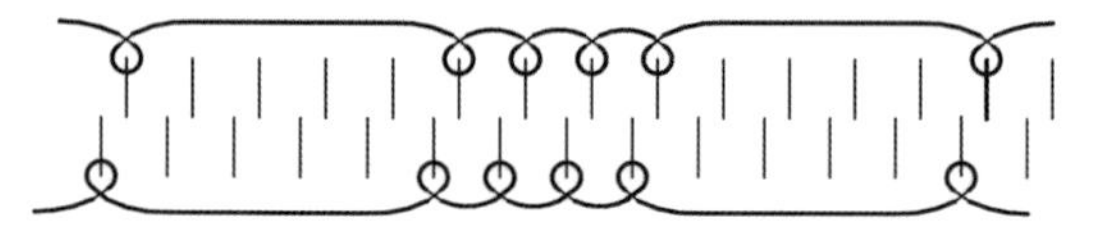

Binding Drawing of Single Jersey Slip Stitch Pattern

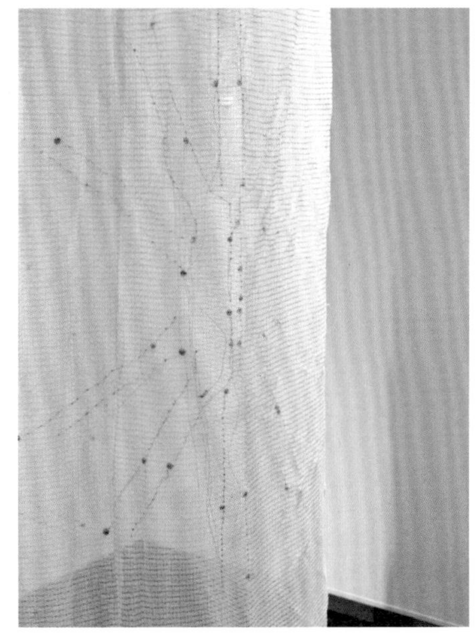

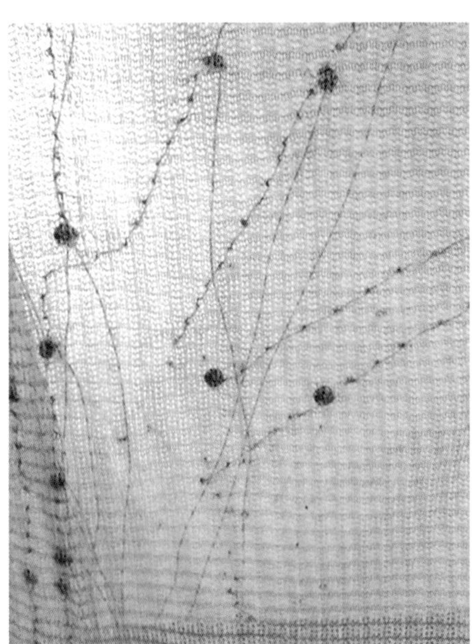

the material, the resistance went up and the authors were not able to activate the tube hanging in the exhibition space. Too much current was required for a response, and the fabric burned. Smaller samples laid upon a table were able to respond. Figure 26.11 shows a sample of the Radiant Daisy textile with a part of the pattern activated. It is possible that this particular knit and structure have a maximum tension limit to perform.

CONTRIBUTIONS OF THE KNITTED HEAT-ACTIVE TEXTILES PROJECT

We have presented the Pixelated Reveal and the Radiant Daisy to demonstrate how electrical current can transform the transparency or opacity of a large knitted textile structure. We have demonstrated that current can be used to activate specific areas in seamless, continuous, tubular structures. We have presented a method for thinking about the time of reaction in the textile pattern by using choreographic temporal drawings. In addition we have presented textile designs that are open to patterning from any designer.

For their help with this project we would like to thank Mika Satomi of Kobakant in Berlin, David Mellis of the High Low Tech Group at the MIT Media Lab, and especially Thomas Martinson and Christian Rodby, both master knitters in the knitting lab at the Swedish School of Textiles.

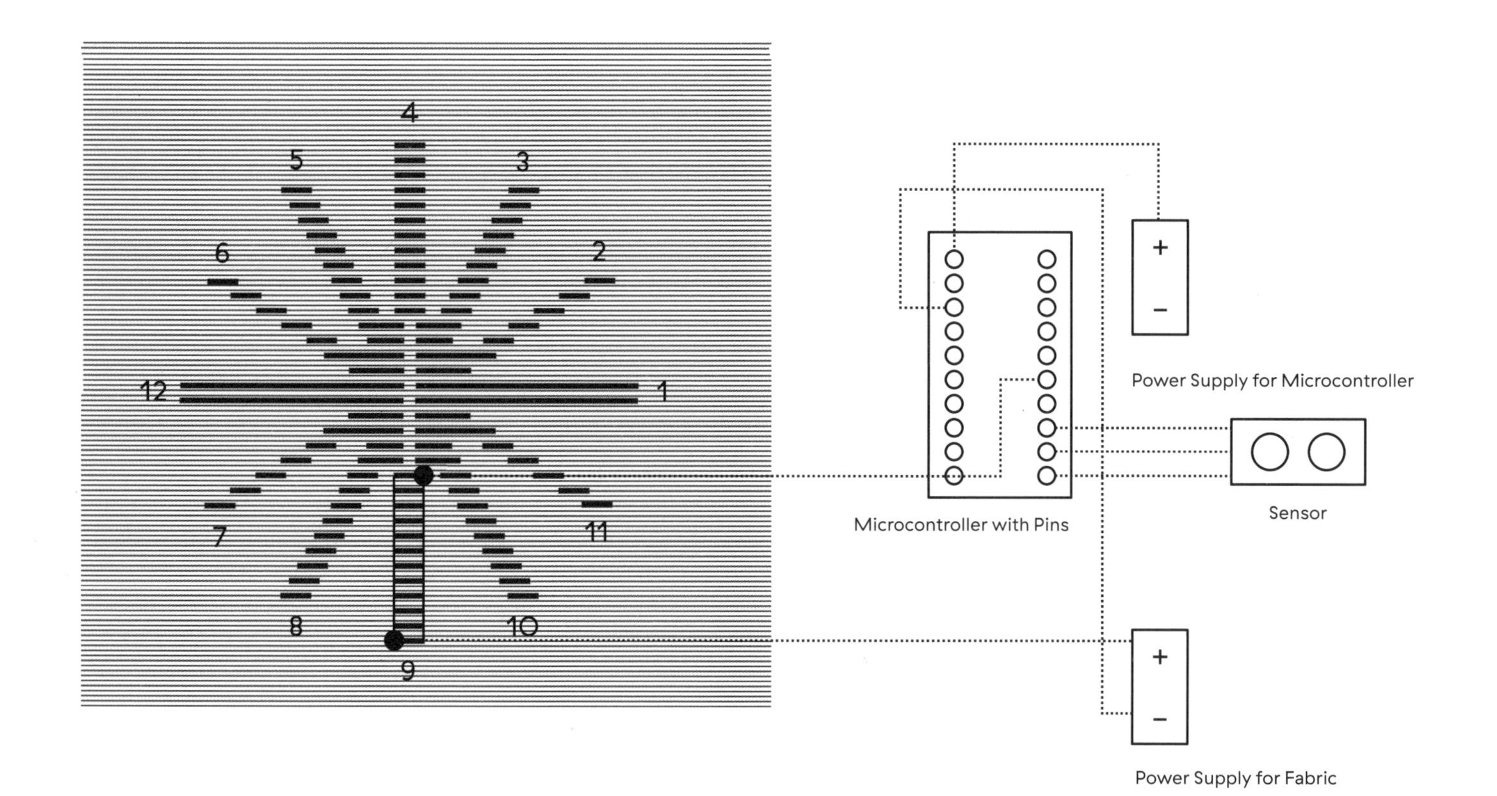

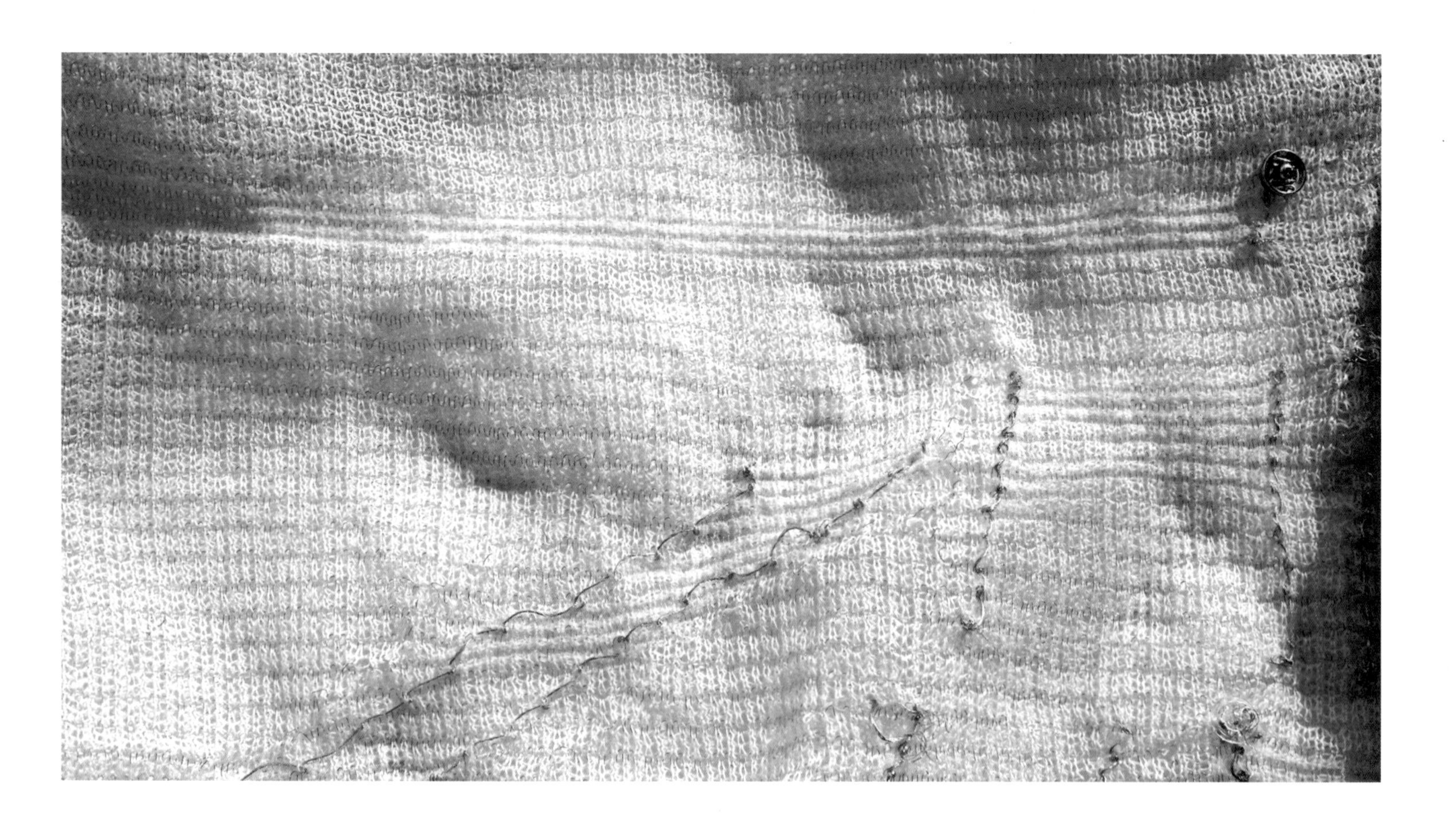

Figure 26.10 (top)
Radiant Daisy choreographic drawing.

Figure 26.11 (bottom)
Radiant Daisy sample activated.

NOTES

1 S. Kennedy, "Responsive Materials," in *Material Design: Informing Architecture by Materiality* (Basel: Birkhäuser, 2011), 118–131.

2 F. Davis and D. Dumitrescu, *Patterning by Heat: Responsive Textile Structures*, exh. cat. (Cambridge, MA: SA+P Press, 2013).

27
Programmable Knitting: An Environmentally Responsive, Shape-Changing Textile System

Jane Scott

27
Programmable Knitting: An Environmentally Responsive, Shape-Changing Textile System

Jane Scott

Programmable knitting presents a new class of behaving textiles, responsive to environmental stimuli, and programmed to change in shape as moisture levels increase. It is a hierarchical system that exploits the inherent functionality of textile fibers, yarns, and knitted fabric structures to integrate shape change behavior into the intrinsic structure of the material. This project documents a collection of three prototypes, Shear, Meander, and Skew, each composed of 100% natural fibers that have been engineered to produce reversible, 2D to 3D actuation.

Programmable knitting was developed using a biomimicry methodology. Insight derived from the structural organization of plant materials used for hygromorphic actuation produced transferable principles for application to responsive textiles. Biological material is structured across many hierarchical levels; this occurs because natural organisms grow material and structure simultaneously.[1] The hierarchies of cellulose, cell walls, cells, and tissue that exist within plant materials provide a model from which to reevaluate the hierarchies of fiber, yarn, knit stitch, and knit structure inherent to a knitted fabric.[2] The principles for programmable knitting have been established through analysis of how these hierarchies interact.

Each of the prototypes Shear, Meander, and Skew illustrates particular characteristics of programmable knitting. These prototypes demonstrate the underlying 2D to 3D actuation, as well as the outcomes of complex interactions between specific yarns and knit structures.

Shear is constructed from 100% linen. It is composed of a balanced 1 × 1 rib edging, with active material configured as three sets of front bed (FB) and back bed (BB) stitches across the fabric width (30 × 30, 40 × 40, 50 × 50). The stitches are transferred to the opposing needle bed after 60 courses. The structure continues for a further 60 courses before repeating the 1 × 1 rib edging.

On actuation, the fabric coils around the central point of interaction between the front and back bed stitches, producing a series of 12 cm spiral peaks on the surface of the fabric. This exploits the opposing directionality within the fabric structure. In addition, the edges of the fabric composed of 1 × 1 rib also transform into an undulating surface. Here, the impact of the deformation in one area transforms the entire surface due to the continuous nature of the material.

In order to design shape-changing fabrics, the constituent materials require the ability to sense and respond to environmental stimuli. In these prototypes, actuation occurs in response to moisture. In nature, hygromorphic behavior can be observed in seed distribution systems including those of pine cones[3] and wheat awns.[4] As a stimulus, moisture is particularly suited to natural fibers, which exhibit exceptional absorption properties. As water is absorbed into an individual fiber, there is little change in the fiber length, but the transverse swelling is extensive (up to 45% in linen).[5]

For shape change to impact at the scale of the knitted fabric, the anisotropic swelling caused by moisture absorption at the level of the fiber must be translated into an actuation motion. This is achieved using twist, the underlying mechanism that holds fibers together within a yarn.[2] Twist can be introduced in two directions, either clockwise (z-twist) or counterclockwise (s-twist), and the level of twist can also vary. In order to generate programmed shape change behavior, the twist direction introduced during the yarn-spinning process is modified through the configuration of knitted stitches within an individual fabric.

Meander demonstrates an alternative iteration of programmable knitting. This prototype explores the properties of two natural fibers, linen and silk. The fabric is constructed to a variable width along its length, increasing and decreasing the number of needles in action as it is knitted. The knit structure is highly complex, using

Figure 27.1 (top left)
Shear, before actuation.

Figure 27.2 (top right)
Shear after actuation, demonstrating programmed shape change behavior.

Figure 27.3 (middle left)
Meander, before actuation. Strips of linen and silk combine with alternating structure and shape.

Figure 27.4 (middle right)
Meander, after actuation. Detail illustrating where structure and yarn interact to enhance or prevent shape change.

Figure 27.5 (bottom left)
Skew.

Figure 27.6 (bottom right)
Skew. Detail of 3D profile after actuation, illustrating scale and direction of programmed shape change.

Figure 27.7
Programmable knitting, detail of environmentally responsive shape change behavior.

a links/links configuration of opposing front bed and back bed stitches. This creates alternative shape change behaviors in the actuated material. The regularity of the spiral peaks and undulating edges observed in Shear are transformed into a rippling surface in Meander. These behaviors are enhanced where the width is reduced and diminished where opposing directionality in both structure and materials interact.

Shear and Meander are small, machine-knitted prototypes. However, to increase the scale and complexity achievable with programmable knitting, Skew is programmed and manufactured using CNC knitting machines. The shape change behavior generated in the fabric on actuation is a series of spiral peaks increasing in size from 1 cm to 7 cm. These peaks form across the fabric, and when the fabric is hanging the peaks form in two directions. This alters the geometries considerably. The fabric structure is composed of repeating sections of FB and BB stitches (5 × 5, 10 × 10, 25 × 25, and 50 × 50), and scale is critical to how the responsive behavior operates. When the scale of pattern across the fabric is changed, the pace of actuation varies, increasing from 3 seconds (10 × 10 stitch repeat) to 10 seconds (50 × 50 stitch repeat). This produces a heightened sense of motion, moving across the surface as the fabric transforms.

Programmable knitting presents a departure from a representation of hygromorphic behavior as a bilayer system. This is achieved by applying a knit logic and introducing twist as a key actuation mechanism to enable shape change. By adapting the inherent hierarchies of a textile material using biomimicry principles, this work situates knitted fabric as a dynamic material interface directly responsive to environmental stimuli. The new behaviors observed using programmable knitting extend the geometries achievable with conventional knitting, whilst producing reversible and repeatable actuation from textiles composed of 100% natural fibers.

NOTES

1 P. Fratzl, and R. Weinkamer, "Nature's Hierarchical Materials," *Progress in Materials Science* 52 (2007): 1265.

2 J. Scott, "Programmable Knitting," PhD Thesis, University of the Arts, London, 2015, 85–90.

3 C. Dawson, J. Vincent, and A. M. Rocca, "How Pine Cones Open Up," *Nature* 390 (1997): 668.

4 E. Reyssat and L. Mahadevan, "Hygromorphs: From Pines Cones to Biomimetic Layers," *Journal of the Royal Society Interface* 6 (2009): 951.

5 W. E. Morton and J. W. S. Hearle, *Physical Properties of Textile Fibres* (Manchester: Textile Institute, 1986), 227.

28
Computational Skins

Marcelo Coelho

Marcelo Coelho

Clothing is one of our earliest technologies. It protects and shelters our bodies from the environment while expressing social, cultural, and economic aspects of our identity.[1] Today's insatiable hunger for information has been driving computers to permeate every aspect of what we wear, supplementing our wardrobes with smart phones, pedometers, electronic jewelry, head-mounted displays, and a whole host of other gadgets. This trend is in large part enabled by an increase in battery life, component miniaturization, and advances in materials science and fabrication, which have allowed electronics to be placed virtually anywhere while conforming to the curvatures and motions of the body. As computers migrate from the desktop to our skins, they will engender a new kind of interface that is primarily wireless, omnipresent, physically rich, and highly contextual. This raises an interesting challenge: how should we design wearable computers to enable a more comprehensive human experience akin to that of our clothing?

In what follows, I will outline a vision for the full integration of computation onto our bodies and describe three separate research efforts that attempt to answer this question.

SOFT MACHINES

Garments are traditionally soft, lightweight, and capable of moving and stretching with the body. Their shape is both a functional tool and a form of expression, and as they become worn their natural form changes over time. As technology improves and garments become full computational extensions of our bodies, it's only natural that they should inherit some of the properties of textiles.

One example of how this could be done is Kukkia: a shape-changing dress developed at XS Labs.[2] Decorated with three animated flowers that frame the neckline, Kukkia is essentially a soft machine that can adapt and respond by modulating its textile properties. Each flower opens and closes through the combined actuation of a custom-designed textile composite made from shape memory alloy and natural felt. While the shape memory alloy deforms the textile in one direction, the felt naturally counteracts this in the opposite direction.

Extending this work, Sprout I/O is a similar textile composite which, in addition to dynamically changing shape, can also capacitively sense when it's being touched.[3] The textile equivalent of a touch screen, this material still looks and feels like fabric, albeit with a whole new set of utilitarian and expressive functionalities.

CROWD NETWORKS

As our garments become dynamic and responsive, they will not only redefine how we access and manage information, but will also dramatically alter how we communicate and relate to each other. Alike is a wearable device that seeks to reinvent how crowds connect, relate, and collaborate when face to face. At its core, Alike is a small circuit board that can be embedded into virtually any article of clothing. It is composed of two primary components: a radio for always-on and continuous device-to-device communication, and an RGB LED that provides real-time feedback to both the wearer and others nearby.

At MIT's first Active Matter Summit, Alike functioned as a distributed and colocated display for attendees to visualize the crowd's networking activity. Embedded into a traditional conference badge, each Alike device kept track of nearby activity. When all conference attendees were stationary, such as when listening to a talk or talking to a single group of people for a long time, their devices would register a low crowd activity and light up a cool, blue color. However, as attendees moved about and interacted with more people, their devices slowly transitioned to a hot, red color. This heat map visualization not only served as a tool for wearers to visualize their behavior, but also provided the feedback and social cues necessary for people to manipulate it.

So far, we have experimented with a variety of crowd dynamics, such as games, crowd-sourced fabrication, and a large-scale performance for the Rio 2016 Paralympics ceremonies, and with a wide range of device form factors, such

Figure 28.1 (top)
Kukkia: electronic textile, a shape-changing garment.

Figure 28.2 (bottom)
Kukkia: electronic textile, a shape-changing garment.

as pendants, bracelets, carnival masks, and even drinking cups. The possibilities for designing new collective human experiences are virtually endless, since each environment and form factor signals, promotes, and precludes a different set of social behaviors. If the body is to become the computer, the relationships between bodies will become the ultimate network.

BIOCOMPOSITES

Looking ahead, it is not hard to imagine how we can go farther and ditch clothing altogether. At LogicINK we are starting to develop chemical and biological functional inks that can be applied directly to the skin to detect both physiological signals from the body and a wide range of environmental stimuli.[4]

One example is an enzyme-based ink embedded into a temporary tattoo that detects wearers' blood alcohol content through their sweat. When applied to the skin, the ink is initially transparent and conceals its design, but as the wearer drinks and their alcohol level goes up, it changes color to reveal a new image and indicate they should not be driving. This biocomposite computational approach has a series of advantages over traditional technologies: it's incredibly low-cost and disposable; its form factor is completely unobtrusive and free of battery and electronics; and since inks can be printed with all kinds of designs and patterns, it provides a completely new outlet for aesthetic expression.

The work described here is for the most part in its infancy, but future developments in biological and chemical engineering will continue to help us design, program, and fundamentally

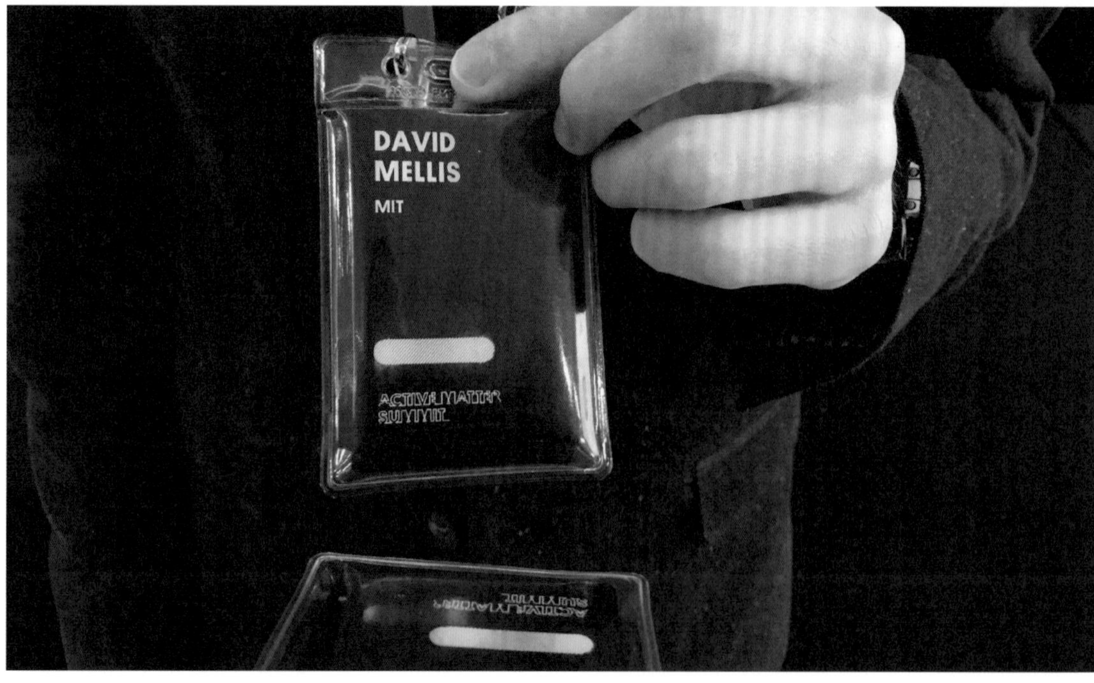

Figure 28.3 (left page, top)
Alike: custom electronics, a social networking wearable.

Figure 28.4 (left page,bottom)
Alike: custom electronics, a social networking wearable.

Figure 28.5 (bottom)
Alike: custom electronics, a social networking wearable.

reinvent our skin's microbiota. Working in tandem with active materials and increasingly small electronics, wearable computers may look a lot more like a sunscreen lotion or spray that we apply to our skins as we are ready to leave the house, rather than the bulky and heavy computers of yesteryear.

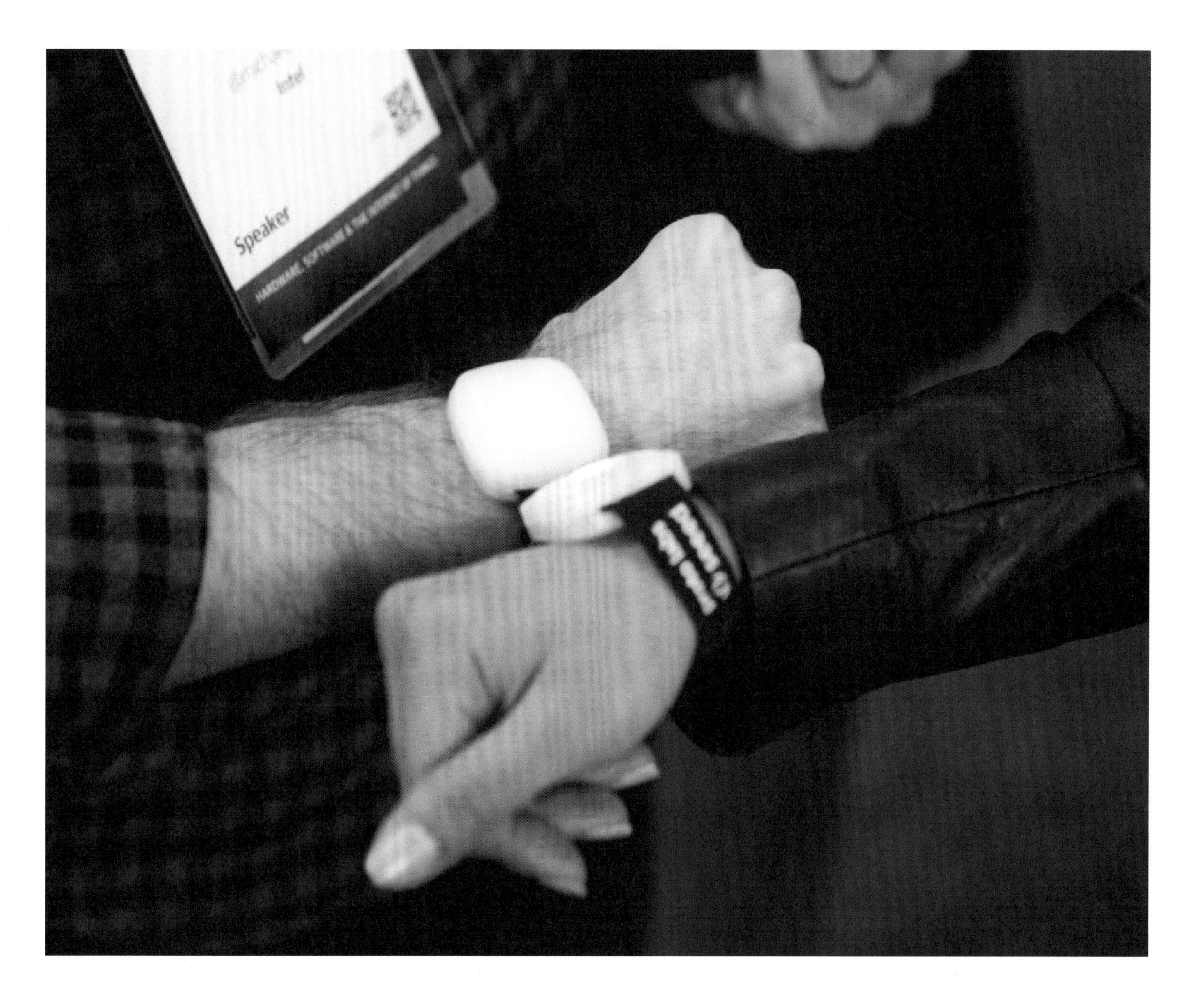

Figure 28.6
Beyond Vision: custom electronics, performance during the Paralympics opening ceremony.

Figure 28.7
Beyond Vision: custom electronics, performance during the Paralympics opening ceremony.

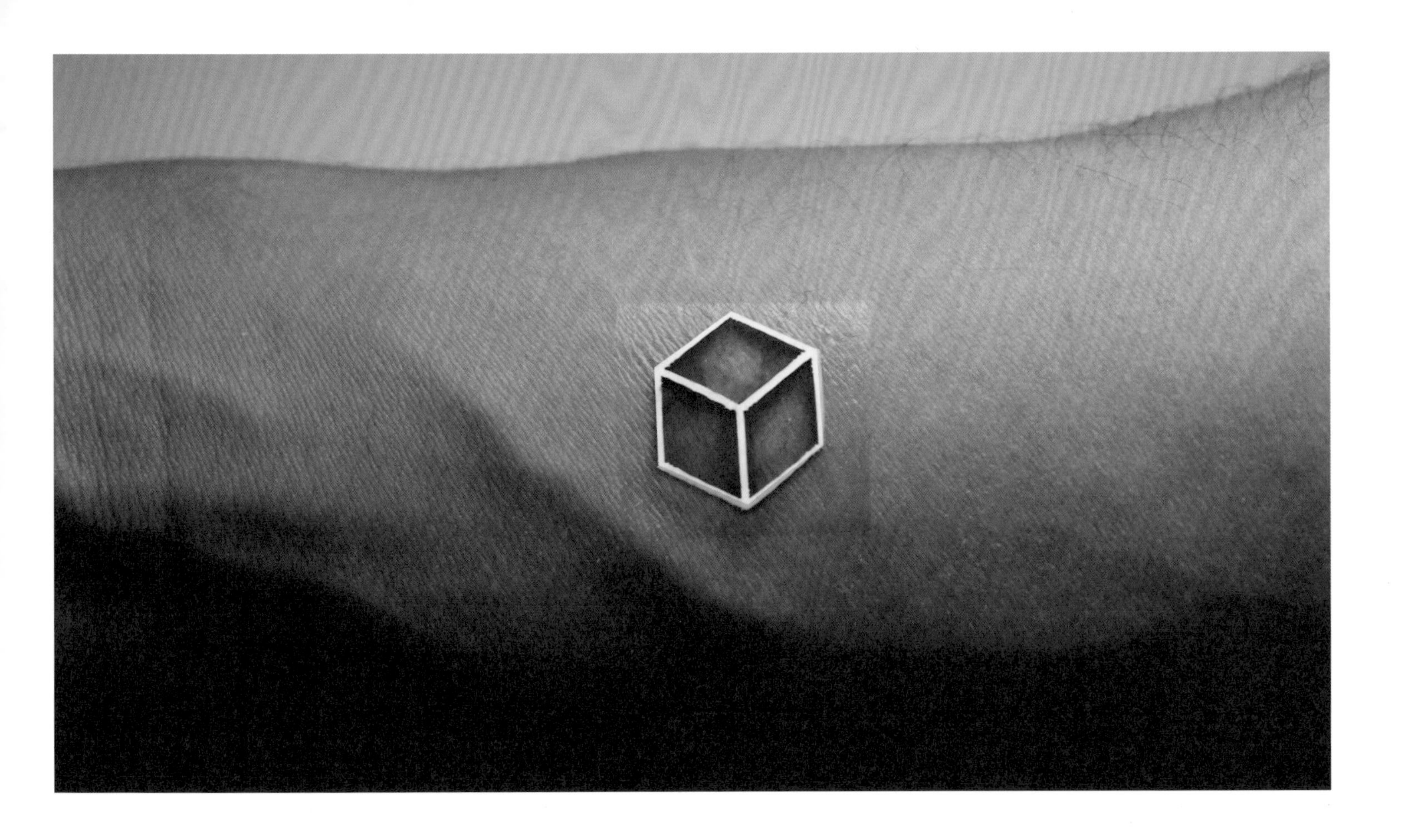

Figure 28.8
Logic INK: programmable ink, programmable temporary tattoo applied on the skin.

NOTES

1 M. Hogenboom, "We Did Not Invent Clothes Simply to Stay Warm," September 19, 2016, http://www.bbc.com/earth/story/20160919-the-real-origin-of-clothes

2 J. Berzowska and M. Coelho, "Kukkia and Vilkas: Kinetic Electronic Garments," International Symposium on Wearable Computers (ISWC'05), IEEE, 2005.

3 M. Coelho et al., "Shape-Changing Interfaces," *Personal Ubiquitous Computing* 15, no. 2 (February 2011): 161–173.

4 LogicINK, http://www.logic.ink

29
Radical Atoms: Beyond the "Pixel Empire"

Hiroshi Ishii

DANCING ATOMS

In a world of static forms, we envision dancing formations as full of life as the pixels on a screen. Our dream is to let atoms dance as pixels do today (figure 29.1).

Physical materials are, at a human scale, seemingly shy, inert, and frozen. In our eyes, these frozen atoms desire to be awoken: to dance, to leap, and to levitate around us. Using the power of computation, we aim to build a new, dynamic, digital and physical medium for expression, design, and communication, to breathe life into static forms. This vision encapsulates the shift from static formations to dancing atoms, which would no longer be confined by their current physical constraints.

ICEBERG RISING

Information can be visualized as water flowing rapidly through the channels of the Internet, connecting billions of machines and users via the cloud. As is true for liquids, we cannot grasp and hold on to the information with our hands, as we would be able to do with wooden blocks or shaping clay. We then must rely on our eyes to make sense of the pixel-based information within the screen, and with remote controllers, such as mice, keyboards, or touchscreens, to indirectly manipulate intangible pixels. This pixel-oriented world, more commonly referred to as a GUI (graphical user interface), allows for interactions via an indirect medium, thus limiting the tangibility and embodied interactions for the user. Given that GUIs currently dominate our interactions, we often refer to the GUI world as the "pixel empire" and to our research as a battle against the far-reaching presence of the ubiquitous pixels (figure 29.2).

Our discontent with the separation between representation (intangible pixels behind 2D screens) and remote controllers led to the conceptualization of Tangible Bits as a pioneering medium. This idea—to partially freeze the ungraspable

Figure 29.1 (left page)
Transform: dancing with red balls.

Figure 29.2
Iceberg metaphor: from GUI (painted bits) and TUI (Tangible Bits) to Radical Atoms.

water, like physically embodying digital information—was born of the desire for people to grasp and manipulate digital information. The tangible user interface (TUI), a next-stage interface, allows for direct interaction and manipulation of information.

In 1997, the Tangible Media Group presented our Tangible Bits vision of the physical embodiment of digital information and computation at the CHI '97 conference.[1] In 2000, Tangible Bits projects were first exhibited at NTT ICC[2] in Tokyo, Japan. Then, between 2001 to 2003, a dozen Tangible Bits projects were presented at the "Get in Touch" exhibition at Ars Electronica Center in Linz, Austria.

However, these frozen forms pose the problem of asynchrony between the ever-changing, dynamic states of the digital and the frozen physical forms—the tip of the iceberg exposed to the physical world. The physical component, which is breaking through the water, is rigid and frozen, yet the digital, which is submerged below the water, is dynamic and ever-changing.[3] To overcome this limitation of frozen atoms, the Tangible Media Group has been working to evolve our original Tangible Bits vision to the next level, resulting in our current vision of Radical Atoms:[4] dynamic, physical, and computational material. We envisioned Radical Atoms[5] to bring atoms to life through their dynamic physical manifestation—malleable and computationally transformable. These advanced new materials composed of radical atoms will be the modern machines that will drive the new paradigm of material user interfaces (MUIs).

As illustrated in figure 29.3, we have developed various streams of Radical Atoms projects, such as actuated tabletop tangibles, kinetic materials, dynamic shape displays, and programmable materials. In the "Radical Atoms" exhibition[6] in 2016 at Ars Electronica Center, we presented

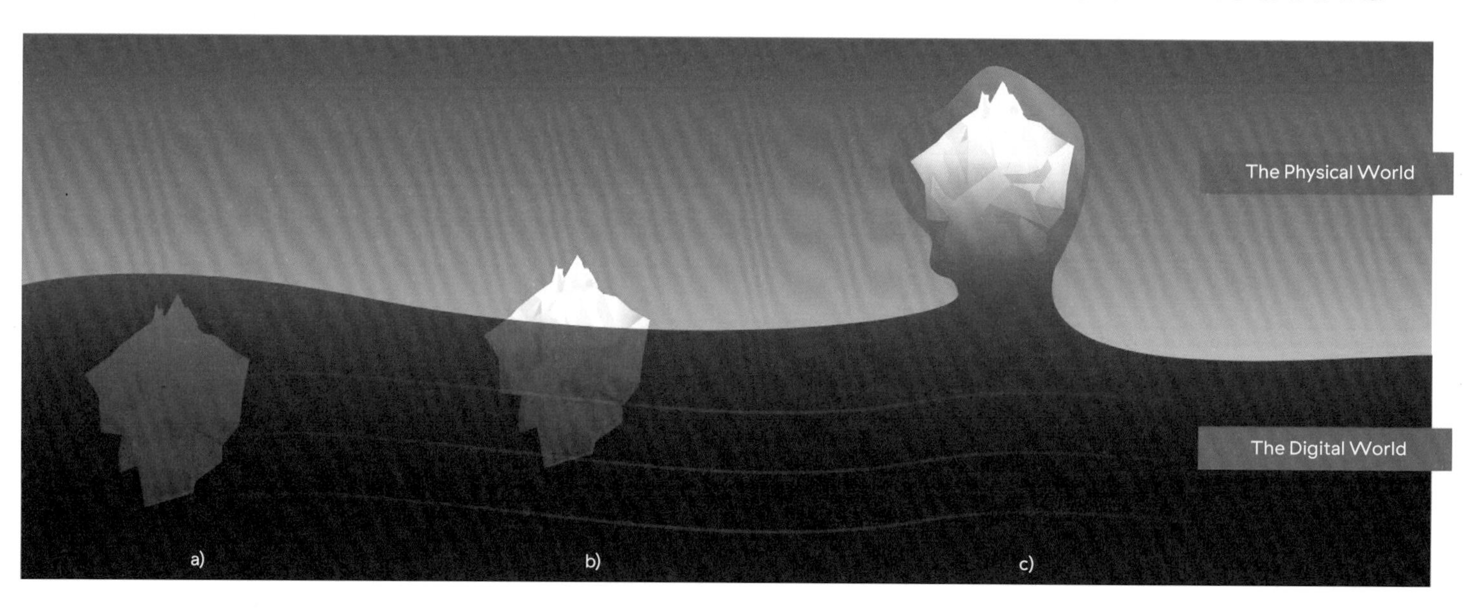

Graphical user interfaces only let users see digital information through a screen, as if looking through the surface of the water. We can interact with the forms below using remote controls such as mice, keyboards or touch screens.

A tangible user interface is like an iceberg: a portion of the digital emerges beyond the surface of the water–into the physical realm–representing the manifestations of computation, allowing us to directly interact with the 'tip of the iceburg'.

Radical Atoms symbolizes our vision for the future of interaction with hypothetical dynamic materials, in which all digital information has a physical manifestation so that we can interact directly with it–as if the iceberg had risen from the depths to reveal its sunken mass.

EVOLUTION

From Tangible Bits to Radical Atoms

STATIC/PASSIVE

DEFORMABLE TANGIBLES

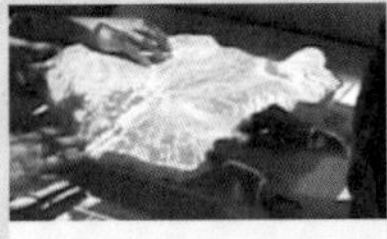

Illuminating clay

CHI '02

SandSacpe

ARS Electronica '02

DYNAMIC SHAPE DISPLAYS

Relief

TEI '09, UIST '11

Recompose

UIST '11

SUB

C

PROGRAMMABLE MATERIALS

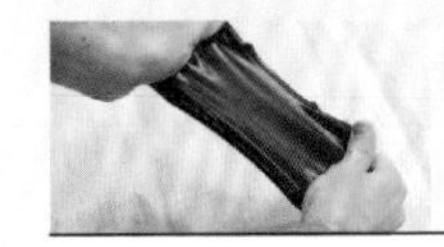

Jamming UI

UIST '12

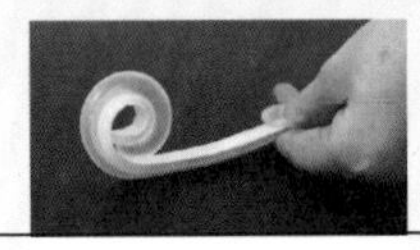

PneUI

UIST '13

jam

T

Sensetable

CHI '01

TABLE TIP TANGIBLES

metaDesk

CHI '97
USIT '97

Urp

CHI '99
SIGRAPH '98

ACTUATED TABLE TOP TANGIBLES

PSyBench

CSCW '98

Actuated Workbench

UIST '12

C

musicBottles

CHI '01

KINETIC MATERIALS

TELESYNCED

inTouch

CSCW '98

RECORD+PLAY

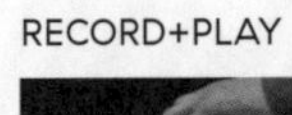

curlyBot

CHI '00

Top

Chi '04,
'06, '08

ACTIVE/KINETIC

inFORM
'13, UIST '14

TRANSFORM
Milano Design Week '14
CHI '15

Cooper Hewitt in FORM
Cooper Hewitt
Museum '15

Kinetic Blocks
UIST '15

Materiable
CHI '16

ptiElastic
UIST '14

bioLogic
UIST '14, CHI '15

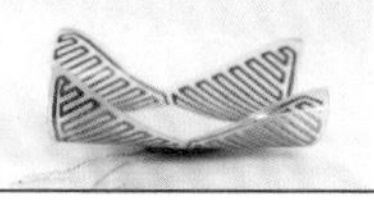
uniMorph
UIST '15

biologic
Media Lab Exhibit '15

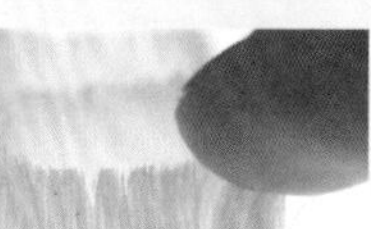
Cillia
CHI '16

LEVITATING MATERIALS

ZeroN
UIST '11

SIGN TOOLKITS

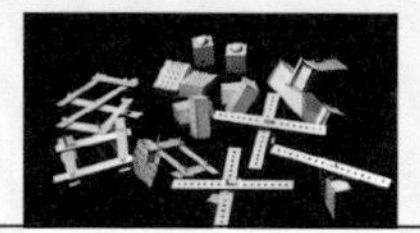
Kinetic Sketchup
DIS '10

Bosu
DIS '10

SELF RECONFIGURABLE

LineFORM
UIST '15

ChainFORM
UIST '16

RADICAL

The Future of Radical Atoms

PHASE SHIFT

PROGRAMMABLE MATERIALITY

BIOLOGICAL SYMBIOSIS

LEVITATION

SELF ASSEMBLY

the evolution from our classic Tangible Bits projects, such as musicBottles[7] (figure 29.4) and SandScape[8] (figure 29.5), to Radical Atoms projects, such as Topobo[9] (figure 29.6), inFORM[10] (figure 29.7), LineFORM[11] (figure 29.8), PneUI[12] (figure 29.9), jamSheets[13] (figure 29.10), and bioLogic[14] (figure 29.11). Through a combination of projects[15] crossing altitudes and domains, we presented our vision of Radical Atoms, where atoms dance across scale and axes and the static becomes alive and active.

MATERIALS TO THINK WITH

This leads to the following key question: Why care about these new materials for the representation of ideas? By using mathematical equations to represent ideas, we can apply the embedded mathematical, logical, and linear operators to manipulate ideas. By drawing on a whiteboard or on sheets of paper, we can apply spatial mental operators to manipulate the idea by rearranging them spatially within our mind's eye.

Tangible Bits and Radical Atoms represent the combined effort to create new physical and digital representations of ideas and concepts for the artists, designers, and communicators of the future. They personify our way to *defy the gravity of the pixel empire* by demonstrating the new possibilities of embodied and kinesthetic relationships between humans and digital information.

We are determined to invent new media to express creators' ideas and to engage body-syntonic interactions with users. Hence our focus on Radical Atoms, those computational, physical, and dynamic materials.

MACHINES BECOME MATERIALS ...

MATERIALS BECOME MACHINES

Blurring the boundary between machine and material is one of our dreams. Radical Atoms could represent the future of digital information and computation, and these, in turn, could provide us with a rich variety of embedded interfaces with which to manipulate them. Thus, machines become materials ... and materials become machines.

While we might need to wait decades before atom hackers like material scientists, biologists, or self-organizing nanorobot engineers can invent the necessary enabling technologies for Radical Atoms, we strongly believe the exploration of artistic expression and interaction design can begin today.

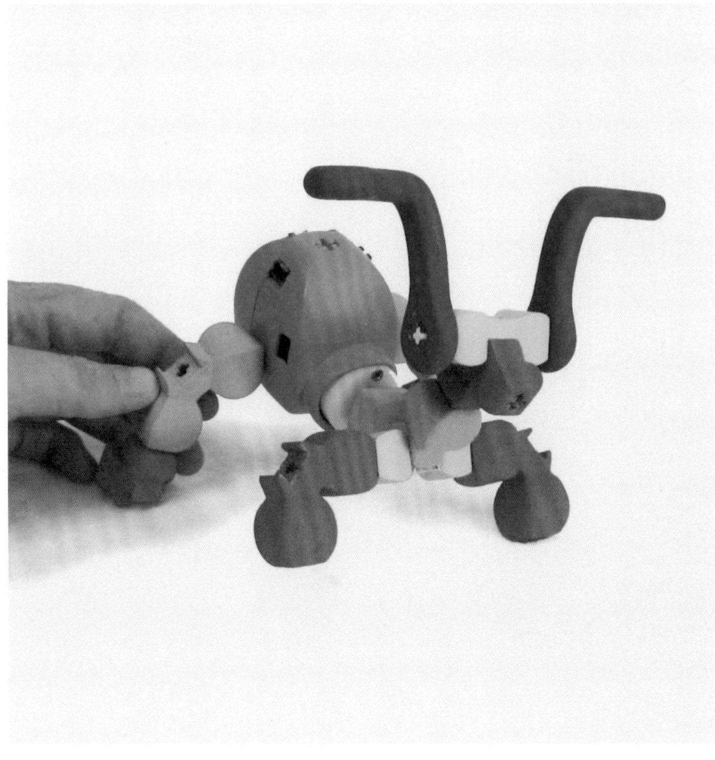

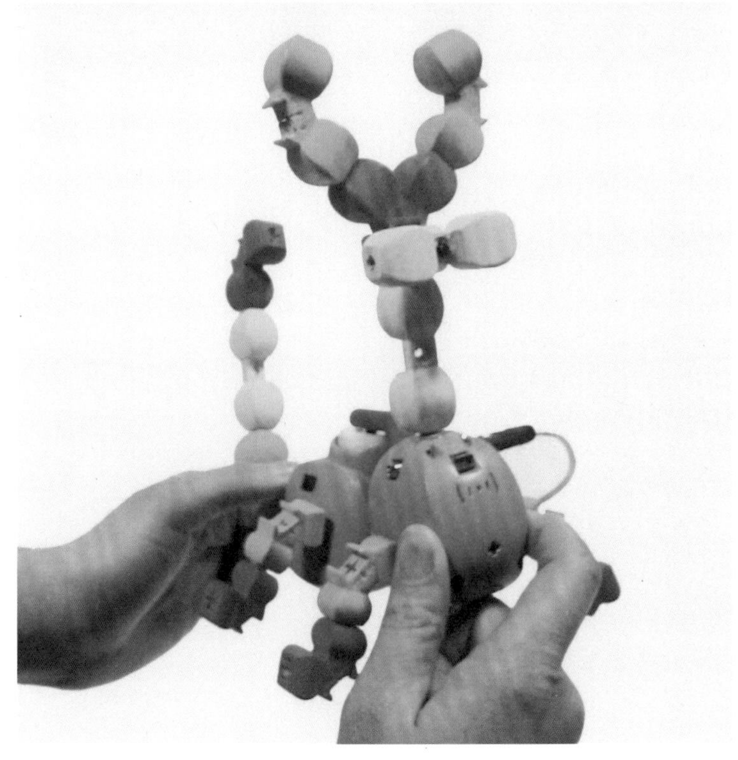

Figure 29.4 (left page)
musicBottles.

Figure 29.5 (top)
SandScape.

Figure 29.6 (bottom left, right)
Topobo.

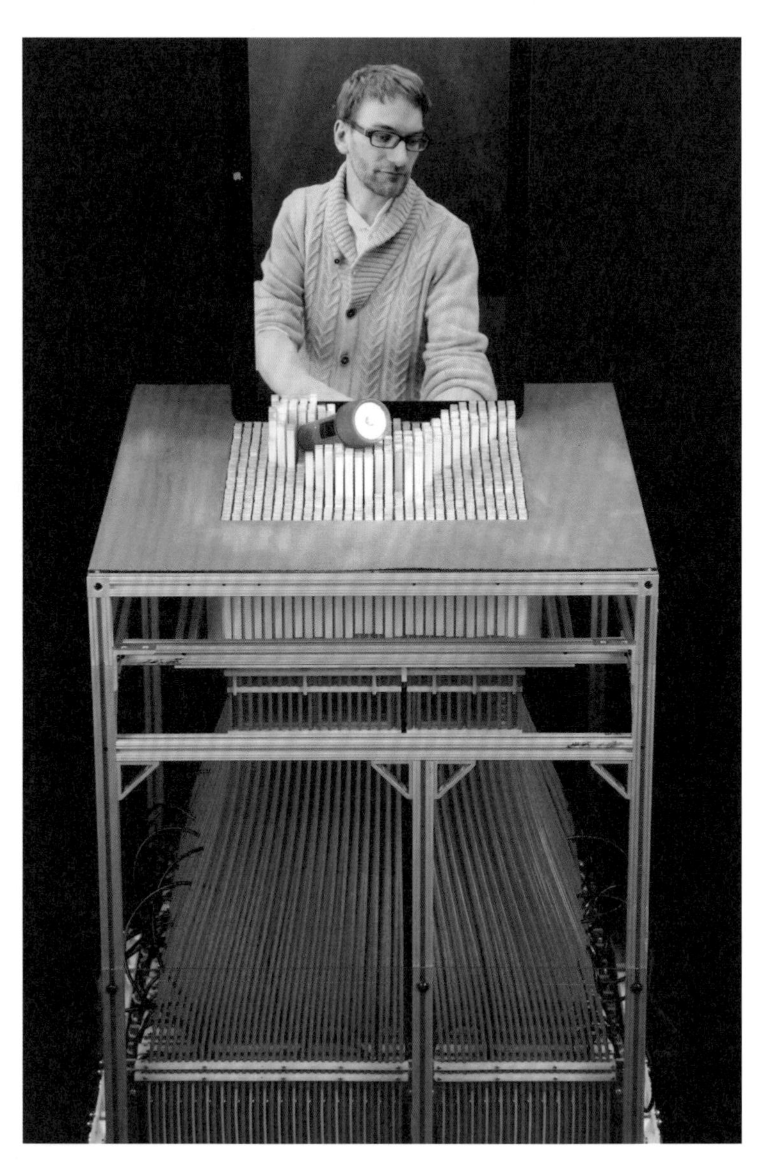

Figure 29.7 (left)
inFORM (telepresence).

Figure 29.8 (right)
LineFORM.

Figure 29.10 (bottom, 1st row)
jamSheets.

Figure 29.9 (bottom, 2nd row)
PneUI.

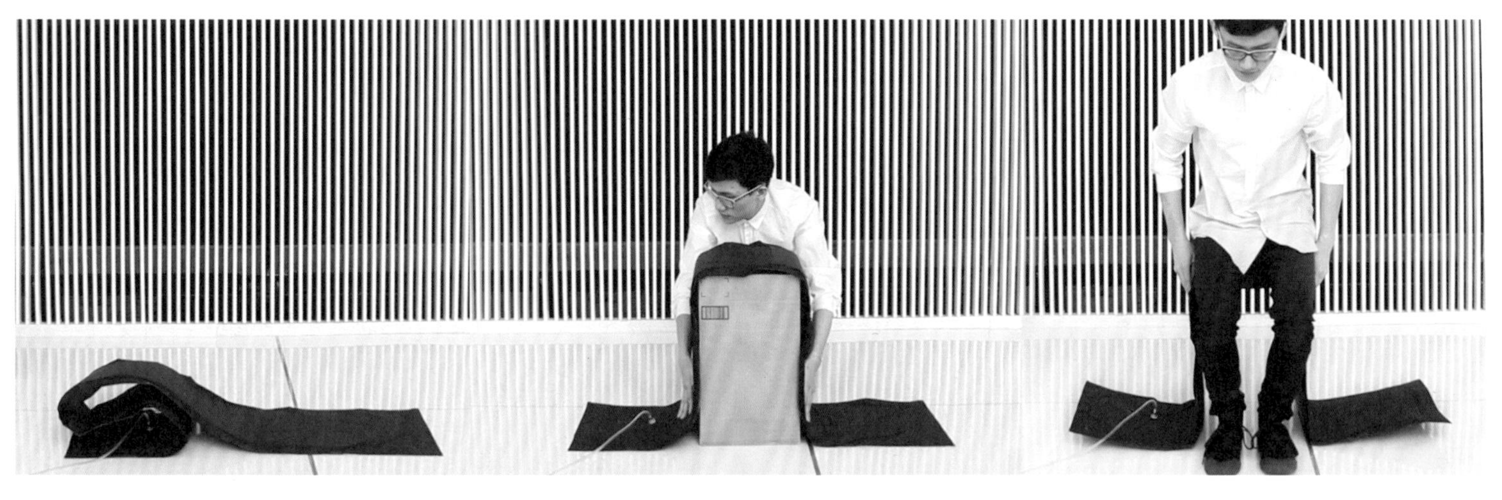

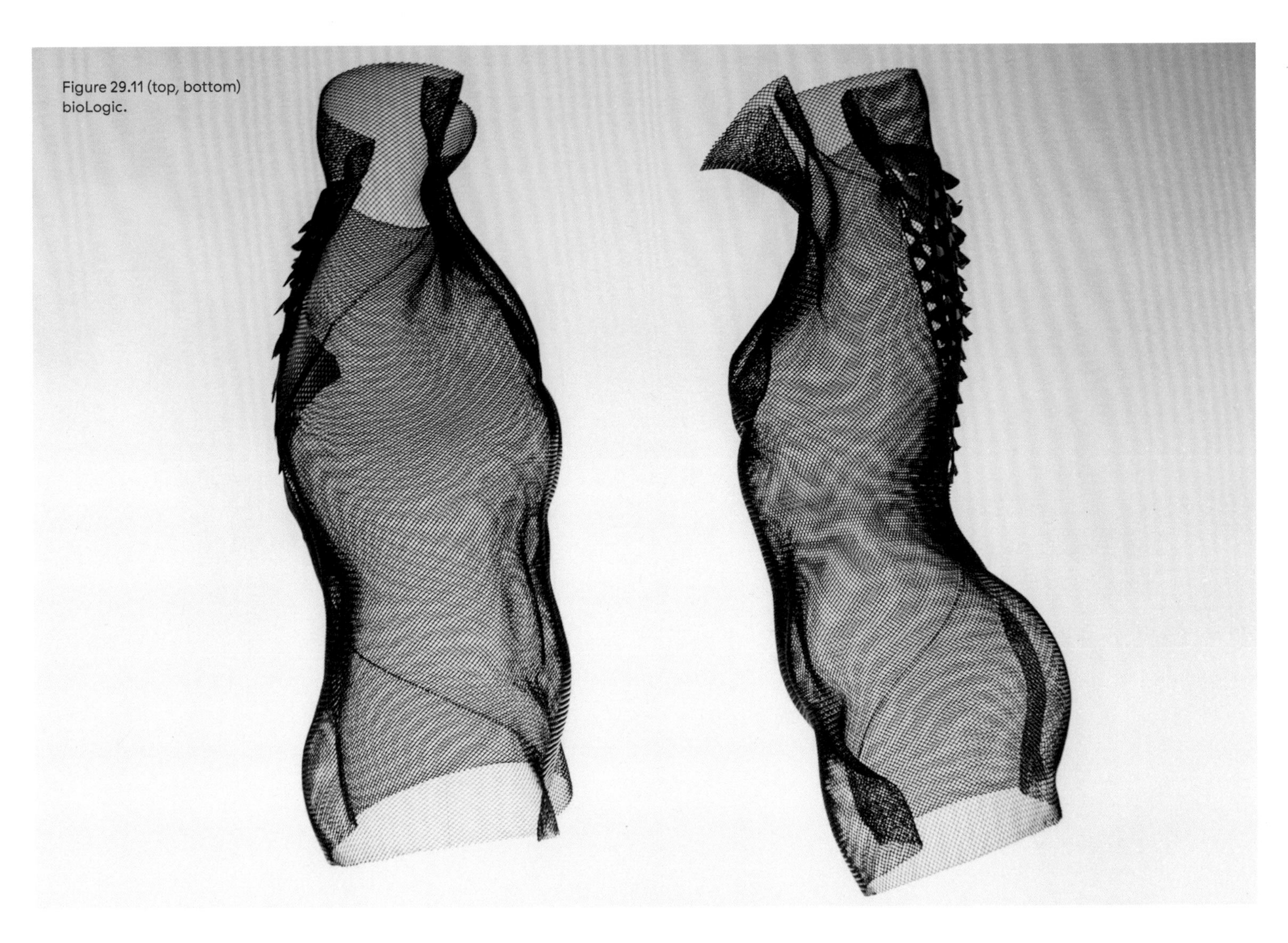

Figure 29.11 (top, bottom) bioLogic.

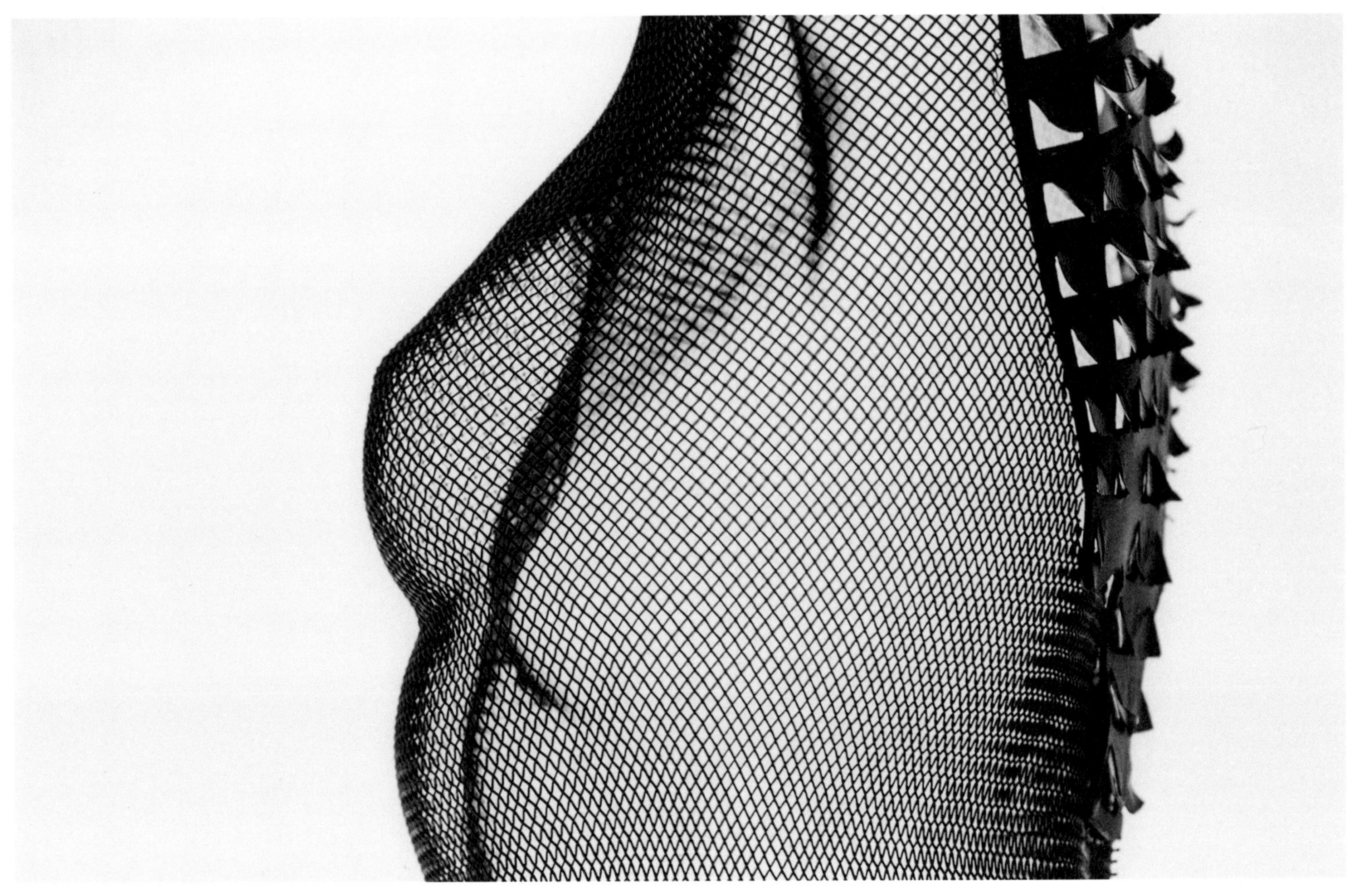

ACKNOWLEDGMENTS

The vision of Radical Atoms has been developed and evolved through a series of design exercises and intensive discussion within the Tangible Media Group since 2008. A large number of past and current members of the Tangible Media Group and colleagues in the MIT Media Lab have contributed to shape the concept and articulate the interaction techniques and possible applications. We would like to thank all of them for their contributions. Special thanks are due to: Brygg Ullmer, John Underkoffler, Daniel Leithinger, Sean Follmer, Lining Yao, Jifei Ou, Wen Wang, Amit Zoran, Ken Nakagaki, Hayes Raffle, Amanda Perks, James Patten, Xiao Xiao, Leonardo Amerigo Bonanni, Jean-Baptiste Labrune, Carlo Ratti, Robert Jacob, Mike Ananny, Catherine Vaucelle, Kimiko Ryokai, Ali Mazalek, Amos Golan, Daniel Levine, Chin-Yi Cheng, Daniel Fitzgerald, Penny Webb, Udayan Umapathi, Luke Alexander Jozef Vink, Virginia (Viirj) Kan, Clark Della Silva, Basheer Tome, Felix Heibeck, Philipp Schoessler, Anna Pereira, Ryuma Niiyama, Peter Schmitt, Sheng Kai Tang, Austin Lee, Samuel Luescher, Dávid Lakatos, Anthony DeVincenzi, Jinha Lee, Keywon Chung, Jamie Zigelbaum, Marcelo Coelho, Adam Kumpf, Angela Chang, Jason Alonso, Jim Gouldstone, Richard Whitney, Vincent Leclerc, Chad Dyner, Bradley Kaanta, Jennifer S. Yoon, Dan Chak, Yao Wang, Dan Maynes-Aminzade, Gian Pangaro, Luke Yeung, Ben Piper, Jay Lee, Phil Frei, Seungho Choo, Sandia Ren, Craig Wisneski, Paul Yarin, Victor Su, Scott Brave, Andrew Dahley, Matt Gorbet, Oksana Anilionyte, Helene Steiner, Daniel Tauber, Nikolaos Vlavianos, Paula Aguilera, Jonathan Williams, and all the research assistants and research staff of the Tangible Media Group[16] and the MIT Media Lab. I also would like to thank a number of visiting students and the students who have taken the Fall Tangible Interfaces course since 1998 and contributed to the development of our visions. We thank the MIT Media Lab for their support of this ongoing project since 1995 when I joined the Media Lab.

NOTES

1 Hiroshi Ishii and Brygg Ullmer, "Tangible Bits: Towards Seamless Interfaces between People, Bits and Atoms," in *Proceedings of the ACM SIGCHI Conference on Human Factors in Computing Systems* (CHI '97) (New York: ACM, 1997), 234–241.

2 http://tangible.media.mit.edu/project/icc-2000-tangible-bits-exhibition/

3 Hiroshi Ishii, "Tangible Bits: Beyond Pixels," in *Proceedings of the 2nd International Conference on Tangible and Embedded Interaction* (TEI '08) (New York: ACM, 2008), xv–xxv.

4 Hiroshi Ishii, Dávid Lakatos, Leonardo Bonanni, and Jean-Baptiste Labrune, "Radical Atoms: Beyond Tangible Bits, toward Transformable Materials," *Interactions* 19, no. 1 (January 2012): 38–51.

5 "Radical" came from the "free radical" in chemistry. A free radical is an atom, molecule, or ion with unpaired valence electrons; this makes free radicals highly chemically reactive toward other substances or even between themselves. https://en.wikipedia.org/wiki/Radical_(chemistry)

6 http://tangible.media.mit.edu/project/radical-atoms-exhibition-at-ars-electronica/

7 Hiroshi Ishii, H. R. Fletcher, J. Lee, S. Choo, J. Berzowska, C. Wisneski, C. Cano, A. Hernandez, and C. Bulthaup, "musicBottles," in *ACM SIGGRAPH 99 Conference Abstracts and Applications* (SIGGRAPH '99) (New York: ACM, 1999), 174–.

8 Ben Piper, Carlo Ratti, and Hiroshi Ishii, "Illuminating Clay: A 3-D Tangible Interface for Landscape Analysis," in *Proceedings of the SIGCHI Conference on Human Factors in Computing Systems* (CHI '02) (New York: ACM, 2002), 355–362.

9 Hayes Solos Raffle, Amanda J. Parkes, and Hiroshi Ishii, "Topobo: A Constructive Assembly System with Kinetic Memory," in *Proceedings of the SIGCHI Conference on Human Factors in Computing Systems* (CHI '04) (New York: ACM, 2004), 647–654.

10 Daniel Leithinger, Sean Follmer, Alex Olwal, and Hiroshi Ishii, "Physical Telepresence: Shape Capture and Display for Embodied, Computer-Mediated Remote Collaboration," in *Proceedings of the 27th Annual ACM Symposium on User Interface Software and Technology* (UIST '14) (New York: ACM, 2014), 461–470.

11 Ken Nakagaki, Sean Follmer, and Hiroshi Ishii, "LineFORM: Actuated Curve Interfaces for Display, Interaction, and Constraint," in *Proceedings of the 28th Annual ACM Symposium on User Interface Software and Technology* (UIST '15) (New York: ACM, 2015), 333–339.

12 Lining Yao, Ryuma Niiyama, Jifei Ou, Sean Follmer, Clark Della Silva, and Hiroshi Ishii, "PneUI: Pneumatically Actuated Soft Composite Materials for Shape Changing Interfaces," in *Proceedings of the 26th Annual ACM Symposium on User Interface Software and Technology* (UIST '13) (New York: ACM, 2013), 13–22.

13 Jifei Ou, Lining Yao, Daniel Tauber, Jürgen Steimle, Ryuma Niiyama, and Hiroshi Ishii, "jamSheets: Thin Interfaces with Tunable Stiffness Enabled by Layer Jamming," in *Proceedings of the 8th International Conference on Tangible, Embedded and Embodied Interaction* (TEI '14) (New York: ACM, 2014), 65–72.

14 Lining Yao, Jifei Ou, Chin-Yi Cheng, Helene Steiner, Wen Wang, Guanyun Wang, and Hiroshi Ishii, "bioLogic: Natto Cells as Nanoactuators for Shape Changing Interfaces," in *Proceedings of the 33rd Annual ACM Conference on Human Factors in Computing Systems* (CHI '15) (New York: ACM, 2015), 1–10.

15 http://tangible.media.mit.edu/projects/

16 http://tangible.media.mit.edu/people/

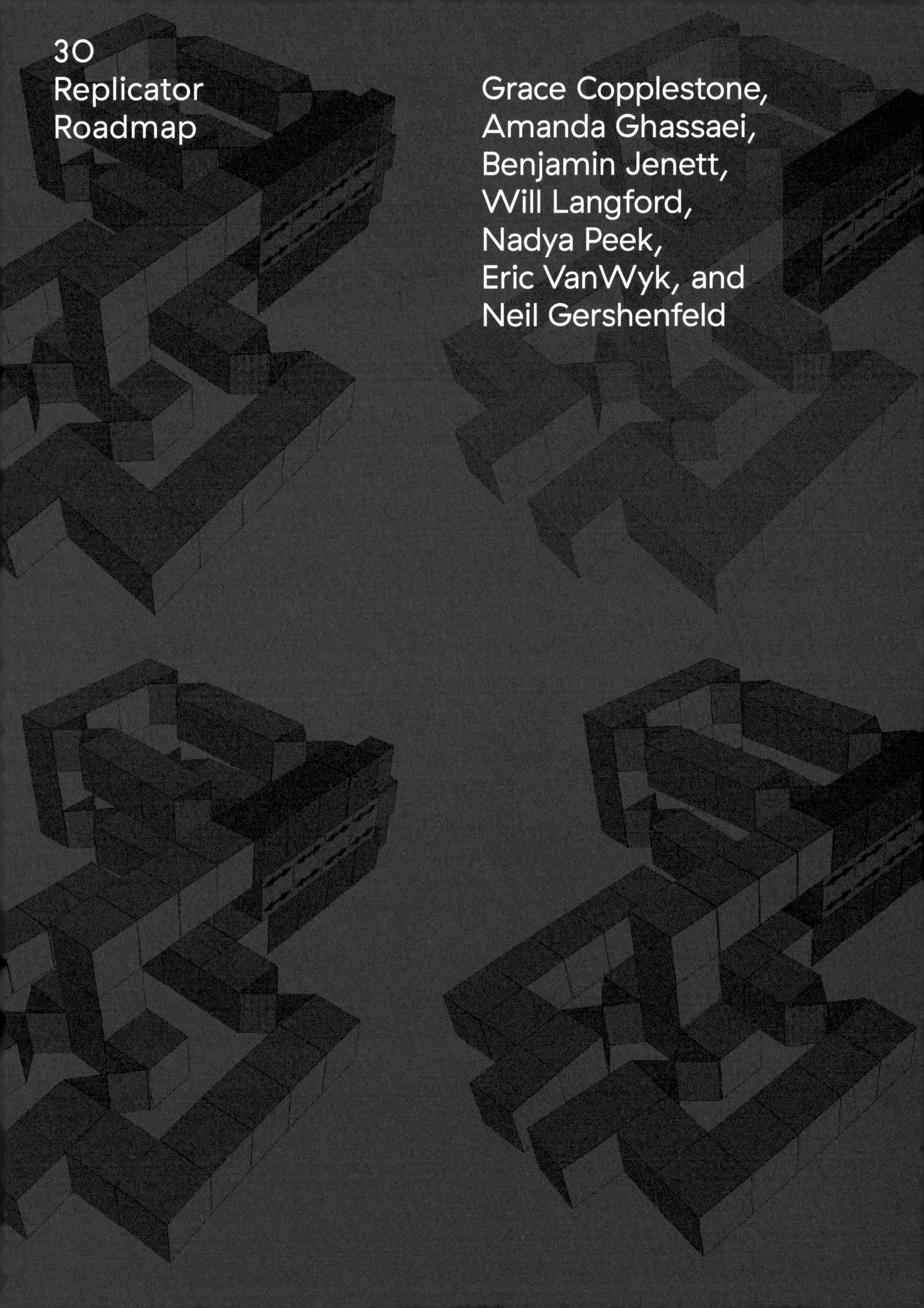

30 Replicator Roadmap

Grace Copplestone, Amanda Ghassaei, Benjamin Jenett, Will Langford, Nadya Peek, Eric VanWyk, and Neil Gershenfeld

Grace Copplestone,
Amanda Ghassaei,
Benjamin Jenett,
Will Langford,
Nadya Peek,
Eric VanWyk, and
Neil Gershenfeld

Programmable matter has been a "desirement": something that exists in research program statements and fanciful renderings but hasn't been realized. There have been approaches based on complex robotic modules that can each do everything independently, resulting in relatively small numbers of large units, and approaches based on larger numbers of simpler units that can't do very much. Neither has come close to the original inspiration of biology's ability to adapt, evolve, and repair itself.

Molecular biology is based on a very different approach. The twenty regular amino acids don't self-assemble and aren't themselves programmable. Rather, they do coded assembly with a careful division of labor. Messenger RNAs bring sequences of what to construct to the ribosome, transfer RNAs bring amino acids that the ribosome elongates into a protein, followed by chaperones guiding folding. This progresses through the hierarchical stages of the primary structure of the amino acid sequence, the secondary structure of their geometrical motifs, the tertiary structure of functional subunits, and the quaternary structure of molecular machines like the myosin molecular motors that move muscles. At the heart of this hierarchy is the recursion of ribosomes making ribosomes. Scalability comes from the exponential increase in massively parallel serial fabrication. And programmability comes from the combination of an active assembler with a feedstock of passive parts. Nothing about this requires biochemistry. MIT's Center for Bits and Atoms is pursuing a roadmap to replicate its essential features in nonbiological materials, in order to obtain properties not available in molecular biology. This roadmap is progressing in the three stages described in the following sections; they are being developed in parallel but emerging with successively longer lead times.

The first stage is developing the architecture for rapid prototyping of rapid-prototyping machines, based on the composition of stateless modular machine-building components controlled by virtual machines over

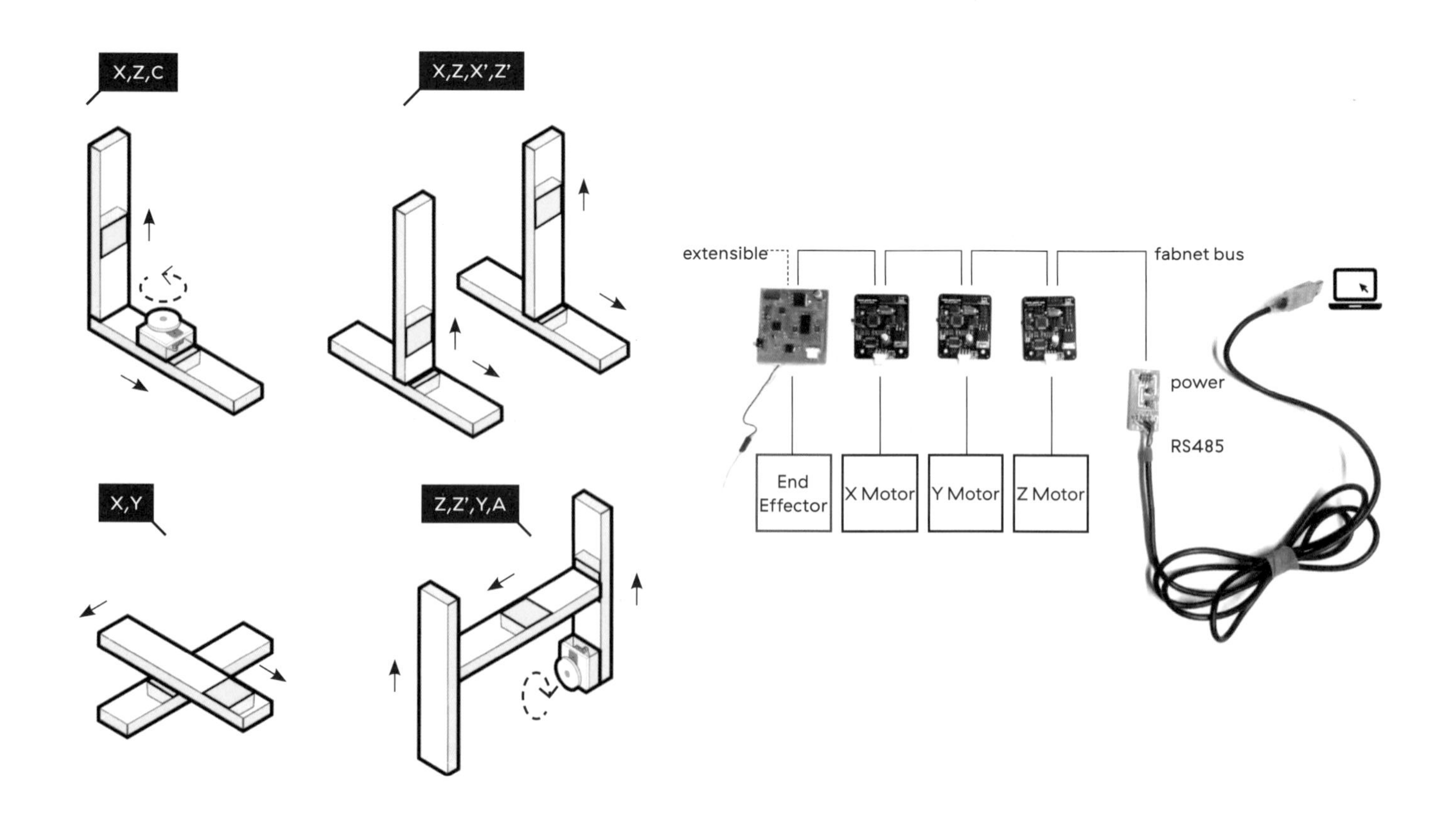

Figure 30.1 (left, right pages) Object-oriented hardware. Top left: Configurations of primitive modules into machines. Bottom left: A network of control nodes that make up the electrical controls for a machine (in this case, a three-axis 3D printer). Right: Machines built to teach machine-building, from top to bottom a four-axis hot wire cutter, an omelet ketchup printer, a lathe, and a 3D scanner. Their primitive motion modules are shown schematically on the right. (Peek et al. 2017.)

real-time networks. The second stage progresses from additive and subtractive rapid-prototyping processes to the assembly and disassembly of digital materials based on robots placing a discrete set of parts with a discrete set of relative positions and orientations relative to the structures that they're constructing. The third stage concludes by assembling an assembler out of the parts that it assembles.

What's remarkable about amino acids is that they're not remarkable. They have twenty properties that are typical but not extremal: basic vs. acidic, hydrophobic vs. hydrophilic, etc. Yet these are sufficient to synthesize the diversity of life. In the same way, the twenty or so properties that we assume for assembling an assembler—conducting vs. insulating, rigid vs. flexural, etc.—are sufficient to synthesize the diversity of technology. In that sense this is a roadmap for a universal replicator, able to construct, reconfigure, and recycle (almost) anything.

MODULAR MACHINES

Rapid-prototyping tools (such as laser cutters or 3D printers) can be instructed to produce any part geometry within their application space. These application spaces, however, are typically fixed when the machine is manufactured. There is an analogy with how early software was written, as monolithic programs, rather than as software objects that can be reused as is done today.[1] And there is an analogy with how early inflexible central-office phone switches connected to dumb telephones, unlike today's end-to-end Internet architecture in which applications are determined by what is connected to the Internet rather than by how the Internet is constructed.[2] Based on these principles, we are developing the practice of rapid prototyping of rapid-prototyping machines by composing objects that simultaneously embody physical degrees of freedom, communicating nodes in real-time networks, and computing nodes in virtual machine controllers.[3]

The first stage in our roadmap begins by partitioning our machines

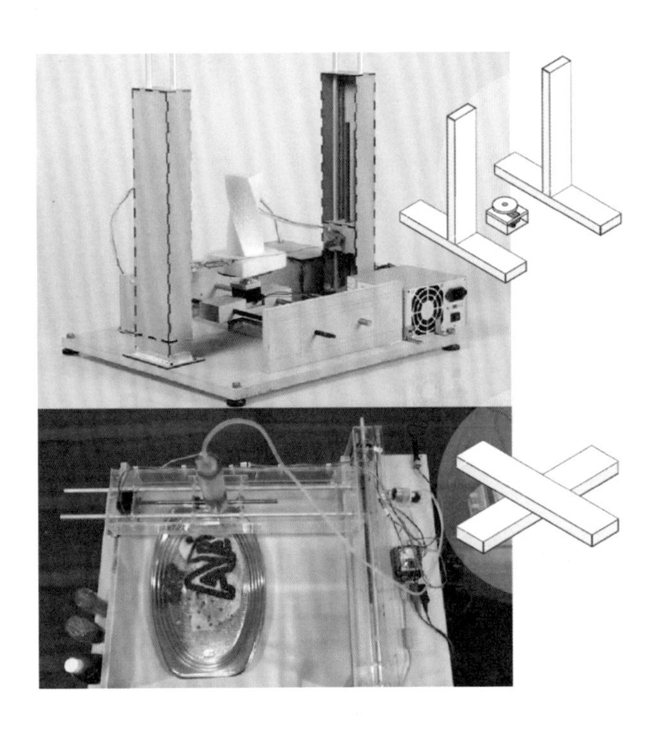

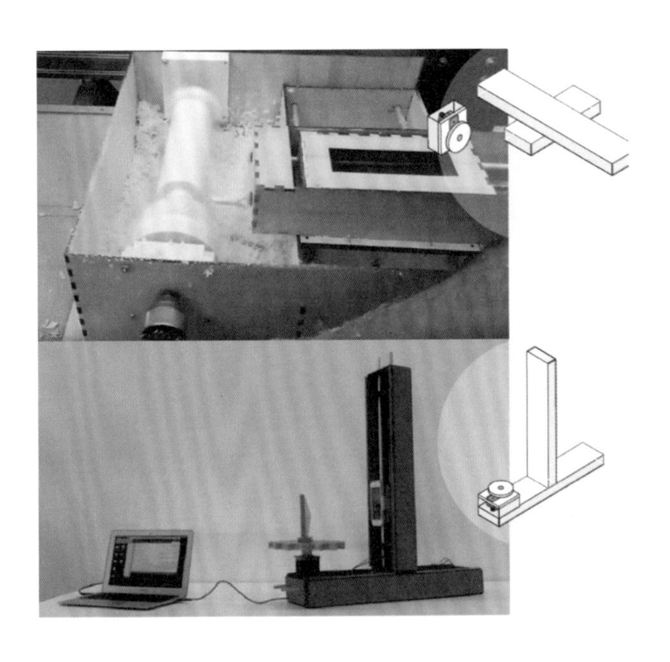

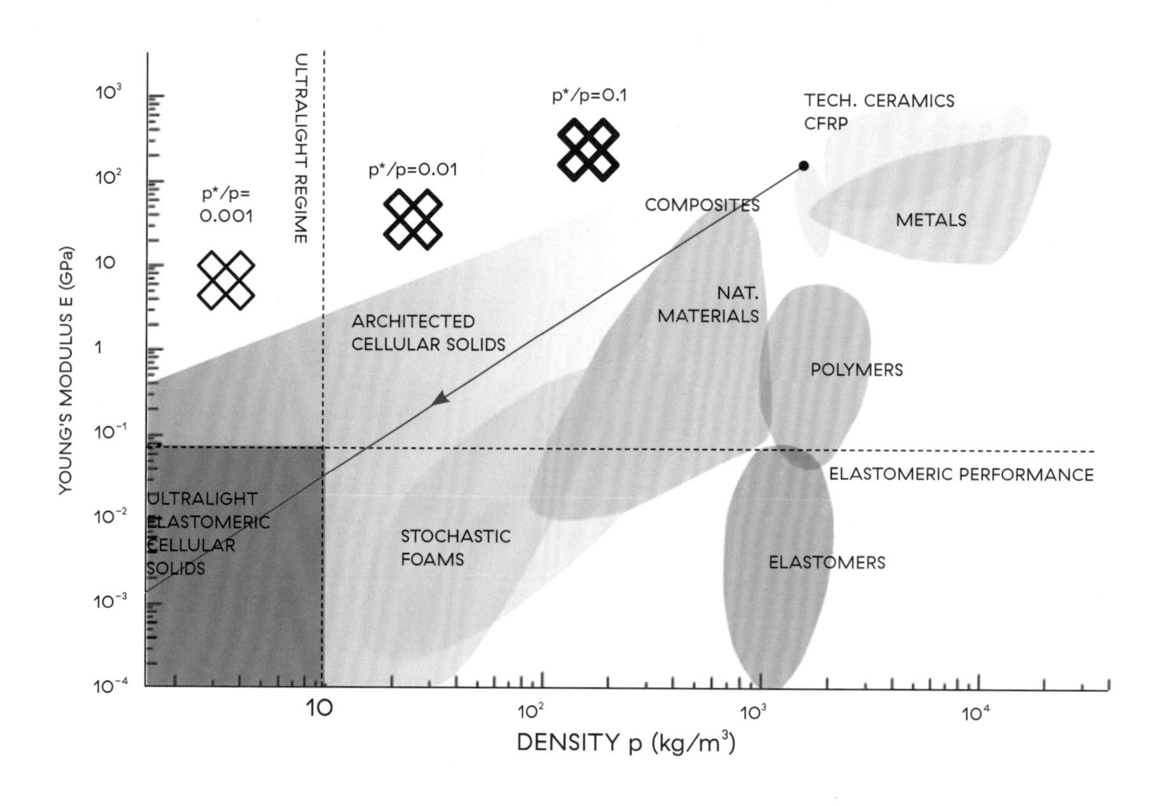

into physical modules, as shown in figure 30.1, that can be thought of as object-oriented hardware.[4] A particular application might require linear or rotary motion, using three or four or five degrees of freedom, and an end-effector with an additive or subtractive process with open-loop or closed-loop control. Rather than these choices being fixed by the static machine construction, they can change as application needs change.

Historically CAD, CAM, machine control, and motion control were developed and done by different people, in different places, at different times. As a result there is internal state at each of these stages, requiring multiple steps of file import and export to translate between them. Applying the end-to-end principle here means letting a design's mathematical model communicate directly with a machine's motors to make it. That's why our machine modules are stateless. A G-code interpreter must be modified if the machine that it is controlling is modified, and that change must also be reflected in the G-code generator. Here, knowledge of physical changes to a machine can be logically located in an application anywhere in the network.

To support scaling the complexity of these machines, we are developing a network protocol, asynchronous packet automata (APA), that is inspired by a spatial computing model.[5] APA combines source routing, network coordinates, and back-pressure flow control, so that naming, routing, and flow control emerge as the system is built rather than having to be provided as an external service. APA can be carried over wired and wireless transports that represent logical token passing.

The modular construction of these machines is mirrored in a modular framework for composing software workflows, the mods project (http://mods.cba.mit.edu). This framework is based on just-in-time compilation

of browser-based modules that pass events in a dataflow graph among parallel software threads. CAD, CAM, machine control, and motion control can all be consistently represented in this common framework.

Taken together, these technologies for rapid prototyping of rapid-prototyping machines have an even greater implication: rapid automation. Rather than distinguishing among machines, robots, and factories, they all become special cases of composing a common set of integrated hardware objects.

MODULAR MATERIALS

The second stage in the roadmap discretizes the construction process with digital materials, which are defined as those reversibly assembled from a discrete set of parts, with a discrete set of relative positions and orientations.[6] These attributes allow the global geometry to be determined by local relationships, so that an assembler can make a structure larger than itself, with errors to be detected and corrected so that part placement is more accurate than that of the assembler; parts with dissimilar properties to be joined in a common process so that bulk properties can be varied by their relative placement; and parts to be disassembled and reused rather than disposed of.

We've used a variety of geometries for the primitive parts, including planar cruciform elements,[7] discretely assembled struts and nodes,[8] and mass-produced injection-molded 3D voxels.[9] The resulting metamaterials behave like cellular solids, accessing previously inaccessible regions of material property space (figure 30.2), such as high strength and stiffness per weight at very low mass density.[10]

A number of parameters can be varied in designing digital materials, such as the building block's constituent material, the overall structure's relative density, and the type of lattice geometry. Highly interconnected lattices

Figure 30.2 (left, right pages) Digital materials performance and construction. Left: Young's modulus versus density for engineering materials and cellular solids. Right: Meso-scale digital material construction (Jenett, Cellucci, et al. 2016).

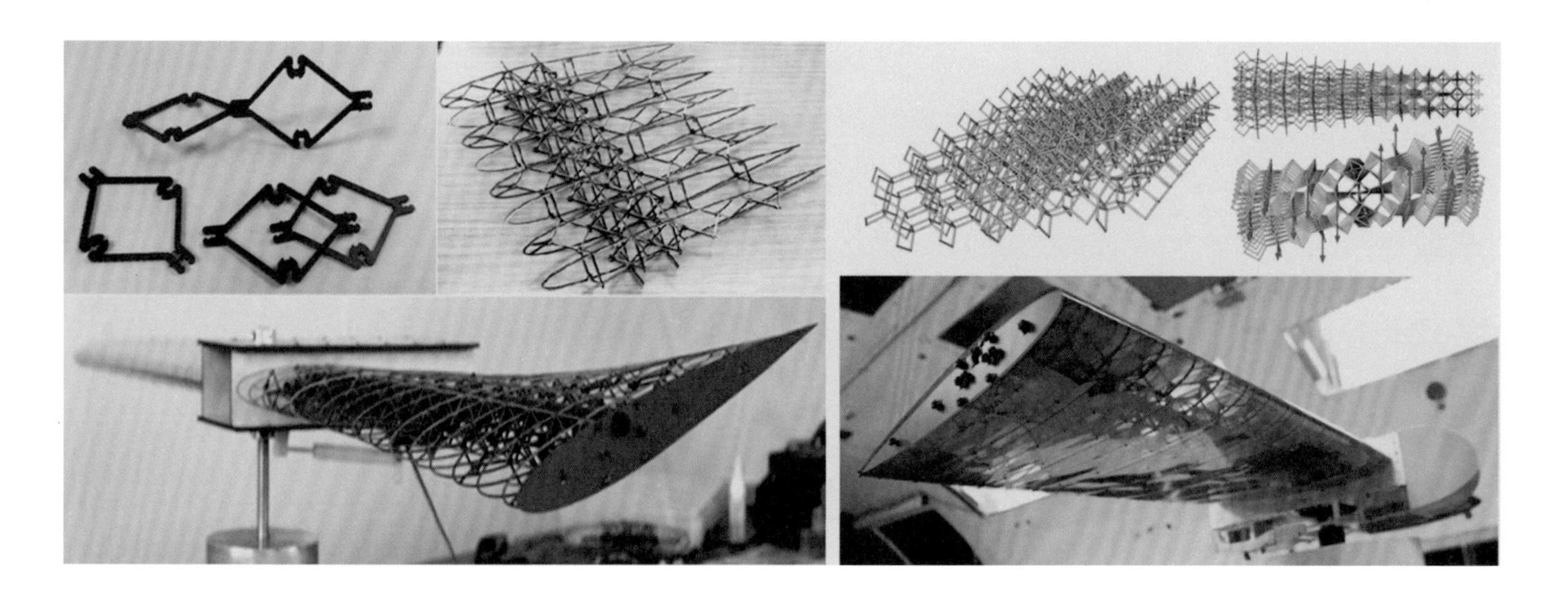

result in stiff structures that transfer loads through axial forces, whereas less connected geometries transfer loads through bending and result in compliant structures. This heterogeneity can be leveraged to design anisotropic structures with tunable and spatially varying properties. We've shown that these degrees of freedom can be used to design shape-changing robotic structures, with aerospace applications such as morphing wings.[11] We design the wing to be stiff in bending but compliant in torsion (figure 30.3), and, compared to traditional means of control using rigid control surfaces, we use a single degree of freedom for global deformation of an ultralight cellular structure. Such pure roll deformation improves both efficiency and agility.

Central to the use of digital materials is automating their construction. Due to the regularity of these structures, task-specific robots can be designed to operate with reference to the lattice rather than to their absolute position. These robots, which we call relative robots, take advantage of this structured environment to simplify controls and reduce the number of degrees of freedom required for the robot's configuration space. MOJO (Multi Objective Journeying rObot) is designed to climb within the lattice,[12] using two pairs of arms to climb linearly and one hip to rotate orthogonally (figure 30.4). This robot can access any point inside the lattice and perform structural health-monitoring by sensing broken areas on the structure using current feedback from its motors.

BILL-E (Bipedal Isotropic Lattice Locomoting Explorer) traverses the outside of the lattice using an inchworm design and mating interfaces[9] (figure 30.4). This robot is designed to assemble and disassemble the structure using an end-effector to add or remove building blocks. We have found that due to the discrete nature of its locomotion based on local metrology, it is possible to reduce error-stacking over long distances,[13] and we can parallelize construction with swarms of robots. These attributes are being investigated for constructing large-scale space structures for transport and habitation.[14]

PROGRAMMABLE MATERIALS

The third stage in this roadmap uses digital materials to assemble an assembler. The model for this is the ribosome. Proteins are synthesized from a set of 20 standard amino acids, yet they exhibit a wide range of morphologies and functions in the cell. We are developing analogs to amino acids for engineering, a set of digital material parts that can form the basis for electronic and mechanical systems. The key insights drawn from biology are that the properties of a protein are encoded in the sequence of its amino acids, and that the amino acids are recycled and

Figure 30.3 (left page)
Digital material morphing wing. Left: Construction of shape-changing structures from discrete lattice-building block elements. Right: Simulation and wind-tunnel-testing of active wing twist (Jenett, Calisch, et al. 2016).

Figure 30.4 (top)
Relative robotic assembly. Left: MOJO (Multi Objective Journeying rObot) can crawl through the lattice (top) and rotate to travel in orthogonal directions (bottom) (Cellucci and Jenett 2017; image credit: NASA ARC CSL). Right: BILL-E (Bipedal Isotropic Lattice Locomoting Explorer) can climb the exterior of the lattice and can place and manipulate parts. Teams of robots enable coordinated assembly and material transportation (Jenett et al. 2017).

Figure 30.5
The hierarchy of assembled assemblers. An assembled assembler (top) uses walking modules to locomote. The walking leg module is in turn assembled from functional parts, including piezo shear actuators. Those actuator elements are ultimately assembled from single-material building blocks including conductive, insulating, rigid, flexible, magnetic, and piezoelectric.

reconfigured after the protein's lifetime has passed. Similarly, our digital material building blocks are programmed by controlling and reconfiguring their spatial arrangement.

Through the relative placement of just three kinds of part (conductive, insulating, and resistive), arbitrary three-dimensional wiring and passive electronic components (inductors, capacitors, and resistors) can be constructed.[15] The addition of four types of semiconducting parts can extend this construction system to include active electronic components such as diodes and transistors.

Furthermore, parts with flexural degrees of freedom can be used to assemble mechanisms and motion platforms. With the addition of magnetic part types, electromagnetic actuators can then be assembled. This leads to the ability to assemble robots that are capable of manipulating and assembling the parts that they are composed of.[16]

The complexity of designing and assembling such recursive robotic systems becomes unmanageable without another principle evident in biology: hierarchy. Just like the primary, secondary, tertiary, and quaternary levels of protein structure, there are natural hierarchical divisions in our assembly architecture. Single-material parts, which exhibit basic material properties like conducting or insulating and stiff or flexible, are built up into functional parts, which perform higher-level behaviors including switching, actuating, and constraining motion. These functional parts are assembled into modules, which in turn perform useful operations like gripping, latching, and large-displacement motion, and modules are finally assembled into higher-level systems like the assembler. We are developing these intermediate levels of description in parallel with their

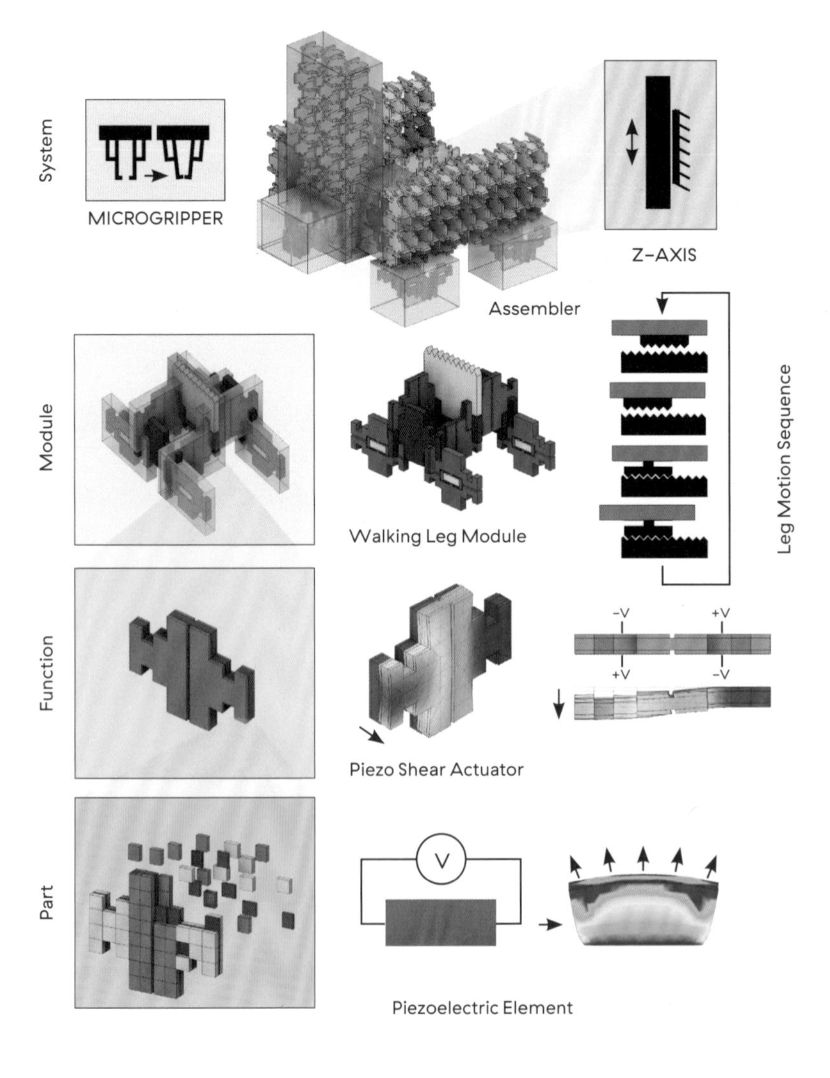

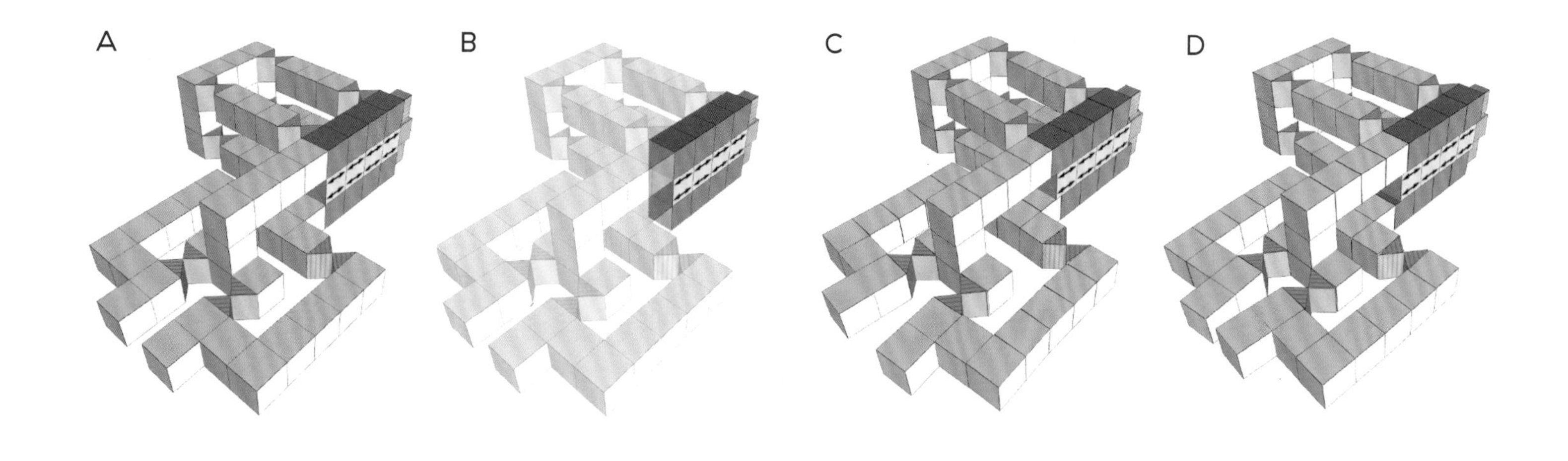

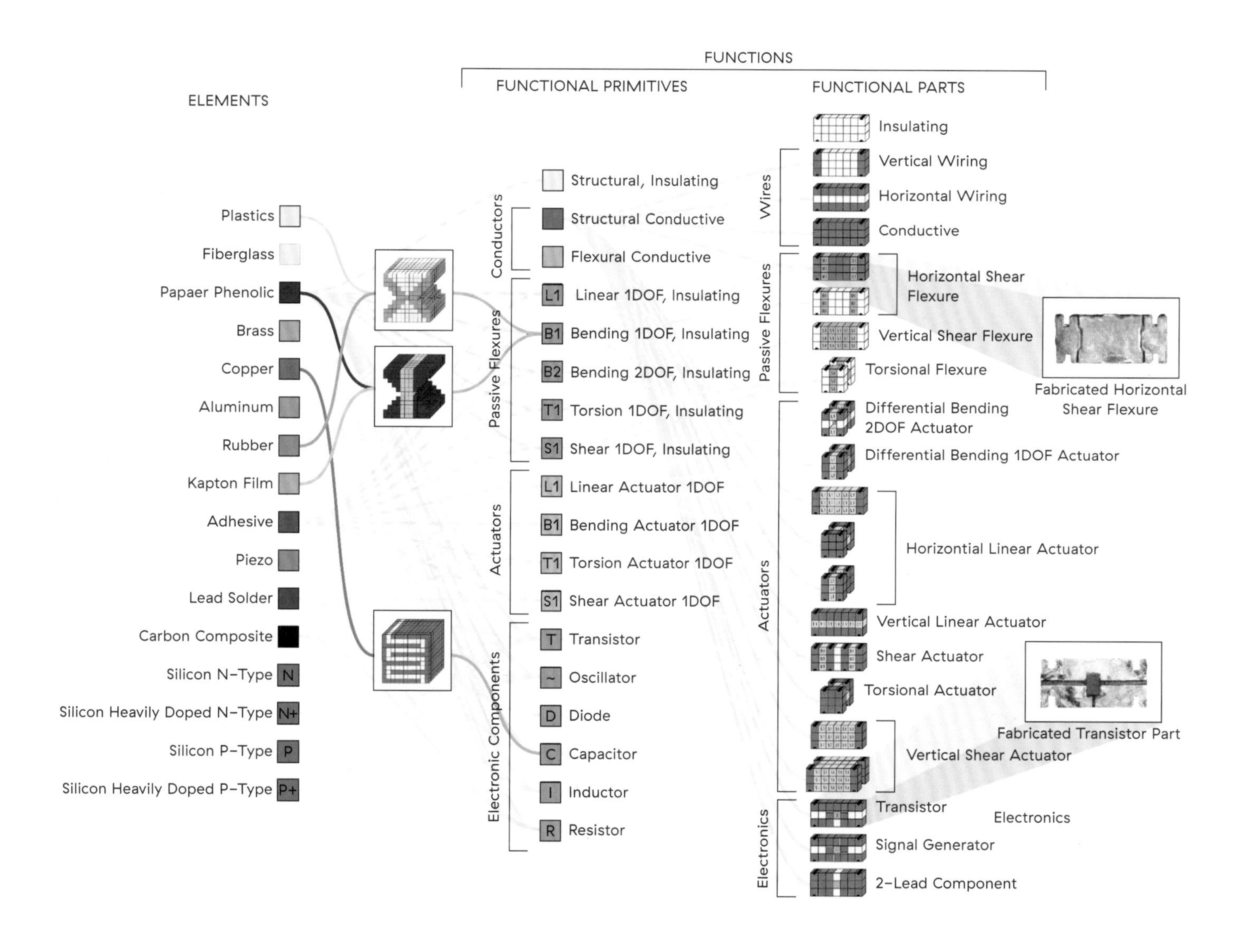

Figure 30.7 (top)
Discrete gripper modeling. (A) With electronic circuitry (highlighted, B). Full stroke of shear actuators results in opening (C) and closing (D) motion of gripper.

Figure 30.8 (bottom)
Digital material function decomposition. The discrete decomposition of active and passive functions into digital material building-block elements.

Figure 30.6
Digital material parts. Left: Structure assembled from conducting digital material parts. Top: Parts with embedded flexural hinges form a parallelogram linkage. Bottom: Intermediate coil and transistor parts, and batch fabrication of passive mechanical parts.

decomposition into primitive parts.

Ribosomes are slow, adding about one amino acid per second, but because ribosomes can make ribosomes their number can increase exponentially. That is why we're focusing on massively parallel serial fabrication. Amino acids, and our building blocks, are not themselves directly programmable; that's a property of the combined system of the parts and the assembler. This division of labor relaxes the requirements on the individual parts, allowing programmable mechanisms to be built out of rather than into them.

CONCLUSION

We've surveyed research on physical discretization in successively finer scales, from machines to materials, until the distinction between them disappears. This ends up looking less like the history of either modular robotics or smart materials, and more like the hierarchical decomposition of complex functions into simple buildings blocks in molecular biology[17] (figure 30.8). The roadmap that we've presented can thus be viewed as forward- rather than reverse-engineering biology, arriving at a similar architecture in inorganic materials. The matter that becomes active is not any one of these elements, it's their integration into a system.

NOTES

1 Tim Rentsch, "Object Oriented Programming," *ACM Sigplan Notices* 17, no. 9 (1982): 51—57.

2 Jerome H. Saltzer, David P. Reed, and David D. Clark, "End-to-End Arguments in System Design," *ACM Transactions on Computer Systems* 2, no. 4 (1984): 277—288.

3 Nadya Peek, "Making Machines That Make: Object-Oriented Hardware Meets Object-Oriented Software," PhD dissertation, Massachusetts Institute of Technology, 2016.

4 Nadya Peek, James Coleman, Ilan Moyer, and Neil Gershenfeld, "Cardboard Machine Kit: Modules for the Rapid Prototyping of Rapid Prototyping Machines," in *Proceedings of the 2017 CHI Conference on Human Factors in Computing Systems* (New York: ACM, forthcoming 2017).

5 Neil Gershenfeld, "Aligning the Representation and Reality of Computation with Asynchronous Logic Automata," *Computing* 93, nos. 2—4 (2011): 91—102.

6 George A. Popescu, Tushar Mahale, and Neil Gershenfeld, "Digital Materials for Digital Printing," *NIP & Digital Fabrication Conference* 2006, no. 3 (2006).

7 Kenneth C. Cheung and Neil Gershenfeld, "Reversibly Assembled Cellular Composite Materials," *Science* 341, no. 6151 (2013): 1219—1221.

8 Benjamin Jenett, Daniel Cellucci, Christine Gregg, and Kenneth C. Cheung, "Meso-Scale Digital Materials: Modular, Reconfigurable, Lattice-Based Structures," in *Proceedings of the 2016 Manufacturing Science and Engineering Conference* (2016).

9 Benjamin Jenett and Kenneth C. Cheung, "BILL-E: Robotic Platform for Locomotion and Manipulation of Lightweight Space Structures," in *Proceedings of AIAA Sci-Tech Conference 2017* (forthcoming).

10 M. F. Ashby, "The Properties of Foams and Lattices," *Philosophical Transactions. Series A, Mathematical, Physical, and Engineering Sciences* 364, no. 1838 (2006): 15—30.

11 Benjamin Jenett, Sam Calisch, Daniel Cellucci, Nick Cramer, Neil Gershenfeld, Sean Swei, and Kenneth C. Cheung, "Digital Morphing Wing: Active Wing Shaping Concept Using Composite Lattice-Based Cellular Structures," *Soft Robotics* 3, no. 3 (2016).

12 Daniel Cellucci and Benjamin Jenett, "A Mobile Robotic Platform for Exploration, Sensing, and Manipulation in a 3D Periodic Lattice Environment," in *Proceedings of the 2017 International Conference of Robotics and Automation* (forthcoming 2017).

13 Matthew Carney and Benjamin Jenett, "Relative Robots: Scaling Automated Assembly of Discrete Cellular Lattices," in *Proceedings of the 2016 Manufacturing Science and Engineering Conference* (2016).

14 Daniel Cellucci, Benjamin Jenett, and Kenneth C. Cheung, "Digital Cellular Solid Pressure Vessels: A Novel Approach for Human Habitation," in *Proceedings of the 2017 IEEE Aerospace Conference* (forthcoming 2017).

15 Will Langford, Amanda Ghassaei, and Neil Gershenfeld, "Automated Assembly of Electronic Digital Materials," *ASME 2016 11th International Manufacturing Science and Engineering Conference* (2016).

16 Will Langford, Amanda Ghassaei, Benjamin Jenett, and Neil Gershenfeld, "Hierarchical Assembly of a Self-Replicating Spacecraft," in *Proceedings of the 2017 IEEE Aerospace Conference* (forthcoming 2017).

17 Amanda Ghassaei, "Rapid Design and Simulation of Functional Digital Materials," MS thesis, Massachusetts Institute of Technology, 2016.

31

Toward Automating Construction with Decentralized Climbing Robots and Environmentally Adaptive, Functionally Specified Structures

Justin Werfel and Paul Kassabian

31 Toward Automating Construction with Decentralized Climbing Robots and Environmentally Adaptive, Functionally Specified Structures

Justin Werfel and Paul Kassabian

Construction is a multitrillion-dollar industry, in which—even in our highly automated age and in striking contrast to other major production industries like manufacturing—automation remains largely absent. Heavy machinery is ubiquitous but always under the close control of an operator, and foremen direct human workers but not robotic builders. Moreover, traditional human construction relies on detailed structural design, careful preplanning and sequencing of assembly steps, and extensive top-down coordination of workers. Other animals work very differently. Builders in nature create large-scale structures through the actions of many independent agents, relying on limited and local information and without the benefit of global knowledge or centralized control.

These considerations inspire the research area of collective construction, whose goal is to create teams of independent robots under decentralized control, which together build large-scale structures of interest. The multiplicity of independent builders provides benefits including speedup through parallelism, scalability of control, and robustness to factors like loss of individuals and variability in the order and timing of their actions.

A past project we build on here is the TERMES system,[1] inspired by mound-building termites. The TERMES robots build with specialized bricks, and create structures that match blueprints they're given, an important capability for engineering human-relevant systems. However, animals again take a different approach. They build forms responsive to features of the local environment and contingent events in the construction process, and they maintain and modify the structures throughout their lifetime (often across the lifetimes of many generations of builders) in response to dynamic conditions, damage, and the changing needs of the group.

The goal of our current project is to develop a framework for automated multirobot construction systems in which independent climbing robots build strut-based structures satisfying desired functional goals. The robots, limited to locally available information and not knowing the overall state of the full structure at any moment, measure local forces on the structure to help determine their actions: e.g., whether local stresses indicate that a part of the structure is stable enough to support further building or additional reinforcement is required. The structures to be built are specified by their purpose rather than by a full blueprint: e.g., to build a bridge across a gap, in ways responsive to the conditions encountered on site, without needing to know the gap geometry or

Figure 31.1
Animals that build large-scale structures as collectives include beavers, sociable weaverbirds, and termites. Individual animals take actions according to local conditions, with no central control or master plan; the resulting structure is built according to the demands of its particular environment. For instance, beavers add material to their dams in response to the sound of flowing water, reinforcing the structure in needed areas and creating an overall result adapted to perform the function of blocking the flow.

Figure 31.2 (top)
The TERMES system, inspired by mound-building termites, comprises independent climbing robots that use specialized bricks to build user-specified structures. The system takes as input a representation of a desired target structure, and the robots follow local rules that provably guarantee correct completion of that structure, regardless of unpredictable variations in factors like the number of robots and the order and timing of their actions. The robots use strictly onboard sensing and only indirect communication, relying on manipulation of the shared environment as an implicit coordination mechanism.

Figure 31.3 (bottom)
The goal of the current project is to develop a framework for collective construction systems in which independent climbing robots build strut-based structures satisfying desired functional goals. Robots will move over struts, measuring local forces as cues to inform their actions. As a structure is built, the changing forces that result on individual elements shape the way in which further construction proceeds. Functional goals may include building a tower of a specified height, or bridging a gap, in initially unknown environments, and maintaining stability throughout the build sequence.

Figure 31.3
The goal of the current project is to develop a framework for collective construction systems in which independent climbing robots build strut-based structures satisfying desired functional goals. Robots will move over struts, measuring local forces as cues to inform their actions. As a structure is built, the changing forces that result on individual elements shape the way in which further construction proceeds. Functional goals may include building a tower of a specified height, or bridging a gap, in initially unknown environments, and maintaining stability throughout the build sequence.

Figure 31.6
Composite structures can be built satisfying multiple functional requirements. A key goal is for the building process to be adaptive to initially unknown environments: e.g., built atop and conforming to uneven terrain not known in advance; taking advantage of an existing rock face to help provide partial support.

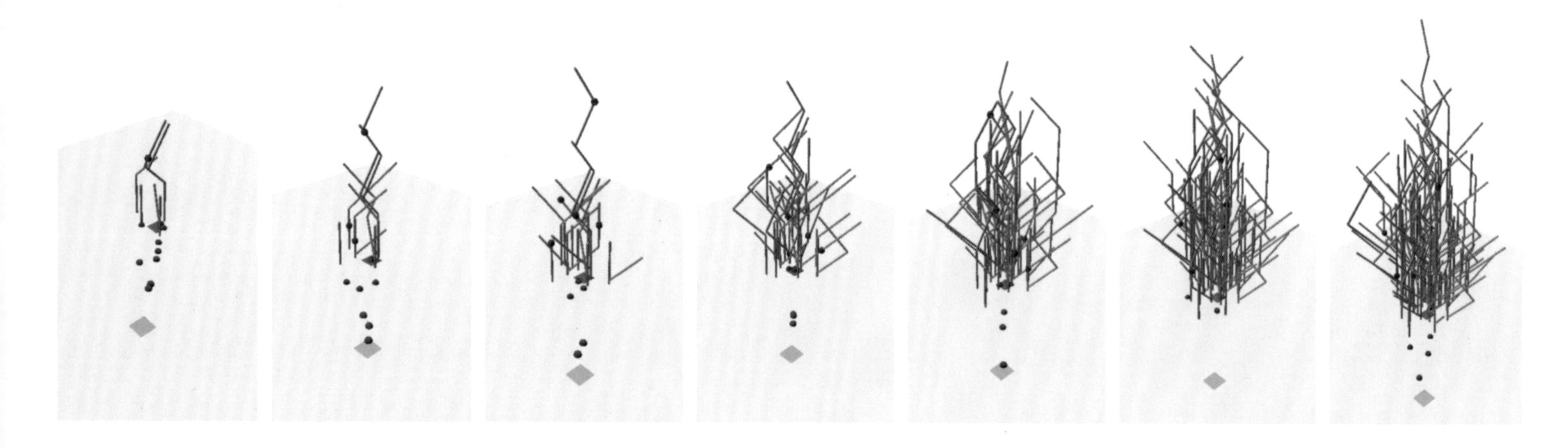

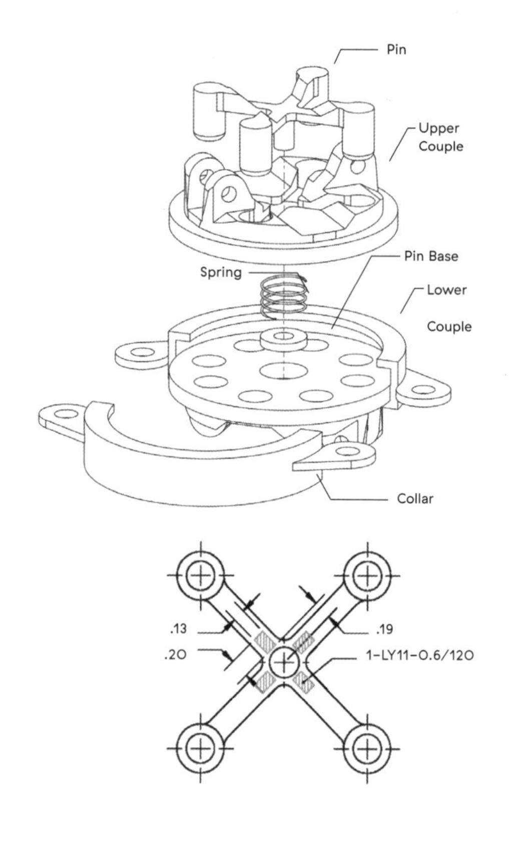

Figure 31.4 (top)
A multirobot simulation environment created for this current project, linked to commercial structural engineering software for high-fidelity physical accuracy in modeling, aids in developing and validating formal analytic approaches. In this sequence, independent agents (blue) retrieve struts from a reservoir (green) and build a tower at a designated construction site (red), adding reinforcing struts as necessary according to local forces they sense as they travel.

Figure 31.5 (bottom)
Special-purpose instrumented clamps could be used with standard scaffolding tubes to provide information about forces at nodes. This prototype design allows attachment of two struts at a discrete set of relative angles. The angle is locked by the act of attaching a strut, reducing the number of separate mechanical operations a robot would need to perform.

environmental details in advance.

This current project has begun to develop a framework and new theories for designing and verifying dynamic systems of interacting agents, reliably achieving collective goals while adapting to each other and to uncertain and changing environments. The research will address the scientific challenges of understanding how high-level results emerge from low-level rules, and the engineering challenges of designing low-level rules to achieve specific high-level results. It will thus advance our ability to robustly achieve desired collective effects using large numbers of simple, cheap, potentially unreliable components. Bringing automation to construction could be transformative, offering benefits like lower costs, reduced accident rates, higher efficiency, and increased ability to build in problematic settings like disaster areas or extraterrestrial environments.

NOTES

1 Justin Werfel, Kirstin Petersen, and Radhika Nagpal. "Designing Collective Behavior in a Termite-Inspired Robot Construction Team," *Science* 343, no. 6172 (2014): 754–758.

32 Living Matter

David Benjamin

David Benjamin

Living organisms have the ability to grow, adapt, regenerate, self-replicate, and biodegrade. For this reason, they are difficult to completely control. And when living organisms are combined with construction materials, they challenge the properties we typically rely on for the built environment: predictability, fixity, and durability. Yet this combination of living organisms and materials—living matter—is irresistible.

Over the past several years, my studio, The Living, has been developing three approaches that utilize living organisms at the intersection of biology, computation, and design. The first approach, biocomputing, uses living organisms as tiny processors to solve design problems. The second, biosensing, uses living organisms to detect conditions of their environment and respond accordingly. The third, biomanufacturing, uses living organisms as tiny factories to produce materials for the built environment.

In a recent open-ended experiment using biomanufacturing and living matter, we designed and built a project called Hy-Fi. The project was a tower made of a new kind of architectural brick, a brick of living matter. To create the bricks, we placed agricultural waste (corn stover) and mycelium (the rootlike part of

Figure 32.1 (top)
Growing bricks, Hy-Fi, New York, 2014. Courtesy of The Living.

Figure 32.2 (bottom)
Composting bricks, Hy-Fi, New York, 2014. Courtesy of The Living.

Figure 32.3
Test at a medium scale, Hy-Fi, New York, 2014. Photo by Amy Barkow, courtesy of The Living.

Figure 32.4
Context of glass and steel Manhattan buildings and traditional clay brick MoMA PS1 building, Hy-Fi, New York, 2014. Photo by Iwan Baan, courtesy of The Living.

Figure 32.5
At once familiar and new, Hy-Fi, New York, 2014. Photo by Charles Roussel, courtesy of The Living.

mushrooms) in a mold. Over five days, the mycelium grew and fused the stover into a solid object. The brick was literally alive.

In several ways, the project was a test of scale: creating a medium-size building (four stories tall) with a medium amount of building blocks (10,000 units). Since it was the first outdoor construction of this size with this material, we faced many unknowns. And since Hy-Fi was commissioned by the Museum of Modern Art and MoMA PS1 as a public installation, we had to make it functional and safe.

So we killed the bricks. After harnessing a living organism to create a new brick—at once familiar and startling, ancient and cutting-edge—we rejoined the traditional architectural process, transforming the material from living and dynamic to dead and inert. This allowed us to prove that the material is viable for architecture.

Yet this project was only the first prototype. The next version might explore more thrilling possibilities for living matter by keeping the bricks alive during and after construction. In the near future, architecture with mycelium bricks may be self-repairing: if the material is cut, it will fuse back together. It may also be self-adhering: if living bricks are stacked, they will bind together, offering a single material that is both brick and mortar. And it may even be self-protecting: if the material is attacked by mold or decay, it will defend itself.

Living matter offers a new world of future possibilities. But in the present, our initial prototype offered one more feature related to active matter: it remained biodegradable. In the ultimate test of Hy-Fi, we disassembled the building at the end of the installation, crumbled the bricks, combined them with worms and food scraps, and composted them in 60 days, ultimately transforming physical matter from earth to crops to construction to compost, and therefore back to earth.

33
Material Computation: Toward Self-X Material Systems in Architecture

Achim Menges

bam
bam
bam
bam

Figure 33.1
ICD/ITKE Research Pavilion 2015–16, Institute for Computational Design, Institute of Building Structures and Structural Design, University of Stuttgart, 2016. The pavilion consists of 151 different shell elements. Their individual shapes are programmed into the material and locked by robotic sewing.

Achim Menges

The notion of active matter profoundly resonates with the exploration of self-x material systems in architecture conducted at the Institute for Computational Design (ICD) at the University of Stuttgart. This research aims at developing material systems that have the capacity to compute form, structure, and space on a material level. The current state and scope of this work are presented through three projects that demonstrate the capacity to self-form during assembly, to self-stabilize based on particle morphology, and to self-adapt through weather-responsive, hygroscopically driven actuation. In the future, we seek to synthesize these various modes of material computation in comprehensive self-x material systems for architecture.

ICD/ITKE RESEARCH PAVILION 2015/16 SELF-FORMING DURING ASSEMBLY

The ICD/ITKE Research Pavilion 2015/16, developed by the ITECH program at the ICD and the Institute of Building Structures and Structural Design (ITKE), is the latest in a line of research projects that explore material self-forming in architecture. Self-forming here refers to physically programming the shape of a building element into its material makeup. The final element shape then unfolds by a relatively simple input force, which is a tension force pulling the ends of strips together. Previous projects had explored the self-forming of linear plywood lamella and conical plywood surfaces through active bending. In this case the building element settles into a particular shape when connected to itself or to other elements. While the material was homogeneous throughout in these initial experiments,[1] and the shape was varied using only different locations of connection points, this research explores additional differentiation on the material level.

Based on interdisciplinary, biomimetic research on the plate skeleton of two species of the order Clypeasteroida (sand dollars) in the class Echinoidea (sea urchins), the structure of the pavilion was developed as a double-layered segmented shell.[2] The shell segments are made from extremely thin, elastically bent wood strips, which not only vary in shape from segment to segment but also change radii within each segment. In order to achieve the various target geometries, the required stiffness distribution was computed and physically programmed into each segment by varying the grain direction of the highly anisotropic veneer layers. The initially planar custom-laminated strips settle into their complex target geometry entirely by themselves when their three open ends are connected and locked into shape by robotic sewing. In this way 151 geometrically different

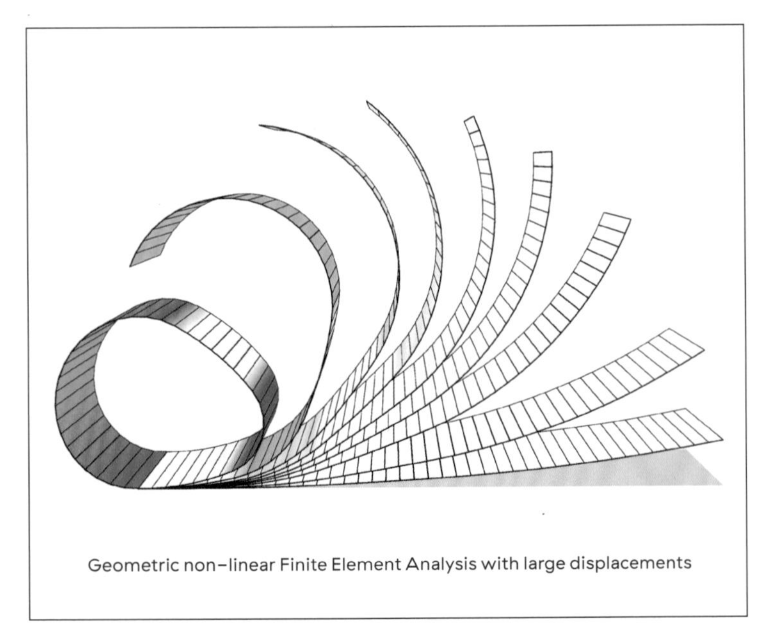
Geometric non-linear Finite Element Analysis with large displacements

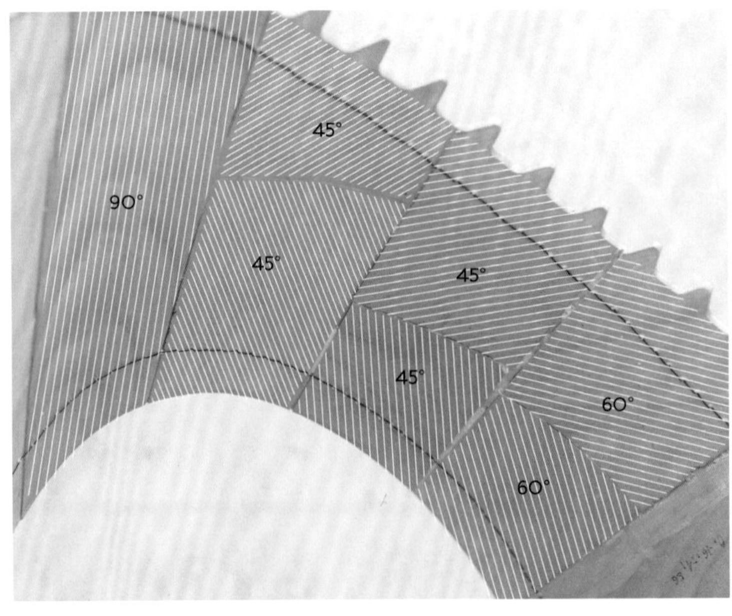

Figure 33.2 (left page, bottom) ICD/ITKE Research Pavilion 2015–16, Institute for Computational Design, Institute of Building Structures and Structural Design, University of Stuttgart, 2016. The required stiffness distribution for a desired target shape is computed and physically programmed into the shell element by varying the grain direction of the veneer layers.

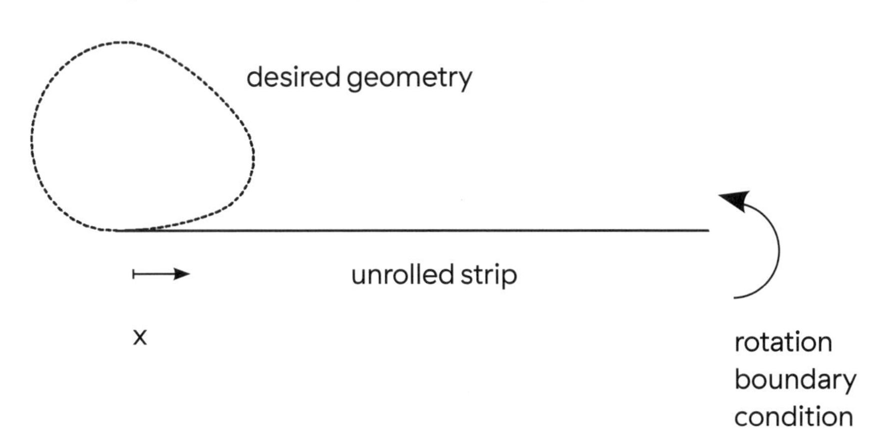

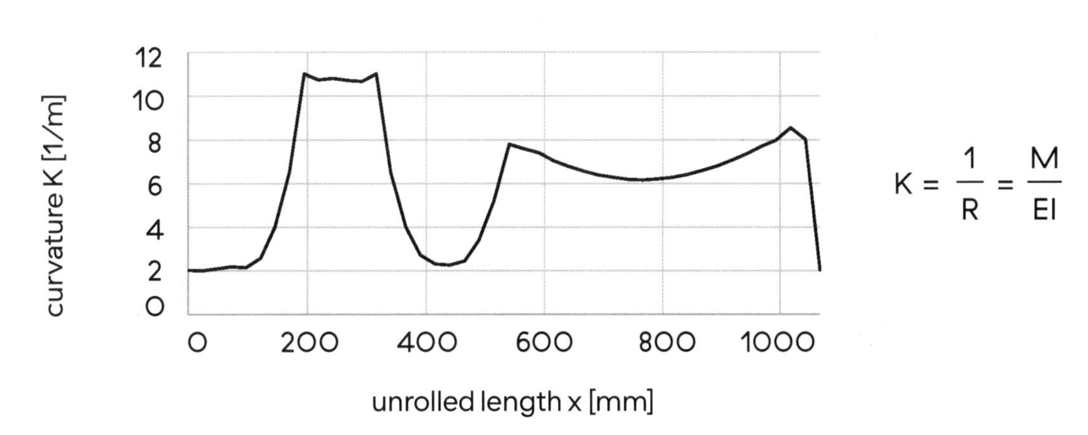

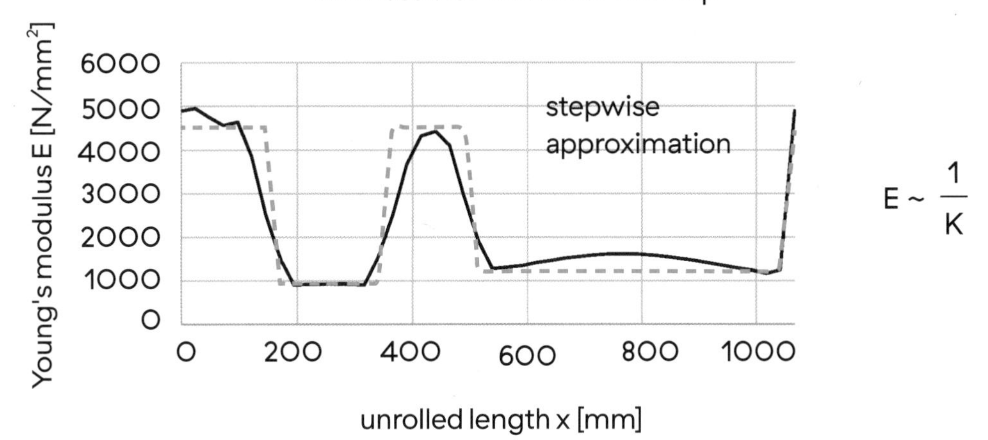

elements can be produced, which result in a stiff doubly curved shell structure when assembled. The entire pavilion weighs only 780 kg and covers an area of 85 m^2 and a span of 9.3 meters. With a resulting material thickness/span ratio of 1/1000 on average, it has a structural weight of only 7.85 kg/m^2.

ICD AGGREGATE PAVILION 2015 SELF-STABILIZING DURING AGGREGATION

The Aggregate Architectures research project, conducted by Karola Dierichs at the ICD, investigates designed granular materials in architecture.[3] Granular materials are defined as large groupings of particles between which only short-range contact forces act. If the individual particle is defined in its geometry and materiality, the overall behavior of the granular material can be calibrated. Key to the design process with granular materials are suitable methods of observation and information collection, which encompass both physical experimentation and numerical simulation.[4] Both types of computation have to be conducted with statistical repetition in order to arrive at reliable predictions of the behavior of a specific granular structure. The relevance of designed granular materials in architecture is threefold: they can be fully recycled, they can adaptively reconfigure as a structure over time, and at the same time they allow for the development of novel architectural characteristics.

The ICD Aggregate Pavilion 2015 is the first full-scale architectural implementation of a designed granular material within the framework of research conducted at the ICD. The pavilion explores the fact that, through the use of highly nonconvex particles, vertical structures can be erected that deviate from the sloped angle of repose usually encountered in a wide range of granular materials. The structures are graded using two to three different particle types to allow for a better load transfer from top to bottom. Based on initial statistical tests of packing densities for different particle geometries, two particle types were injection-molded in higher numbers using recycled plastics. Construction with a cable robot was tested first within a confined framework. The eventual construction setup on site used four trees as fixing points between which the robot's cables were suspended. Precise calibration allowed

Figure 33.3 (left page)
ICD Aggregate Pavilion 2015, Karola Dierichs and Achim Menges, Institute for Computational Design, University of Stuttgart, 2015. The project explores vertical structures made from designed granular materials.

Figure 33.5 (bottom)
3D Printed Hygroscopic Programmable Material System, David Correa, Steffen Reichert, and Achim Menges, Institute for Computational Design, University of Stuttgart, 2014. 3D-printed hygroscopic responsive aperture (left) compared to previously developed wood veneer and fiberglass composite aperture (right) with shape change actuation in response to relative humidity changes.

for the positioning and repositioning of the designed granular material in situ.

3D PRINTED HYGROSCOPIC PROGRAMMABLE MATERIAL SYSTEMS SELF-ADAPTING TO ENVIRONMENTAL CONDITIONS

The field of bio-inspired 3D-printed hygroscopic material systems has been investigated by David Correa at the ICD. His research looks at biological models of material organization to achieve autonomously responsive material systems that react to weather changes without the need to supply any operational energy. Various natural plant movements are based on the multihierarchical structuring of the material that enables a passive response to adapt to the changing environmental conditions. The opening of pine cones is one example of such systems, where external environmental changes induce shape change through the hygroscopic differential expansion of microscopic multilayer structures. Without electromechanical components, these compliant mechanisms function as smart material assemblies capable of responding without the need for discrete controllers, sensing units, or external energy input.

The first case study of the 3D Printed Hygroscopic Programmable Material System project presents the technical transfer from previous research instrumentalizing wood tissue for shape-changing wood veneer composites to purposefully designed custom-3D-printed wood/polymer composites.[5] The wood veneer composites previously developed for climate-responsive architectural applications, as shown in the HygroScope Installation and the HygroSkin Pavilion,[6] provided great insight into the micro- and macrostructural relation of wood grain and hygroscopic expansion in relation to atmospheric changes. These findings were then transferred to multimaterial and multihierarchical structures using fused filament fabrication to generate 3D-printed composites capable of programmable shape change deformation with differentiated functional regions.[7] Enabled by digital simulation and fabrication tools, it is possible to develop components with unique performance properties that can integrate predefined and preprogrammable functional performance. This project provides the first proof of concept in the innovation and development of custom-material embedded-environmental responsiveness for architectural applications.

PROJECT CREDITS

ICD/ITKE Research Pavilion 2015–16: ITECH students with Simon Bechert, Oliver David Krieg, Tobias Schwinn, Daniel Sonntag, Jan Knippers, Achim Menges, Institute for Computational Design, Institute of Building Structures and Structural Design

ICD Aggregate Pavilion 2015: Karola Dierichs, Achim Menges, Institute for Computational Design

3D Printed Hygroscopic Programmable Material Systems: David Correa, Steffen Reichert, Achim Menges, Institute for Computational Design

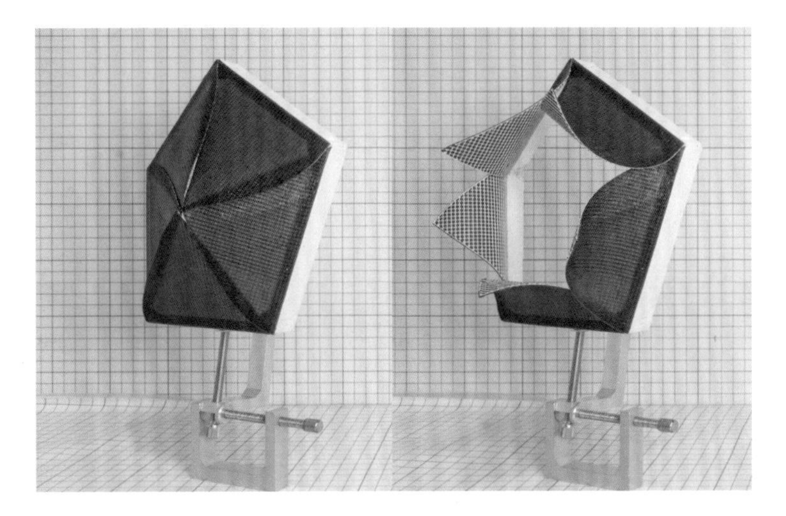

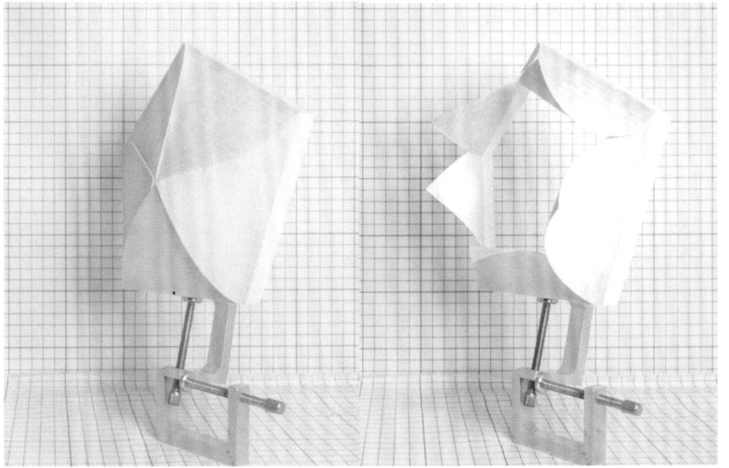

Figure 33.4
ICD Aggregate Pavilion 2015, Karola Dierichs and Achim Menges, Institute for Computational Design, University of Stuttgart, 2015. The pavilion is made from hexapodal and tetrapodal particles that aggregate into stable 4-meter-tall structures.

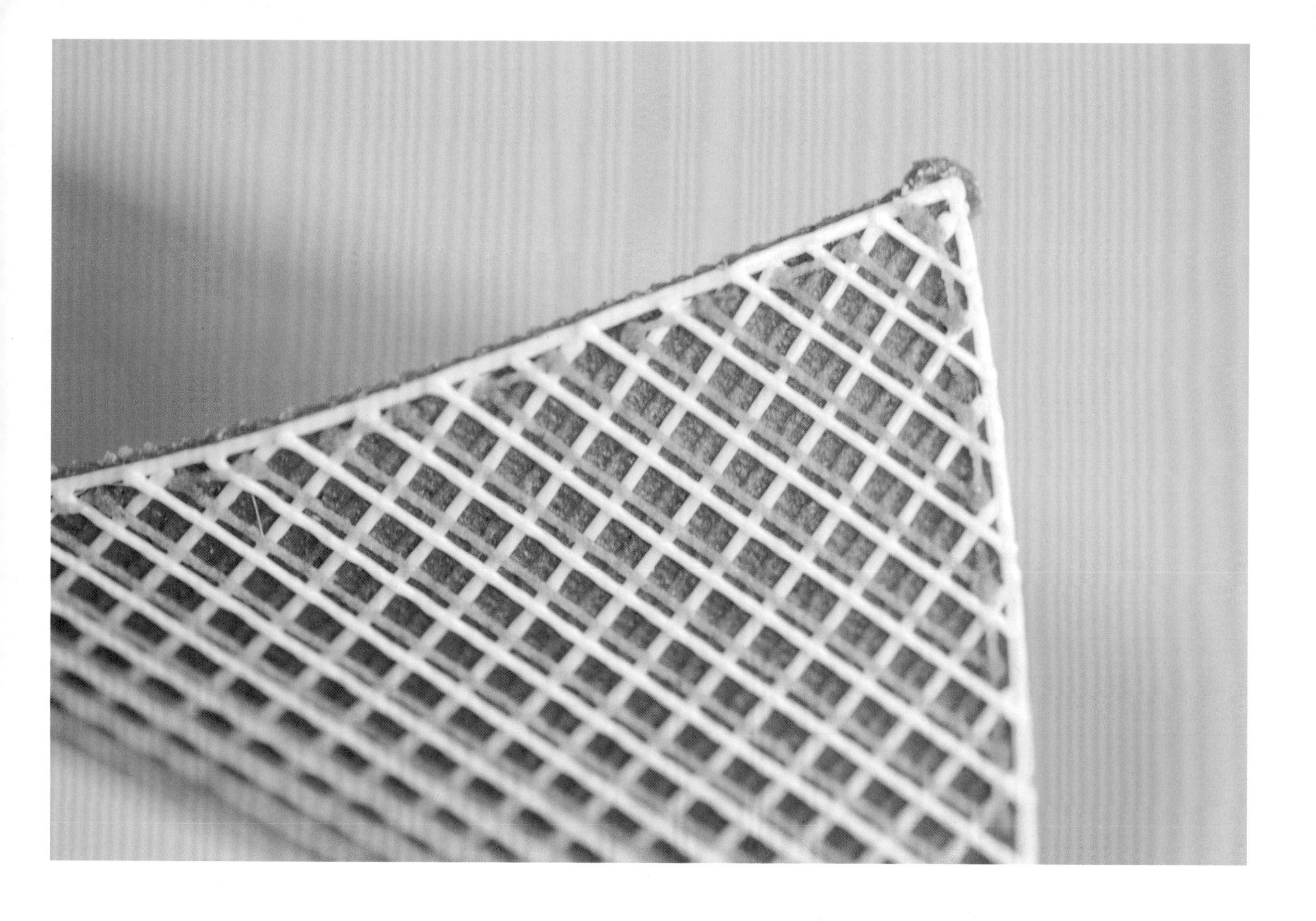

Figure 33.6
3D Printed Hygroscopic Programmable Material System, David Correa, Steffen Reichert, and Achim Menges, Institute for Computational Design, University of Stuttgart, 2014. Multimaterial 3D printing structure close-up of responsive aperture with ABS (white) and wood composite (brown).

NOTES

1 Moritz Fleischmann, Jan Knippers, Julian Lienhard, Achim Menges, and Simon Schleicher, "Material Behaviour: Embedding Physical Properties in Computational Design Processes," *Architectural Design* 82, no. 2 (2012): 44–51.

2 Simon Bechert, Jan Knippers, Oliver David Krieg, Achim Menges, Tobias Schwinn, and Daniel Sonntag, "Textile Fabrication Techniques for Timber Shells," Proceedings of Advances in Architectural Geometry (AAG), Zurich, Switzerland, September 9–13, 2016.

3 Karola Dierichs and Achim Menges, "Towards an Aggregate Architecture: Designed Granular Systems as Programmable Matter in Architecture," *Granular Matter* 18, no. 25 (2016).

4 Karola Dierichs and Achim Menges, "Granular Morphologies: Programming Material Behaviour with Designed Aggregates," *Architectural Design* 85, no. 5 (2015): 86–91.

5 David Correa, Athina Papadopoulou, Christophe Guberan, Nynika Jhaveri, Steffen Reichert, Achim Menges, and Skylar Tibbits, "3D-Printed Wood: Programming Hygroscopic Material Transformations," *3D Printing and Additive Manufacturing* 2, no. 3 (2015): 106–116.

6 Achim Menges and Steffen Reichert, "Performative Wood: Physically Programming the Responsive Architecture of the HygroScope and HygroSkin Projects," *Architectural Design* 85, no. 5 (2016): 66–73.

7 David Correa and Achim Menges, "3D Printed Hygroscopic Programmable Material Systems," in Proceedings of Material Research Society, vol. 1800, San Francisco, April 6–10, 2015.

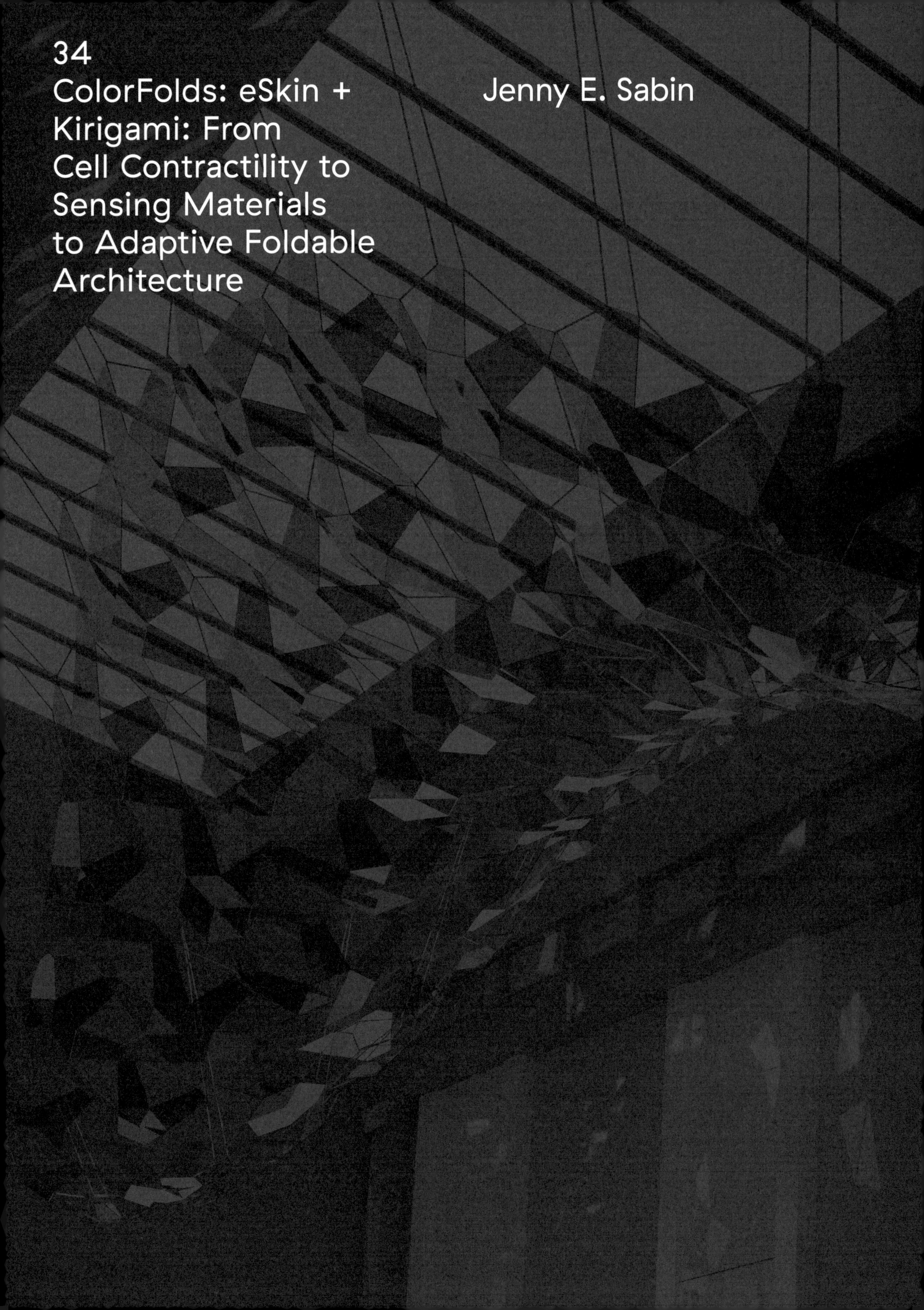

34

ColorFolds: eSkin + Kirigami: From Cell Contractility to Sensing Materials to Adaptive Foldable Architecture

Jenny E. Sabin

Figure 34.1
Rendering of ColorFolds suspended in the architecture building at Cornell University. Color change within the assembly is wavelength-dependent (what is also known as structural color change).

34
ColorFolds: eSkin + Kirigami: From Cell Contractility to Sensing Materials to Adaptive Foldable Architecture

Jenny E. Sabin

As part of two collaborative projects in the Sabin Design Lab at Cornell University, titled eSkin and KATS (Cutting and Pasting—Kirigami in Architecture, Technology, and Science), ColorFolds is one area of ongoing transdisciplinary research spanning the fields of cell biology, materials science, physics, electrical and systems engineering, and architecture. Among other things, the eSkin project aims to explore materiality from nano to macro scales based on understanding of complex cell behaviors on patterned substrates, and to understand how these features and effects may operate at the architectural scale.[1] ColorFolds incorporates two parameters that the team is investigating: optical color and transparency change at the human scale, based on principles of structural color at nano to micro scales (figures 34.1 and 34.2).

In addition to optical color and transparency change, ColorFolds features a lightweight, tessellated array of interactive components that fold and unfold in the presence or absence of people. From architecture to chemistry, from chalkboards to micrographs, and from maps to trompe-l'oeil, we strive to communicate 3D geometry, structures, and features using 2D representations. These drawings have allowed us not only to communicate complex information, but also to create 3D objects, from the act of folding a paper airplane to the construction and digital fabrication of entire buildings. ColorFolds follows the concept of "interact locally, fold globally" that is necessary for deployable and scalable architectures. Each face of the tessellated and interactive components features a novel wavelength-dependent film. This material research builds on the latest prototypes within the eSkin project, including the optical simulation and application of geometrically defined nano/microscale substrates that display the effects of nonlinear structural color change when deployed at the building scale. Not only does this film align

Figure 34.2 (left page)
Based on ongoing collaborative work with Dr. Shu Yang, ColorFolds incorporates two parameters that the team is investigating: optical color and transparency change at the human scale based on principles of structural color at a nano to micro scale. The color change is wavelength-dependent. Many examples of structural color change are found in nature, such as the wings of the blue morpho butterfly or the feathers of hummingbirds. Image courtesy Shu Yang group at University of Pennsylvania.

Figure 34.3 (top)
Precision kirigami based on topology and geometry of defects in sheets. See Castle et al. (note 2).

Figure 34.4 (middle)
The generative design process for ColorFolds began with an examination and study of kirigami processes as a means of creating doubly curved surfaces through a simple implementation of gradient folding conditions.

Figure 34.5 (bottom)
Physical study models of kirigami folding behaviors. Kirigami is similar to origami but with the addition of cuts and holes.

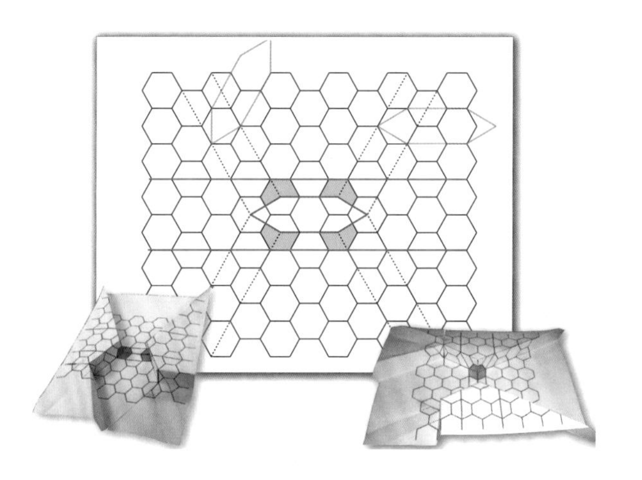

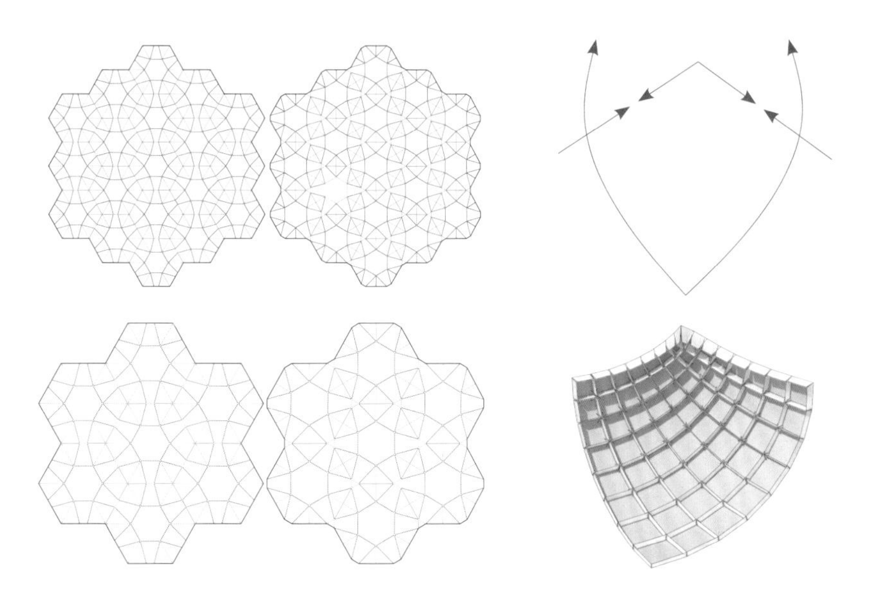

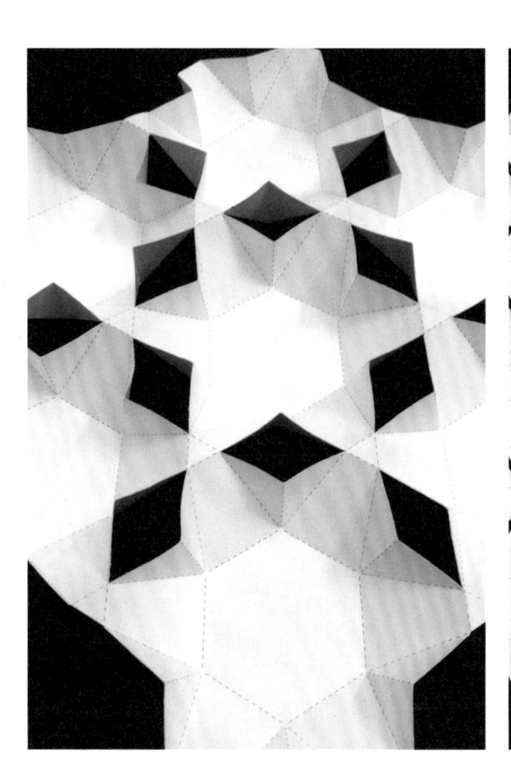

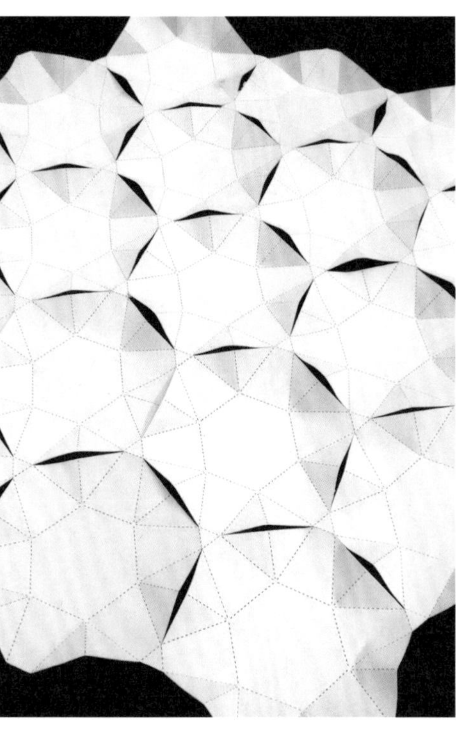

Figure 34.6
The same wavelength-dependent material produces radically different color effects based on viewer orientation and the dynamic quality of the folding/unfolding component.

with our investigations into structural color, but it also allows for room-scale investigations of these nano- to micromaterial effects and features. An array of sensors detects the presence or absence of people below, which in turn actuates a network of nitinol spring systems that open or close the folded components.

The generative design process for ColorFolds began with an examination and study of kirigami processes as a means of creating doubly curved surfaces through a simple implementation of gradient folding conditions. Kirigami is similar to origami but adds cuts and holes. The word comes from the Japanese *kiru*, “to cut,” a geometric method and process that brings an extra, previously unattainable level of design, dynamics, and deployability to self-folding and self-unfolding materials from the molecular to the architectural scale. Our tools and methods were greatly informed by close collaboration with Randall Kamien, a principal investigator and theoretical physicist on the team, based at the University of Pennsylvania (figure 34.3).[2] Through these studies, we developed a series of algorithms for generating an adaptable spatially aware geometry that formally responds to site-based geometries with the added capacity, through its geometry and actuation system, to adapt to various user groups within the installation space (figures 34.3—34.5).

Comprised of a network of low-cost sensors and wavelength-dependent responsive materials, ColorFolds is conceived to be generic and homogeneously structured upon installation, but readily adaptable to local heterogeneous spatiotemporal conditions and user interaction. Our approach to kirigami-based construction will bring a new level of motifs, portability, and nuanced design to recently established techniques to form intricate structures chemically, biologically, elastically, and architecturally. Using mathematical modeling, architectural elements, design computation, and controlled elastic response, ColorFolds showcases new techniques, algorithms, and processes for the assembly of open, deployable structural elements and architectural surface assemblies.

ACKNOWLEDGMENTS

This project is funded by the National Science Foundation and the CCA and is jointly housed at Cornell University and the University of Pennsylvania.

Design research team: Martin Miller (senior personnel and design lead), Daniel Cellucci and Andrew Moorman (mechatronics lead), Giffen Ott (production lead), Max Vanatta, David Rosenwasser, Jessica Jiang. Kirigami / Jenny E. Sabin (co-principal investigator) and Martin Miller (senior personnel) (architecture), Dan Luo (co-principal investigator) (biological and environmental engineering), Cornell University; Shu Yang (co-principal investigator) (materials science), Randall Kamien (principal investigator) (physics and astronomy), University of Pennsylvania.

NOTES

1 Jenny Sabin, Andrew Lucia, Giffen Ott, and Simon Wang, “Prototyping Interactive Nonlinear Nano-to-Micro Scaled Material Properties and Effects at the Human Scale,” in *Symposium on Simulation for Architecture & Urban Design (SimAUD 2014) 2014 Spring Simulation Multi-Conference (SpringSim '14); Tampa, Florida, USA, 13—16 April 2014*, ed. D. Gerber (Red Hook, NY: Curran, 2014), 7—14.

2 T. Castle, Y. Cho, X. Gong, E. Jung, D. M. Sussman, S. Yang, and R. D. Kamien, “Making the Cut: Lattice Kirigami Rules,” *Phys. Rev. Lett.* 113 (2014): 245502.

REFERENCES

Castle, T., Y. Cho, X. Gong, E. Jung, D. M. Sussman, S. Yang, and R. D. Kamien. “Making the Cut: Lattice Kirigami Rules.” *Phys. Rev. Lett.* 113 (2014): 245502.

Sabin, Jenny, Andrew Lucia, Giffen Ott, and Simon Wang. “Prototyping Interactive Nonlinear Nano-to-Micro Scaled Material Properties and Effects at the Human Scale.” In *Symposium on Simulation for Architecture & Urban Design (SimAUD 2014) 2014 Spring Simulation Multi-Conference (SpringSim '14); Tampa, Florida, USA, 13—16 April 2014*, ed. D. Gerber, 7—14. Red Hook, NY: Curran, 2014.

Thompson, L. F. *Introduction to Microlithography*. 2nd ed. Washington, DC: American Chemical Society, 1994.

Xia, Younan, and George M. Whitesides. “Soft Lithography.” *Annual Review of Materials Science*, no. 28 (1998): 153—184.

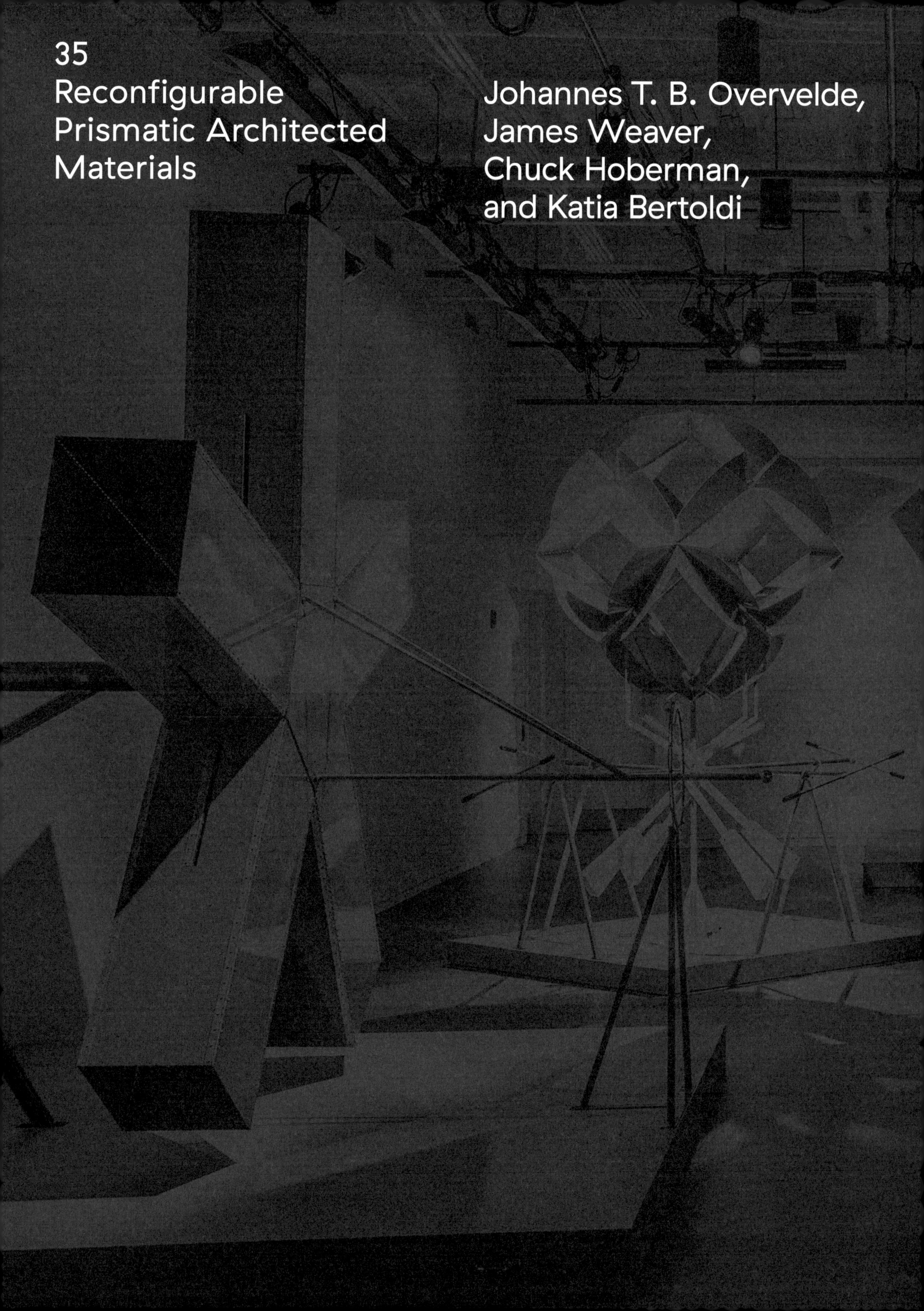

35

Reconfigurable Prismatic Architected Materials

Johannes T. B. Overvelde,
James Weaver,
Chuck Hoberman,
and Katia Bertoldi

35
Reconfigurable Prismatic Architected Materials

Johannes T. B. Overvelde,
James Weaver,
Chuck Hoberman, and
Katia Bertoldi

Advances in fabrication technologies are enabling the production of architected materials with unprecedented properties. While most of these materials are characterized by a fixed geometry, an intriguing avenue is to incorporate internal mechanisms capable of reconfiguring their spatial architecture, therefore enabling the creation of materials that have tunable functionality. Inspired by the structural diversity and foldability of the prismatic geometries that can be constructed using the snapology origami technique, we recently introduced a robust design strategy based on space-filling polyhedra to create 3D reconfigurable materials comprising a periodic assembly of rigid plates and elastic hinges.[1,2] Guided by numerical analysis, cardboard models, and 3D-printed prototypes, we systematically explored the mobility of the designed structures and identified a wide range of qualitatively different deformations and internal rearrangements.[2]

Importantly, since the underlying principles are scale-independent, our strategy can be applied to design structures over a wide range of length scales, ranging from transformable meter-scale architectures to nanoscale tunable photonic systems. For example, Chuck Hoberman has recently realized four large kinetic sculptures that can be transformed by viewers through hands-on play. This sculpture has been shown in a recent exhibit curated by Hoberman at Le Laboratoire Cambridge. At the exhibit, visitors were invited to explore the changes in form of the sculpture through touch and movement, continuously remaking the installation whether individually or with other visitors. By turning a lever, visitors move sculptures that are 12 to 14 feet high and that weigh hundreds of pounds. Large volumes seem to hover above their supports as they are reconfigured. Viewers remark that the sculptures remind them of being underwater, or of floating in outer space.

Here are few visitor quotes:

> "It's such a collaborative experience. Your body interacts with the sculpture, and you have to work with other people to move them."
>
> "I didn't expect such organic movement from something so angular."
>
> "It's like a breathing lung."
>
> "It's soothing, and I could watch it all day long."

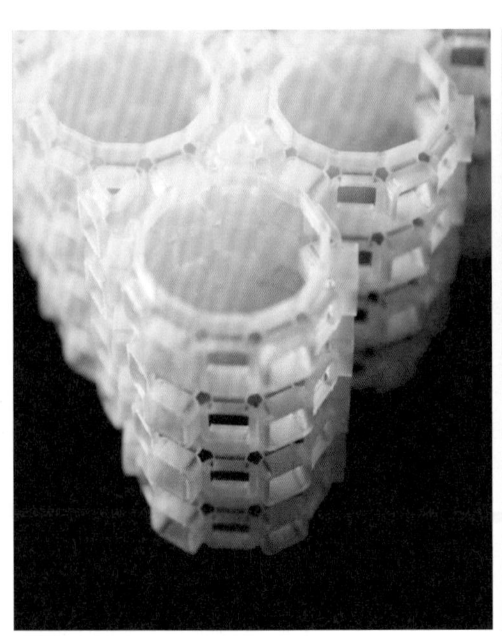

Figure 35.1
Prismatic structure arrays,
3D-printed resin.

Figure 35.2 (left page, left)
Prismatic structure arrays,
3D-printed resin.

Figure 35.3 (left page, middle)
Prismatic structure arrays,
3D-printed resin.

Figure 35.4 (left page, right)
Prismatic structure arrays,
3D-printed resin.

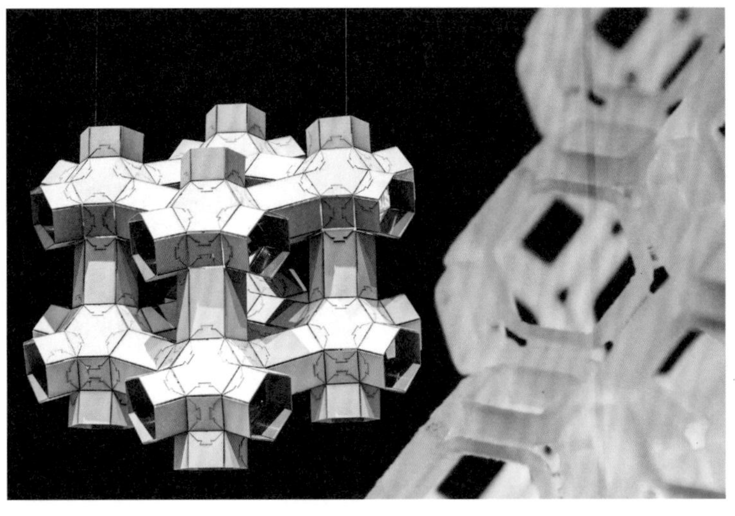

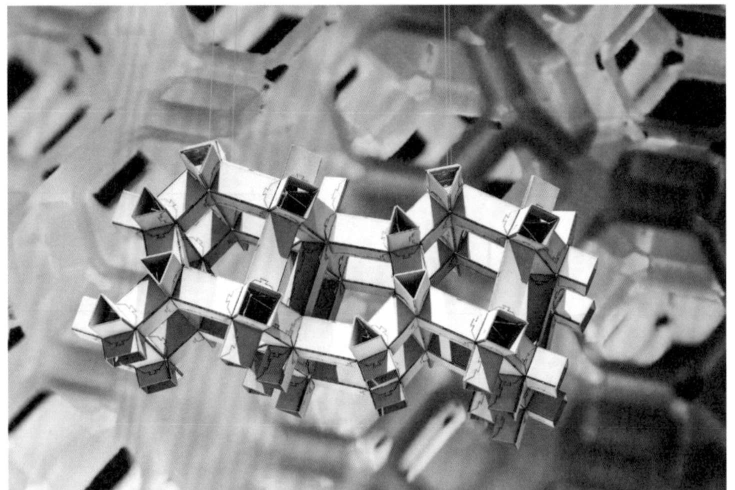

Figure 35.5 (left)
Prismatic sculptures, 10° exhibition,
Le Laboratoire Cambridge.

Figure 35.6 (right, top)
Prismatic model array, 10° exhibition,
Le Laboratoire Cambridge.

Figure 35.7 (right, bottom)
Prismatic model array, 10° exhibition,
Le Laboratoire Cambridge.

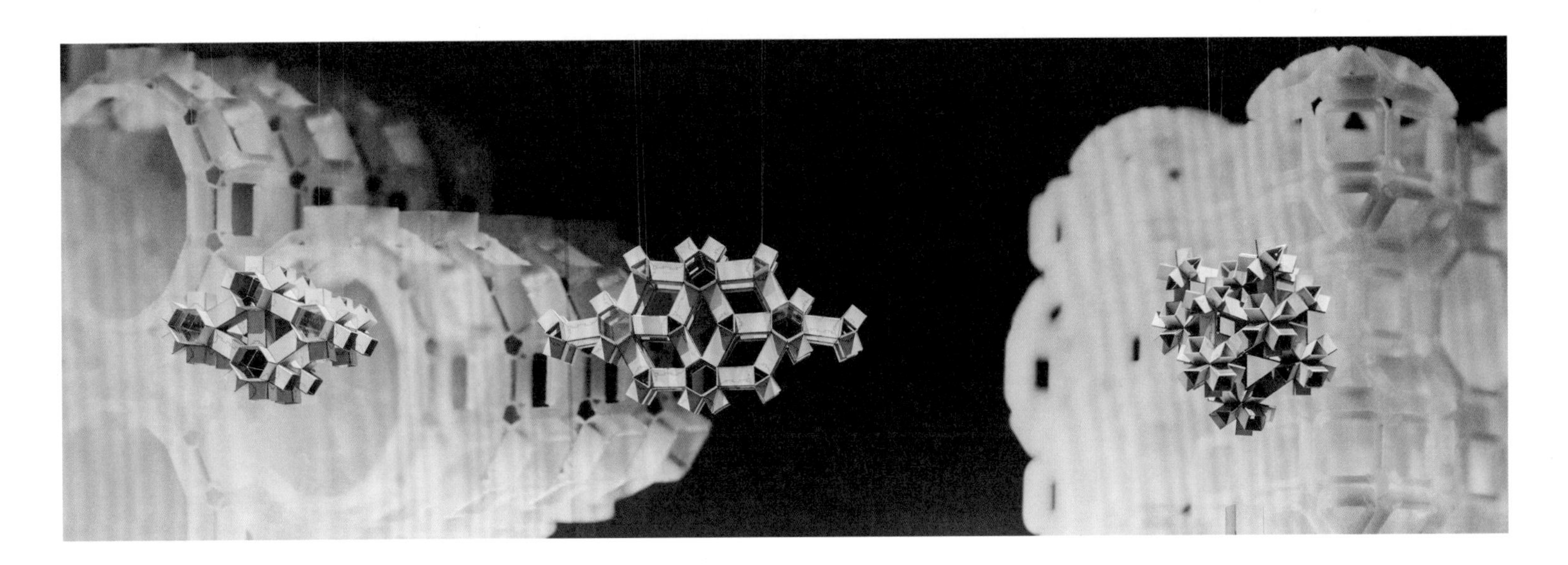

Figure 35.8 (top)
Prismatic model array, 10° exhibition, Le Laboratoire Cambridge.

Figure 35.9 (bottom)
Prismatic model array, 10° exhibition, Le Laboratoire Cambridge.

Figure 35.10
Prismatic sculptures, 10° exhibition, Le Laboratoire Cambridge.

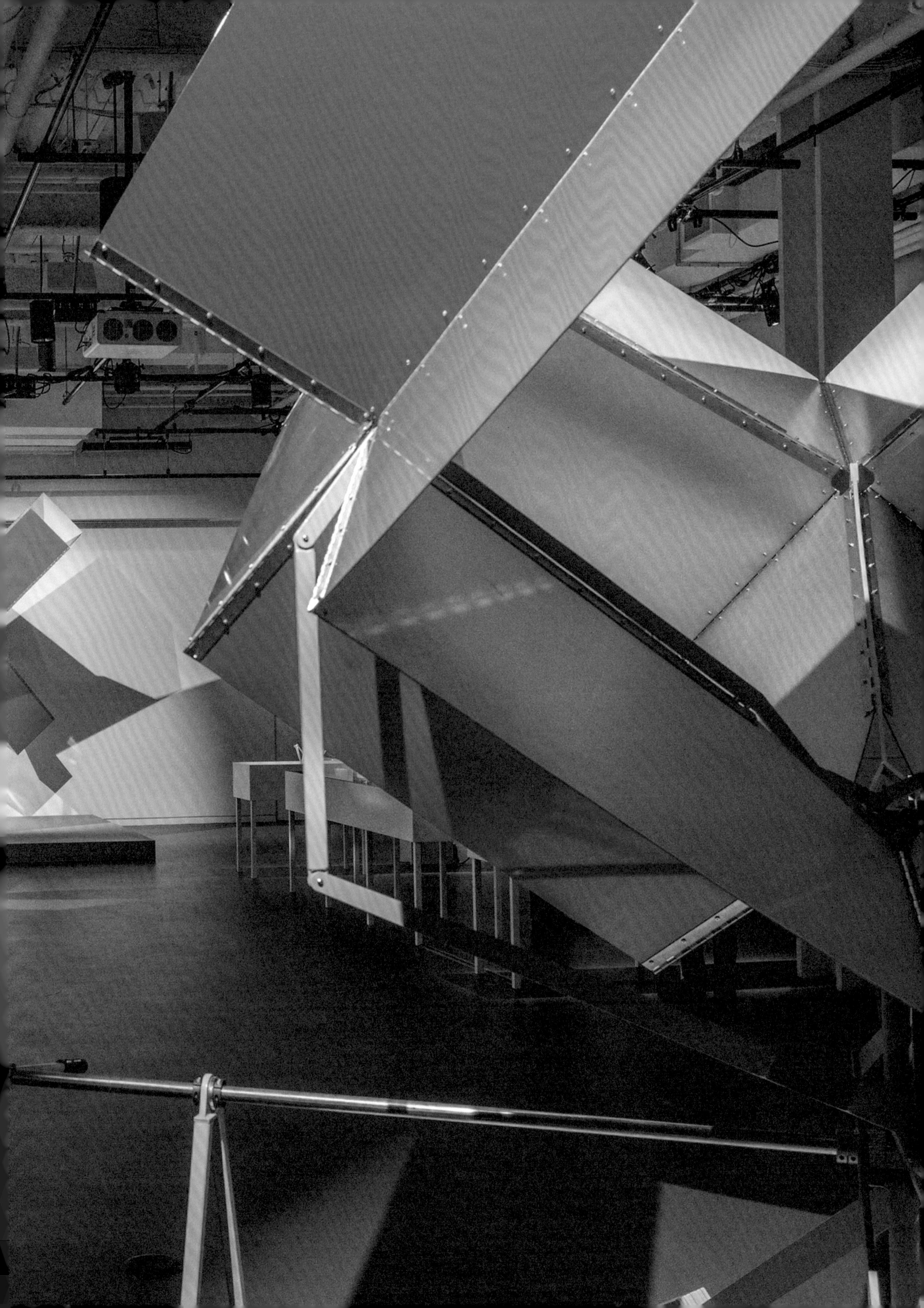

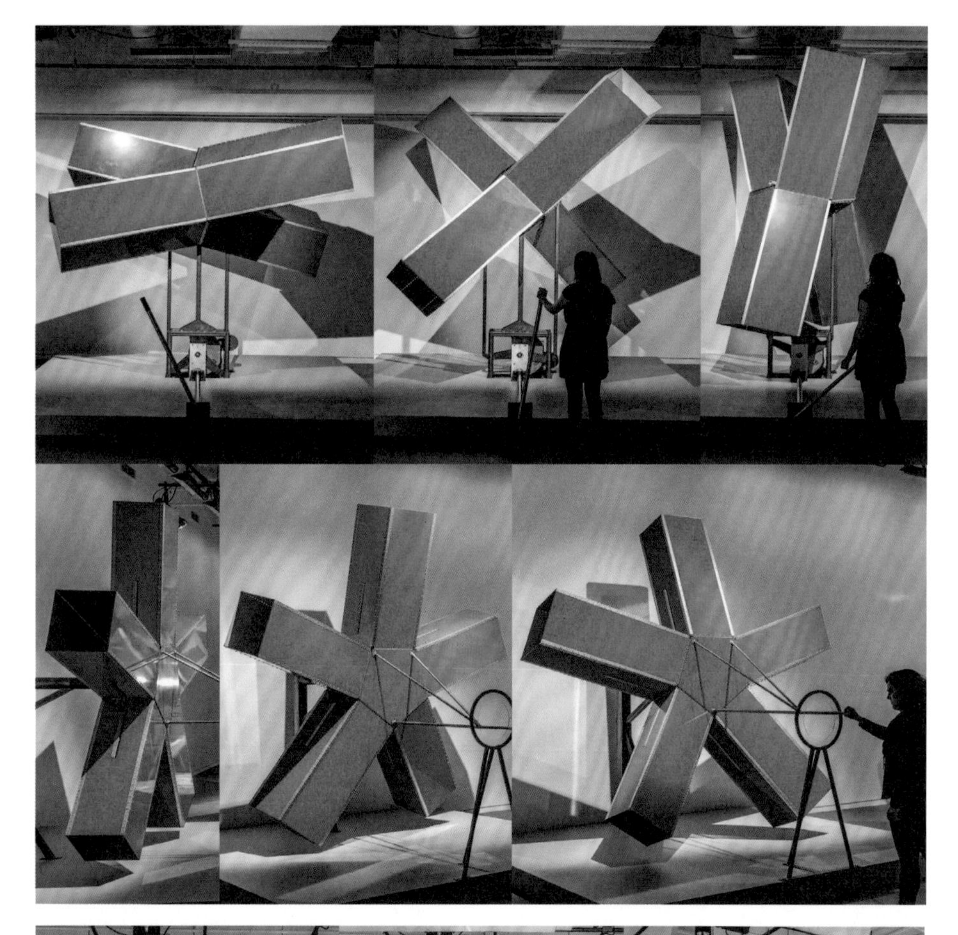

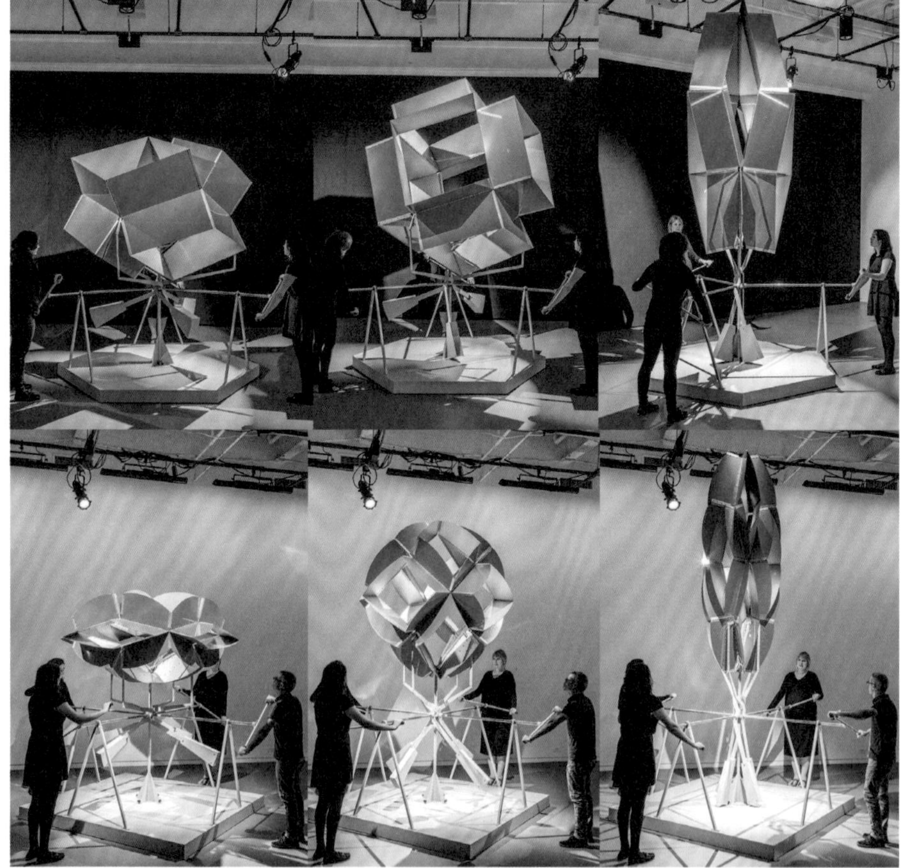

Figure 35.11 (1st row)
Prismatic sculpture movement sequence, 1° ("X").

Figure 35.12 (2nd row)
Prismatic sculpture movement sequence, 2° ("Asterisk").

Figure 35.13 (3rd row)
Prismatic sculpture movement sequence, 3° ("Cube").

Figure 35.14 (4th row)
Prismatic sculpture movement sequence, 4° ("Sphere").

NOTES

1 J. T. B. Overvelde, T. A. de Jong, Y. Shevchenko, S. A. Becerra, G. Whiteside, J. C. Weaver, C. Hoberman, and K. Bertoldi, "A Three-Dimensional Actuated Origami-Inspired Transformable Metamaterial with Multiple Degrees of Freedom," *Nature Communications* (2016).

2 J. T. B. Overvelde, J. C. Weaver, C. Hoberman, and K. Bertoldi. "Rational Design of Reconfigurable Prismatic Architected Materials," *Nature* 541, no. 7637 (2017): 347–352.

36

Adaptive Granular Matter

Kieran A. Murphy,
Leah K. Roth, and
Heinrich M. Jaeger

36
Adaptive Granular Matter

Kieran A. Murphy,
Leah K. Roth, and
Heinrich M. Jaeger

Granular materials are large aggregates of individually solid particles that interact when they come into direct contact. We are familiar with these materials as bulk commodities such as grains, seeds, powders, or pills, all of which are ubiquitous and important to a wide range of industrial processes. Perhaps surprisingly, these seemingly simple materials can exhibit highly complex behavior. Harnessed and optimized, this opens up exciting avenues for creating new types of smart, adaptive response when stress is applied to these materials.

Key to the richly complex behavior of granular matter is the disordered configuration of the constituent particles (figure 36.1). In contrast to ordered, crystalline solids, this enables the granular aggregate to transition between flowing and rigid states easily. Contacts between neighboring particles can form, dissolve, and quickly reform somewhere else, and this gives rise to an intricate network of paths along which forces are transmitted (figure 36.2). The network dynamically reconfigures as the applied load is changing; as a result, granular matter can adapt to applied stresses by morphing between different overall shapes and by self-healing through cooperative particle movement. All of this occurs in real time inside the material and without the need for sensors or computational feedback, features that recently have led to new applications of granular matter ranging from shape-shifting soft robotics to reconfigurable aleatory architectures.[1, 2]

Particle shape is one of the critical parameters controlling the overall aggregate behavior of granular matter and its ability to adapt to changes in the load environment. For example, hook- or Z-shaped particles can create vertical columns that not only are stable under their own weight, but also can sustain significant compression without collapse (figure 36.3), something that would be impossible with most other particle types. Z—shaped particles furthermore provide a balance between column stability under load, flowability of particles during pouring, and easy particle manufacture because they are planar. This makes them a promising candidate particle for aleatory architectural structures.[2,3] Still, in most cases the particle shapes suitable for any given design task are found by laborious trial and error. What has remained elusive is the ability to identify, via an automated process, the shape that can realize a given task.

Our laboratory at the University of Chicago recently has made first steps in this direction by developing a computer-aided approach that uses ideas from artificial evolution to find optimized particle shapes. Arbitrary shapes are approximated by sets of bonded spheres (figure 36.4). We

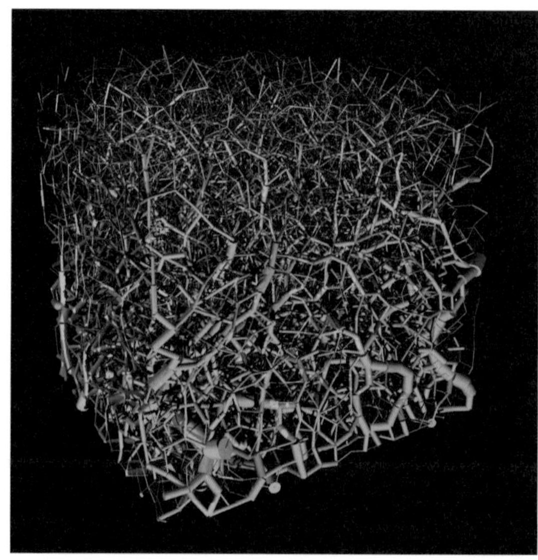

Figure 36.1 (left)
Granular matter. Granular matter comprised of densely packed particles forms a highly disordered aggregate that can transition between overall solid-like and liquid-like behavior. The particles shown here are poppy seeds.

Figure 36.2 (right)
Force chain network. Simulation of the complex network of paths that transmit forces through granular matter, shown here for a three-dimensional aggregate of densely packed spheres that is compressed by its own weight. The cylinders indicate the presence and spatial orientation of a force between any two contacting particles, while the cylinder thickness gives the force magnitude.

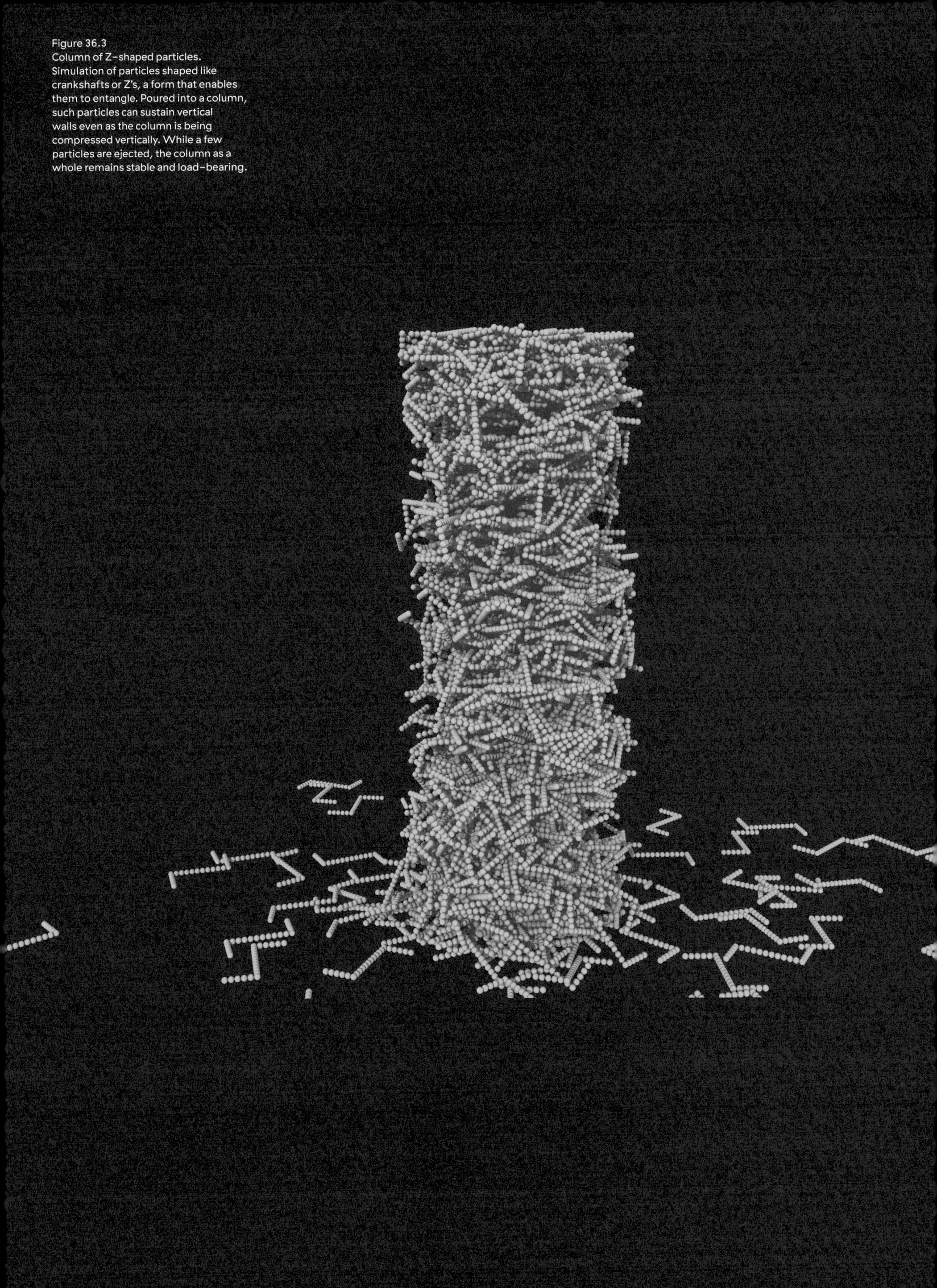

Figure 36.3
Column of Z-shaped particles.
Simulation of particles shaped like crankshafts or Z's, a form that enables them to entangle. Poured into a column, such particles can sustain vertical walls even as the column is being compressed vertically. While a few particles are ejected, the column as a whole remains stable and load-bearing.

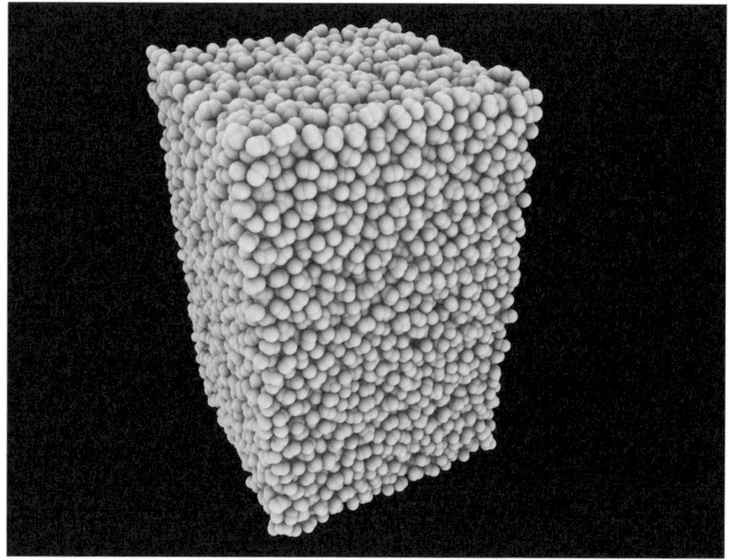

call these sets “granular molecules” in analogy to their atomic-scale cousins. We can construct granular molecules of arbitrary shape by varying the number of constituent spheres, their size, and the degree of overlap among neighbors.[1, 3] Letting the computer optimize the granular molecule configuration with respect to a targeted goal for the aggregate behavior then represents a path toward identifying the best shape. An example outcome is shown in figure 36.5, where the task was to find the shape that produces the densest packing when particles are poured into a container and then the container is lightly tapped to settle the particles. For this task, the best-performing shape turned out to be a planar triangular particle;[4] but, just as with speciation in flora and fauna, the artificial evolution process can lead to strikingly different granular molecules for different tasks or different processi conditions.[1] The real power of using an evolutionary strategy, however, goes far beyond the ability to identify specific molecule shapes as optimal solutions for specific tasks. From the many shapes tested and then discarde along the evolutionary pathway, we ca extract general design rules that make possible to find appropriate solutions f whole classes of related tasks.

Figure 36.4 (left)
Granular molecules. Examples of compound particles formed by bonding a set of spheres. The granular molecules shown were created by 3D printing.

Figure 36.5 (right)
Particle shape optimized for densest packing. The planar, triangular granular molecule shown here, composed of three overlapping spheres of the same size, emerged through artificial evolution as the shape that produces the densest packing when poured into a container and settled by tapping.

NOTES

1 H. M. Jaeger, “Toward Jamming by Design,” *Soft Matter* 11 (2015): 12–27.

2 S. Keller and H. M. Jaeger, “Aleatory Architectures,” *Granular Matter* 18, no. 29 (2016): 1–11.

3 K. A. Murphy, N. Reiser, D. Choksy, C. E. Singer, and H. M. Jaeger, “Freestanding Loadbearing Structures with Z-Shaped Particles,” *Granular Matter* 18, no. 26 (2016): 21–29.

4 L. K. Roth and H. M. Jaeger, “Optimizing Packing Fraction in Granular Media Composed of Overlapping Spheres,” *Soft Matter* 12 (2016): 1107–1115.

37

Rock Print: An Architectural Installation of Granular Matter

Petrus Aejmelaeus-Lindström,
Andreas Thoma,
Ammar Mirjan,
Volker Helm,
Skylar Tibbits,
Fabio Gramazio, and
Matthias Kohler

Figure 37.1
Rock Print at "The State of the Art of Architecture," the inaugural Chicago Architecture Biennial, 2015.

37

Rock Print: An Architectural Installation of Granular Matter

Petrus Aejmelaeus-Lindström,
Andreas Thoma,
Ammar Mirjan,
Volker Helm,
Skylar Tibbits,
Fabio Gramazio, and
Matthias Kohler

Rock Print: a towering mass supported by four legs, a counterintuitive configuration of rock and string (figure 37.1). Rock Print was designed to display the potential of jammed structures in architecture at the inaugural Chicago Architecture Biennial 2015, "The State of the Art of Architecture,"[1] where it was exhibited as a 4-meter-high centerpiece in one of the Chicago Cultural Center's largest rooms.

Rock Print is distinguished by its material system,[2] which exploits the power of contemporary digital design and fabrication tools while using bulk raw materials as cheap and sustainable building material (figure 37.2). The structure is fabricated solely out of rock and string, two ordinary materials with opposite structural behavior. Combining the compressive strength of the gravel, the tensile strength of the string, and the digital control of the amalgamation process of the materials allows for the creation of solid structures with unique material properties. Furthermore, the material system can be fully reversed to its initial state of raw bulk material by simply unwinding the string reinforcement.

The design principle of the artifact accentuates the specific characteristics of the material system. The base of the structure consists of four slender legs, indicating the structural capacities of the system, which meet and form the massive, star-shaped and cantilevered upper body of the structure. The legs start with an almost round profile, growing toward their neighbors until they merge and extend to the boundary of the robotic operational range. The upper part accommodates a higher amount of mass than the lower, to increase the compression of the bottom part and thereby assure a stronger surface strength on the parts more exposed to visitors. Leftover material from the fabrication process is left surrounding

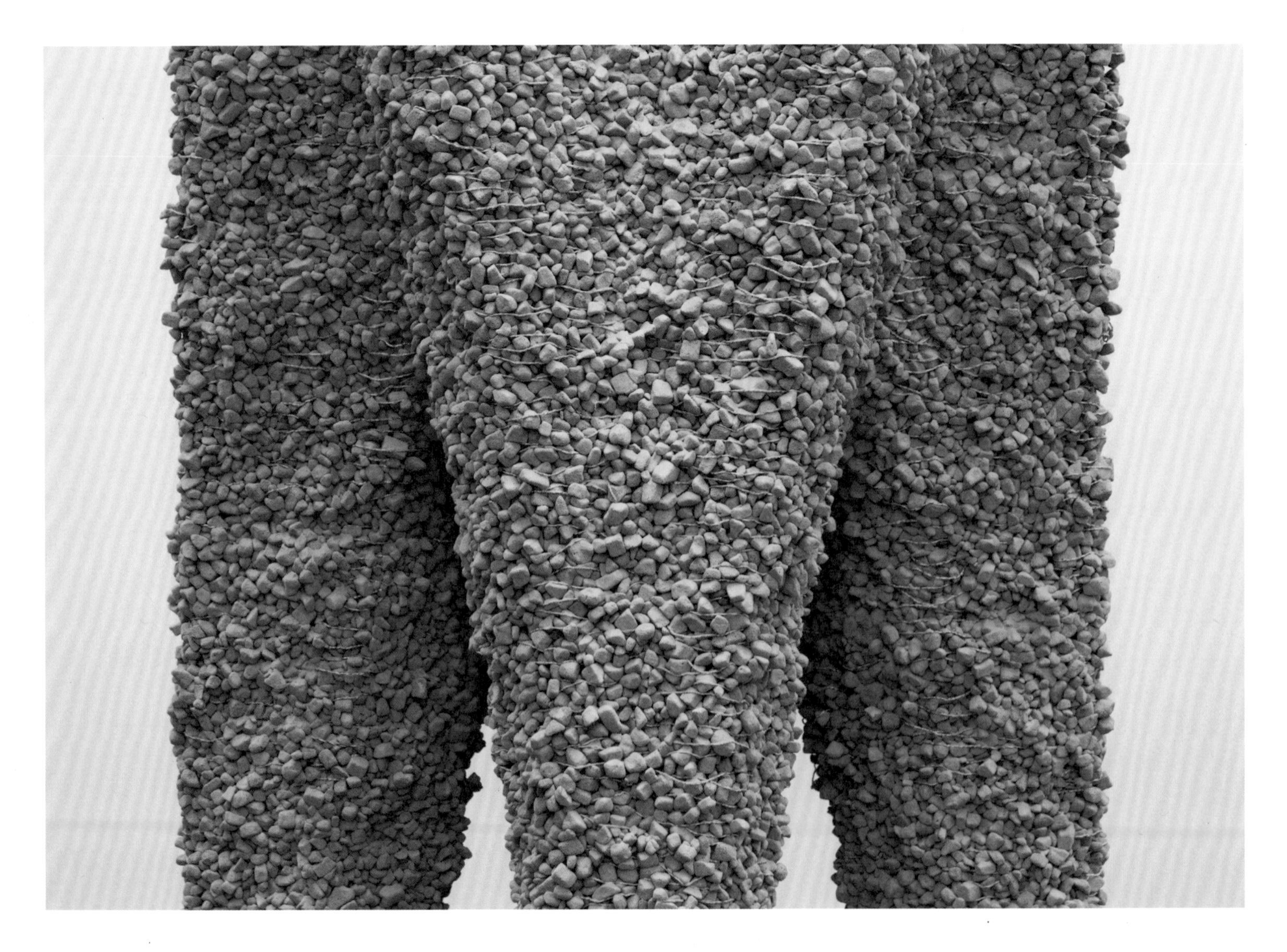

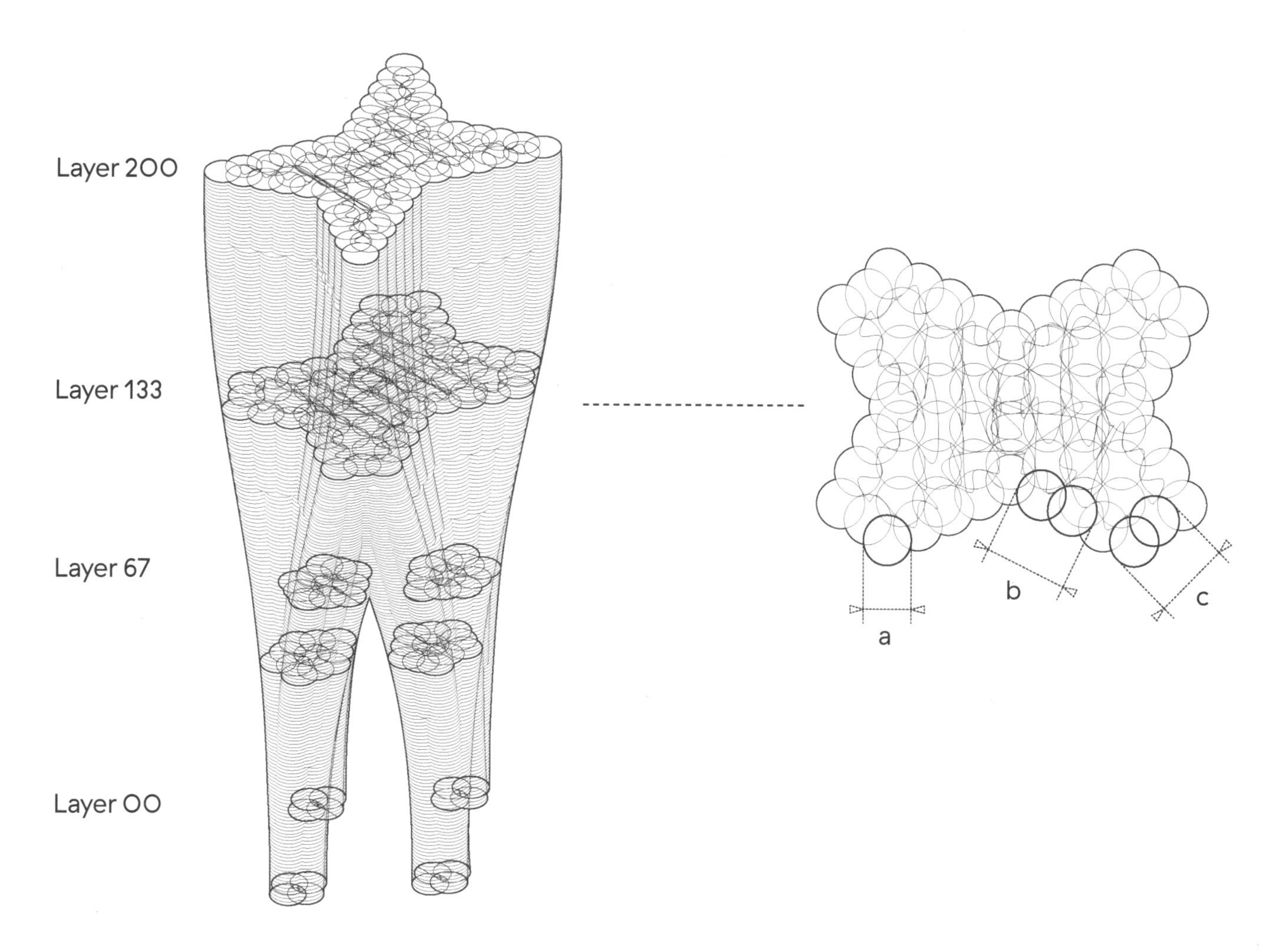

Figure 37.2 (left page)
Close-up of Rock Print revealing the surface texture of the rock-string amalgam.

Figure 37.3 (top)
Isometric drawing of the digital blueprint for the string reinforcement. (a) Diameter of the string pattern circles. (b) Smaller offset of the pattern circles at vertical surfaces.
(c). Larger offset of the pattern circles at cantilevering surfaces.

Figure 37.4 (bottom, 3 images)
Iterative experiments focusing on surface texture: (a) Double string pattern and no packing, resulting in clearly visible pattern layers and weak surface texture. (b) Double string pattern touching the walls of the box and packing, resulting in surface texture and geometry from the box while still clearly showing the pattern layers. (c) Single string pattern not touching the walls of the box, dissolving the pattern layers and resulting in an articulated surface texture.

Figure 37.5 (right page)
(a) the initial segment of the box is assembled on the frame. (b) A thin layer of aggregates (20 mm, which is the average particle size of the 16—32 mm Misapor) is manually placed inside the fabrication bed. (c) The robotic arm equipped with string laying tool deploys the string according to a digital blueprint. This procedure is repeated until the next segment of the box is assembled. (d) Every second layer of aggregate is manually packed with a 10 kg concrete compacter. As a result, the aggregates are compacted by approximately 20%. This has to be taken into account when designing the structure. To accommodate for this, every layer is measured by the robot after building it, to assure a constant build and to allow the self-compacting of the aggregates. The glass foam aggregates are light enough that a manual layer placement can be conducted at a very fast pace. The fabrication of Rock Print took 45 hours; approximately 75% of this time was related to the robotic string laying, while the rest was for measuring, placement of aggregates, and the assembly of the frame. (e) When the box is fully assembled and the whole structure is fabricated, the robotic setup is removed. (f) The box can be dismantled, leaving only the top segment, middle segment, and bottom segment of the frame to stabilize the structure and assure that it will not collapse on the builders. Once the structure has been visually inspected, the last part of the frame can be dismantled. (g) Glass foam aggregates have a tendency to jam even where the string is not placed. Therefore, the structure has to be gently brushed, similar to removing the support material from powder-based 3D prints, revealing the final artifact (h).

the base, to root the exhibit to the space as well as create a natural barrier for the visitors.

JAMMED STRUCTURES

The construction system of Rock Print uses jamming, the physical phenomenon by which granular matter becomes rigid under certain conditions, for example when the free volume per particle is decreased, thereby increasing the strain between the aggregates, making them more and more constrained.[3] The "jamming transition" allows not only a change from a nonsolid, liquid-like state into a solid state, but also for the inverse transition.[4] Controlling constraints such as density and strain between the aggregates allows us to guide the process by which granular matter either jams or remains loose. In the context of architectural materialization, this means manipulating geometry and structural stability. Here the structure is fabricated by compacting the aggregates into a limited volume, containing the rigid particles so they do not move and making the granules jam. Interlacing the granular matter with string and confining the aggregates prevents the outer layer from buckling and falling off. In other words, the volumetric boundary conditions are defined by the string placement and are responsible for the jamming phenomenon. Removing the string would lead to a chain reaction causing the complete collapse of the structure. The placement of the string guides the overall geometry and structural behavior of the material. Because imprecisions in the laying of the string can lead to an altered structural behavior, creating a domino effect and making the structure collapse, it is necessary to design and build using the precision of computational tools. On the one hand, this process informs the pattern of the string in relation to the jammed volume; and, on the other hand, robotic fabrication makes it possible to link the digital model of the string accurately and efficiently with its physical realization.

DESIGNING WITH STRING

Since the clustering of the aggregates is difficult to predict, one cannot precisely estimate the distribution of stress along the force chains. To handle this uncertainty, redundancy is an important factor for the layout of the string pattern. As a basic established rule, the string should always be in tension, to assure the structural integrity of the fabricated artifact, with as little string as possible. Therefore, the string pattern is based on circles due to the their optimal proportions between area and circumference. If a circular string loop is placed on a layer of aggregates and then compressed, the aggregates will try to spread equally in all directions. The string holds the aggregates in place. The string circles are deployed in a layer-based arrangement, forming horizontal string patterns. In fabricating the tower, the most important parameters were the diameter of the circles (figure 37.3a), the planar distribution of the circles, and the distance between the layers (figure 37.3b—c).

Iterative physical experiments (figure 37.4) made it possible to predict and assess the probability of buckling and failure at the outer surface and to react with a corresponding string pattern and tensile reinforcement. There is a direct correlation between the distance between layers and the circle radius: smaller circles require decreasing the distance between layers. The diameter of the circle further informs the possible geometries, particularly in convex forms. The curvature of a convex outer surface cannot be smaller than the radius of the respective string circles, and the smallest circle diameter found to be efficient was approximately eight times the average particle size, while the maximum distance between individual layers also relates to the average particle size.

Varying the layer-based string pattern from layer to layer allows for complex volumetric geometries (figure 37.3). This variation, however, always has to take into account the pattern of the adjacent, upper and lower, layers in order to guarantee a continuous

MISAPOR

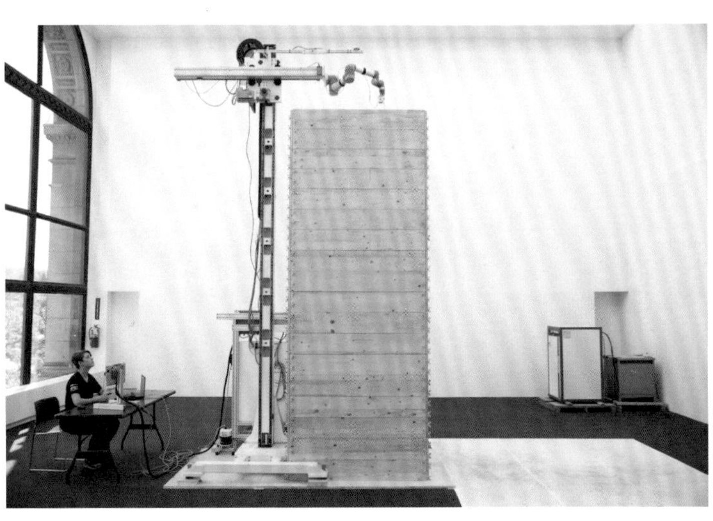

reinforcement in all directions. For example, a cantilever requires decreasing the distance between its layers, whereas a conical shape requires adding circles with an increasing layer area. Having a constant number of circles for all layers, while compensating for geometrical variation only by enlarging the overlap between the circles, results in a string pattern that is too dense, cutting off the connections between the aggregates and thus leading to a weaker structure.

The experiments conducted revealed that the larger the distances between circles, and the subtler the curvature of the boundary condition and the cantilever of the outer form, the greater the structural integrity. Depending on the global shape, the string pattern may have fewer circles in the center of a single layer for flat, vertical geometries or a denser string pattern close to its outer boundary, whereas geometries with high curvatures and cantilevers require a dense string pattern throughout.

TYING STRING AND ROCK

The string pattern is not alone in affecting the overall structural performance of the fabricated structures; the raw material also has a major impact. The aggregate size affecting the resolution of the string pattern has already been mentioned. Factors such as geometry and surface friction of the aggregates, as well as material characteristics of the string, also had to be evaluated since they influence the way the composite binds. Since Rock Print was in an indoor exhibit, it was necessary to exchange heavy crushed rock for lighter aggregates. Glass foam aggregates for self-insulating concrete from Misapor[5] proved to be a good choice in terms of their form and coefficient of friction. By choosing a size range of 16—32 mm, it was possible not only to achieve a suitable resolution for the structure but also to balance the aggregate size and the layer height with an appropriate fabrication speed. Just as with the size and the friction attributes of the aggregates, the string also has to have certain properties. It has to be flexible to be able to form into the desired pattern, and it has to be inelastic to fulfill the function of tensile reinforcement. Rock Print used a 3 mm cord[6] made of a combination of polyester fibers and recycled material from the textile industry.

Rock Print was robotically fabricated in situ, within the exhibition space. The setup consisted of a modular, two-axis gantry system coupled with a lightweight robotic arm, a string-laying end-effector, and an incrementally assembled box. The gantry was used to position the robot at the desired location, whereas the robot was used to lay the string according to the pattern blueprint. The size of the box was equal to the work envelope of the robot, measuring 1.2 × 1.5 × 4 m, allowing use of the full reach of the machine. The box consisted of a frame onto which 20-cm-high segments were attached consecutively. The fabrication process was split into two phases. The first consisted of fabricating the structure inside the box. During the assembly (figure 37.5a—d) the robot placed a layer of string, then aggregates were manually deposited by the team. A concrete compactor, applied at every second layer, assured a correct packing of the aggregates during the construction. New frame elements were added to the box after every 20th layer. This procedure was repeated until the full structure was assembled. The second phase consisted of releasing the box and brushing the support material to reveal the final structure (figure 37.5e—h).

PULLING THE STRING

Rock Print was designed and built to allow the deconstruction of the structure by simply reversing the fabrication process, separating the string from the aggregates. Since the structure is fabricated by layer, from bottom to top, the structure was simply dismantled by unwinding the string network with a pulley system and an electrically driven spool (figure 37.6). For an uninterrupted deconstruction, the string layout has to be continuous during the entire

Figure 37.6
Dismantling of Rock Print by unwinding the string reinforcement with a spool and a power drill.

fabrication process. For Rock Print in Chicago a string for each leg was thus connected to a single string for the upper part of the structure. The speed of the unwinding was adjusted so the aggregates flew off the structure without breaking the string. Because the unwinding happens fast, the artifact remains stable during the whole unravelling, so the jammed structure doesn't collapse and tangle the string. After 10km of string was unwound, layer after layer, for two hours, the first Rock Print was restored to its original raw material: a pile of rock and a spool of string.

CONFIGURING AND RECONFIGURING WITHOUT WASTE

Rock Print demonstrates the fascinating potential of jammed architectural structures. With a growing understanding of the interrelation of aggregates and string, jammed structures can be shaped into nonstandard structures. Furthermore, the material phenomenon works with a large variety of aggregates and gravel types, making it possible in the future to use local materials more widely in fabrication. Since jammed architectural structures can be fully returned to their original state, this construction system presents a shift to a granular architectural material approach, and enables the infinite reconfiguring of the composite into different architectural forms. This idea is supported by a closed life cycle that begins by assembling readily available and cheap building material. As such, Rock Print formulates a striking response both to the ideal of a consumer-oriented (customized) architecture and to the call for sustainability. It simply produces no waste and is 100% recyclable.

ACKNOWLEDGMENTS

Collaborators: Sara Falcone, Jared Laucks, Lina Kara'in, Michael Lyrenmann, Carrie McKnelly, George Varnavides, Stéphane de Weck, Hannes Mayer, Dr. Jan Willmann.
Selected experts: Prof. Dr. Hans J. Herrmann and Dr. Falk K. Wittel (Institute for Building Materials, ETH Zurich), Prof. Dr. Heinrich Jaeger and Kieran Murphy (University of Chicago).

Selected consultants: Walt + Galmarini AG.

Sponsors: Pro Helvetia Swiss Arts Council, swissnex, MISAPOR Beton AG.

Photography: Gramazio Kohler Research, ETH Zurich, and Self-Assembly Lab, MIT.

NOTES

1 Chicago Architecture Biennial 2015, "The State of the Art of Architecture," October 3, 2015—January 3, 2016, guidebook (2015).

2 P. Aejmelaeus-Lindström, J. Willmann, S. Tibbits, F. Gramazio, and M. Kohler, "Jammed Architectural Structures: Towards Large-Scale Reversible Construction," *Granular Matter* 18, no. 2 (2016).

3 H. M. Jaeger, "Celebrating Soft Matter's 10th Anniversary: Toward Jamming by Design," *Soft Matter* 11 (2015): 12—27.

4 C. Song, P. Wang, and M. A. Makse, "A Phase Diagram for Jammed Matter," *Nature* 453 (2008): 629—632.

5 Glass foam aggregates: MISAPOR 16/32, accessed August 31, 2015, http://www.misapor.ch/EN/cellularglass/

6 Polyester-reinforced string: Usacord AG, accessed August 4, 2016, http://usacord.ch/1485/Schn%C3%BCre/Deutsch/Mischfaser.html/#Diverseschnuere

38
Institute of Isolation

Lucy McRae with
Lotje Sodderland

Figure 38.1
Sound-absorbent foam, The Institute of Isolation, Anechoic Chamber.

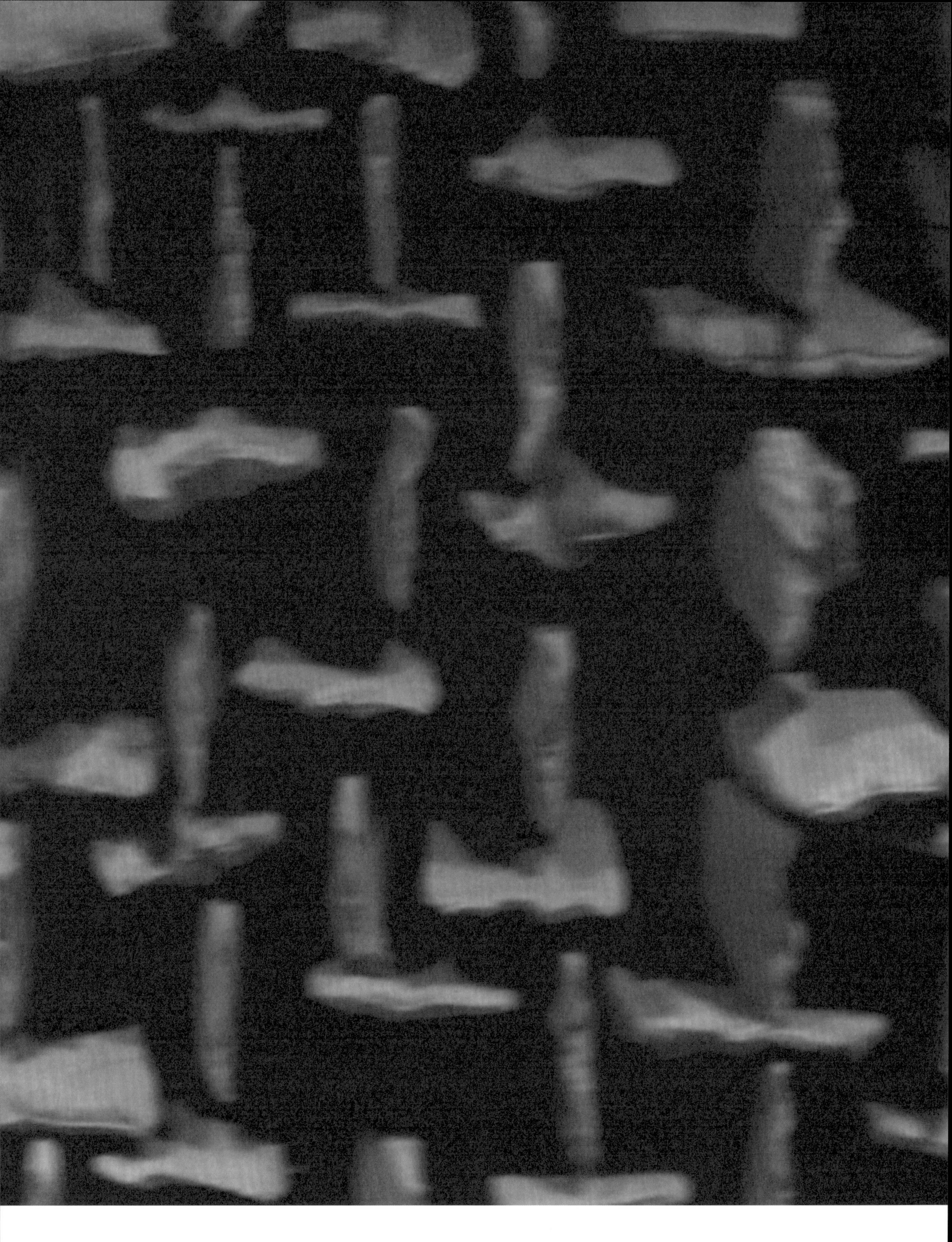

38
Institute of Isolation

Lucy McRae with
Lotje Sodderland

The Institute of Isolation imagines a research and training ground testing ways to prepare the body for far-future scenarios, like space travel. This observational documentary contemplates whether isolation, or more broadly "extreme experience," could be used as a gateway to train human resilience, or our ability to bounce back from the unknown.

The fictional institute is set in a near-future reality offering individuals the chance to optimize their bodies and test the effects that extreme experience could have on evolving fundamental aspects of human biology. Referencing developments in genetic engineering and medical technology, McRae examines how fundamental aspects of the brain and body might be shaped through intentional experiences of sensory deprivation and extreme isolation.

The film contemplates the design of isolation and the body's changing relationship with technology. The protagonist moves through a series of sensory chambers, spending time in an anechoic chamber examining the psychoacoustics of silence or in a microgravity trainer conditioning the body for possible life in space. These fictional locations inquire into the role buildings could have on altering human biology on an evolutionary scale.

As active matter technologies converge, how do we apply this thinking to a time when (1) we live light-years away from our original birthplace or (2) we have been designed from scratch in a petri dish and never experienced the hug of a mother?

Will hardwired senses like touch be genetically engineered? And what does this say about the mind? Perhaps we will need to evolve in very unusual ways and receive more intense experiences taking people beyond the expectations of themselves. Could programmable "isolation" sense and actuate human consciousness? We know that art doesn't give immediate answers, but presents conditions of possibility.

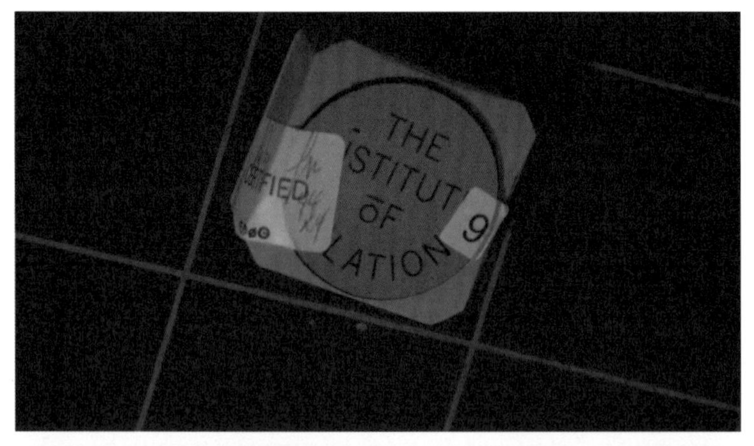

Figure 38.2 (top left)
The Institute of Isolation.

Figure 38.3 (top right)
Increased atmospheric pressure and oxygen saturation, The Institute of Isolation, Hyperbaric Chamber.

Figure 38.4–5 (bottom left and right)
Orthotic harness, The Institute of Isolation, Microgravity Trainer.

Figure 38.6
Orthotic harness, The Institute of Isolation, Microgravity Trainer.

Figure 38.7 (top)
Orthotic harness, The Institute of Isolation, Microgravity Trainer.

Figure 38.8 (bottom)
The Institute of Isolation, Training Circuit.

39

Sentient Spaces and Active Architectures

Meejin Yoon and Eric Höweler

Meejin Yoon and
Eric Höweler

On May 1, 1972 the US Army (in US patent 3766539) described an "automatic personnel intrusion alarm," utilizing infrared radiation as a sensor mechanism. The application described the need to protect strategic establishments from attack by infiltrators in various military and civilian contexts. Passive infrared sensors, operating in the spectral wavelength regions of 3—5 and 8—14 microns, were proposed for military and security applications, overcoming shortcomings of human "operators" who were subject to fatigue. In a civilian context, automatic doors were first introduced in the 1940s but did not utilize motion detection technologies until the 1980s.

Supermarkets were among the first building types to use active infrared sensing to allow shoppers to enter and exit a supermarket with their "hands free," in order to push shopping carts laden with groceries. This early "hands free" concept was intended to improve hygiene as well. Automatic flush urinals became common in public restrooms in the 1990s to consistently provide a "hygienic flush" without requiring the user to touch the fixture. In each case, sensor technologies were introduced into architectural spaces for security, convenience, or hygiene. These technologies augmented spaces and devices, replacing a human guard with a security system, assisting the operation of a door, and ensuring proper flushing without spreading germs. The sensor can enhance an experience, performing faster, safer, or more cost-effectively. Sensors have been introduced into architectural spaces for decades; they have moved beyond utilitarian exploration of sentient spaces, and active architectures are still evolving.

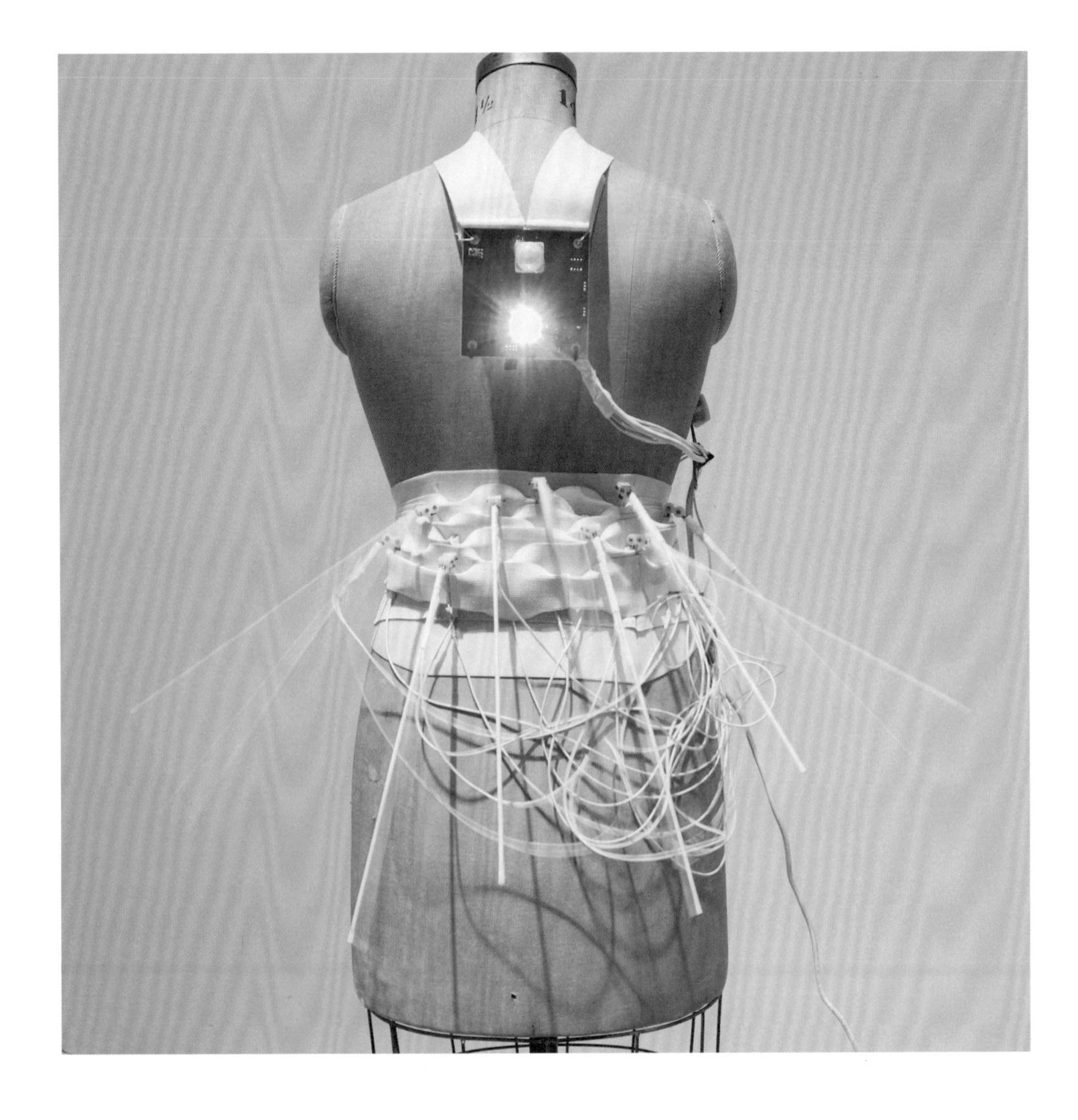

Figure 39.1 (left page)
Defensible Dress.

Figure 39.2
Defensible Dress soldering.

The Defensible Dress, designed and fabricated in 2001, is an active architecture for the body. The project uses sensor technologies and custom microcontrollers to define and protect one's personal space. The wearable is adaptable to inputs that reflect personal comforts and/or cultural norms. It was an early application of passive infrared (PIR) sensors for spatial definition but also provided a commentary on spatial encroachment in urban public spaces.

The Defensible Dress employed PIR, a microcontroller, and a series of actuated "quills" to define a personal space around the body. Each quill consisted of a hollow metal tube with a shape memory alloy (SMA) actuator within it. The wires, made of nickel titanium, dynamically contracted due to changes in the material's internal structure at the heated temperature. When the PIR sensor of the Defensible Dress detected the presence of another individual within the wearer's personal space, an electric current was sent to the SMA wire, causing a contraction like a muscle. Any personal space intrusion was met with a quill-bristling response. In performing this physical transformation, the Defensible Dress was an instrument that defined territories of private space. It signaled the potential for an active architecture to respond to social, gendered, and environmental contexts and stimuli

As a discipline, architecture has been concerned with the design and discourse of the built environment and its materials, technologies, systems, and histories. Formal analysis, proportioning systems, material use, and stylistic innovation are a few of architecture's discourses. Little has been written on the topic of the *use* of the building over time and the *behaviors* of its occupants.

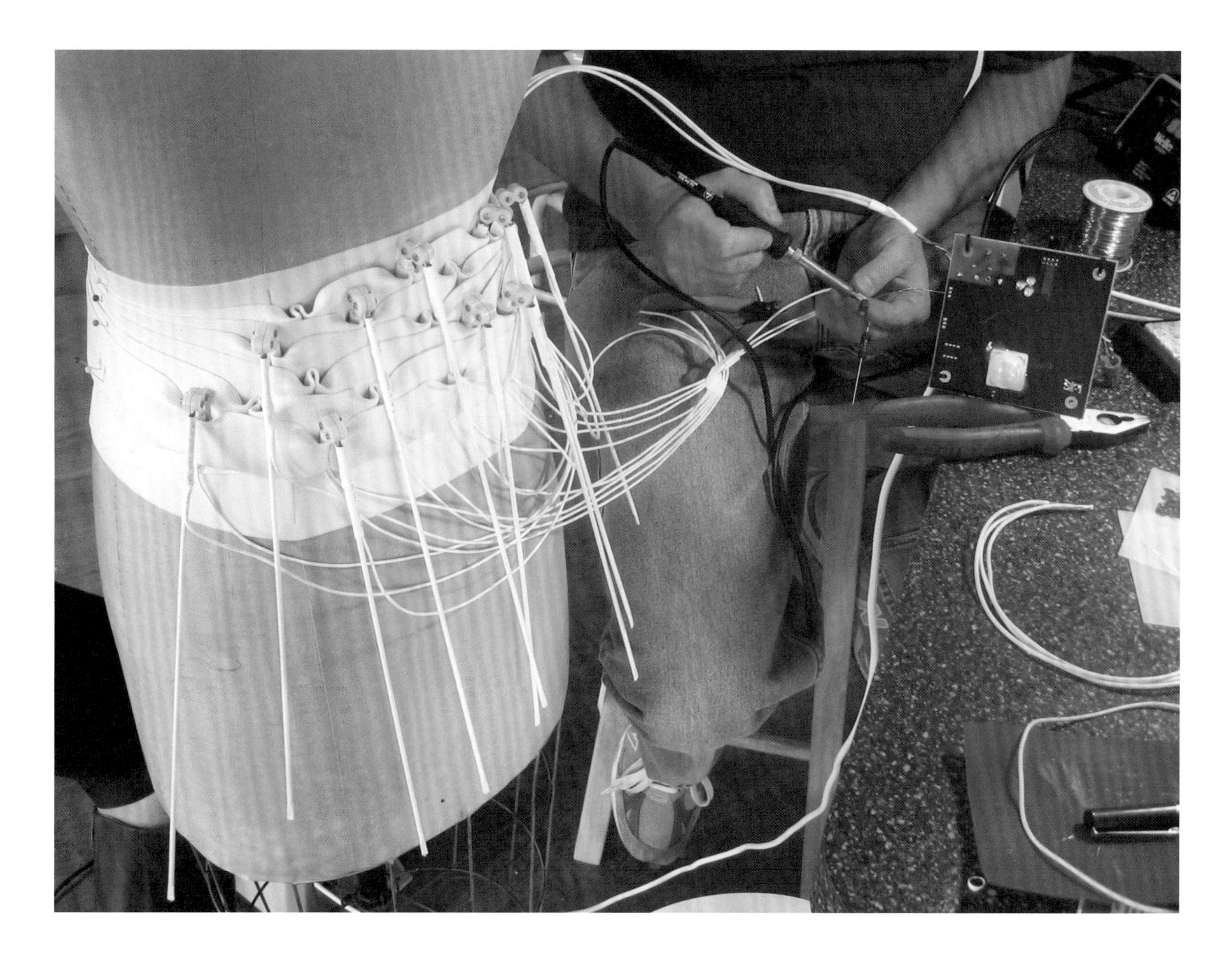

Figure 39.3 Aviary.

The so-called "post-occupancy" of a building, after the project has been published, posted, liked, and shared, is underinvestigated. How do users, occupants, and passersby interact with a building? What behaviors do they exhibit? How does architecture condition behavior? How are a public realm and its associated behavioral codes conditioned by the architecture that defines it? Could a sentient and active architecture promote certain behaviors and discourage others?

Two contemporary projects test these questions and explore the shift in emphasis from buildings to behavior. Aviary is a kinesthetic surround-sound speaker installation consisting of 30 touch-sensitive glass tubes. Aviary relies on integrated sensors programmed to produce a series of responses to light and sound. The installation consists of a radial array of glass tubes that form a circular "grove" of lighting elements. Capacitive-sensing film integrated within the tubes records "touch events." A casual touch creates a vertical burst of light and sound, while a sustained hold slowly fills the column with light and sound. Sliding up and down the pole causes the sounds to be blended in a unique and dynamic sound effect. A quick slide up the pole causes a burst of light to float up to the top and migrate to adjacent poles. A downward slide creates a surround-sound experience where the bird call is played through the networked speakers to create the impression of a bird flying around its users. The project explores touch gestures familiar through smartphones and tablets at a larger scale in public space.

Each of the 30 extruded glass tubes houses a speaker, a domed sound diffuser, a networked computer, a four-way LED core, and a transparent layer of conductive film (TCF) to act as a capacitive sensor. The film enables the interactivity of the piece, as it is specifically reactive to the electric current present in the human body. As

Touch Duration

Aviary responds to the touch with a display of light and sound effects that evokes a bird in flight or a bird's natural habitat. A casual touch creates a vertical burst of light, while a sustained hold slowly fills the column with light

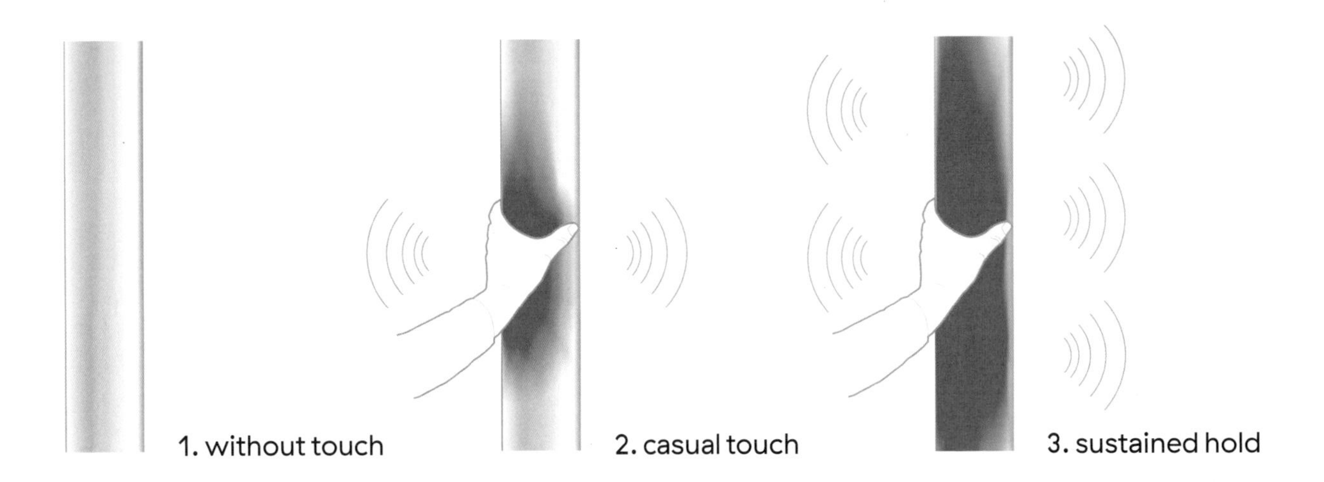

Touch Duration

Each pole has a unique set of related sounds based on a specific bird, with bird songs at the top of the pole and abstracted musical interpretations at the bottom. Sliding up and down the pole creates a dynamic blend of the three channels of distinct audio content

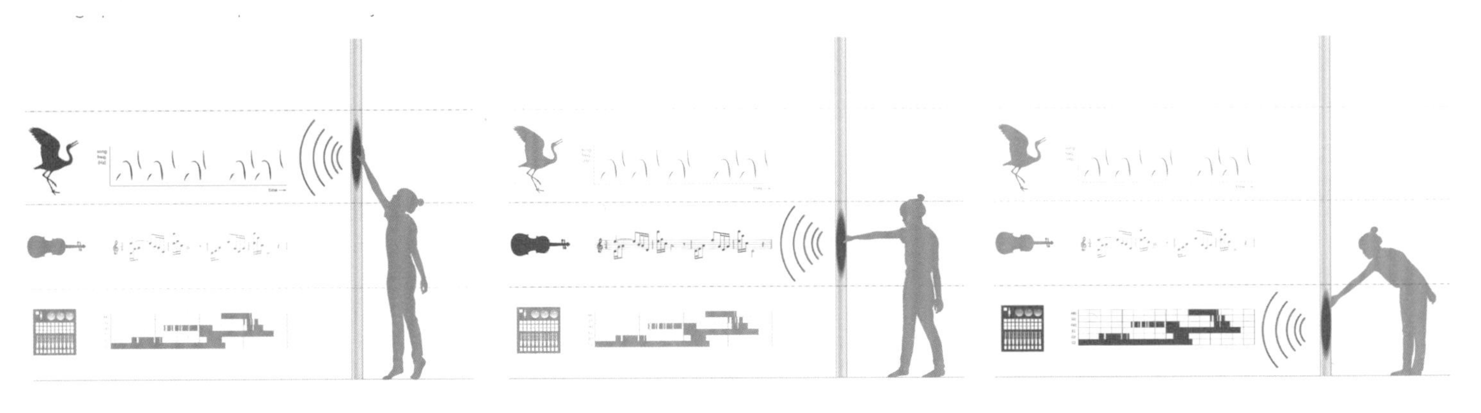

1. Touching top – Bird call

2. Touching middle – Blended Sound

3. Touching bottom – Abstracted musical interpretations

Figure 39.4 (left page, left)
Aviary close-up.

Figure 39.5 ((left page, right)
Aviary process image.

Figure 39.6
Aviary touch-sensing diagram.

Figure 39.7 Swing Time.

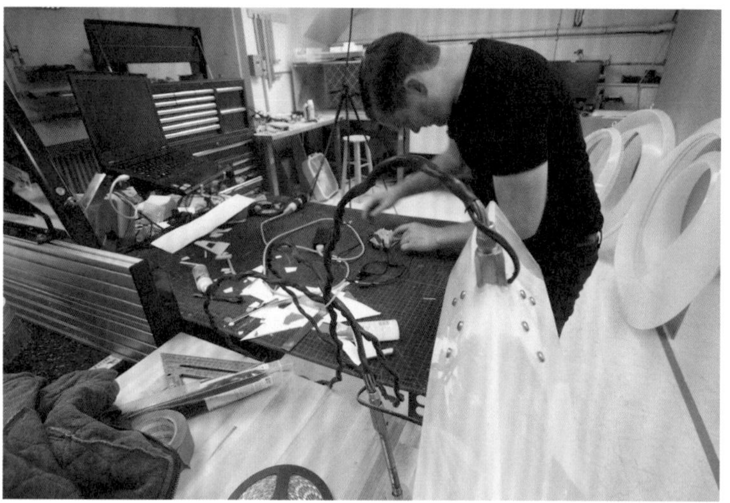

Figure 39.8 (top)
Swing Time roto-molding process.

Figure 39.9 (bottom left)
Swing Time electronics.

Figure 39.10 (bottom right)
Process of making Swing Time.

Figure 39.11 (right page)
Swing Time energy production
and consumption.

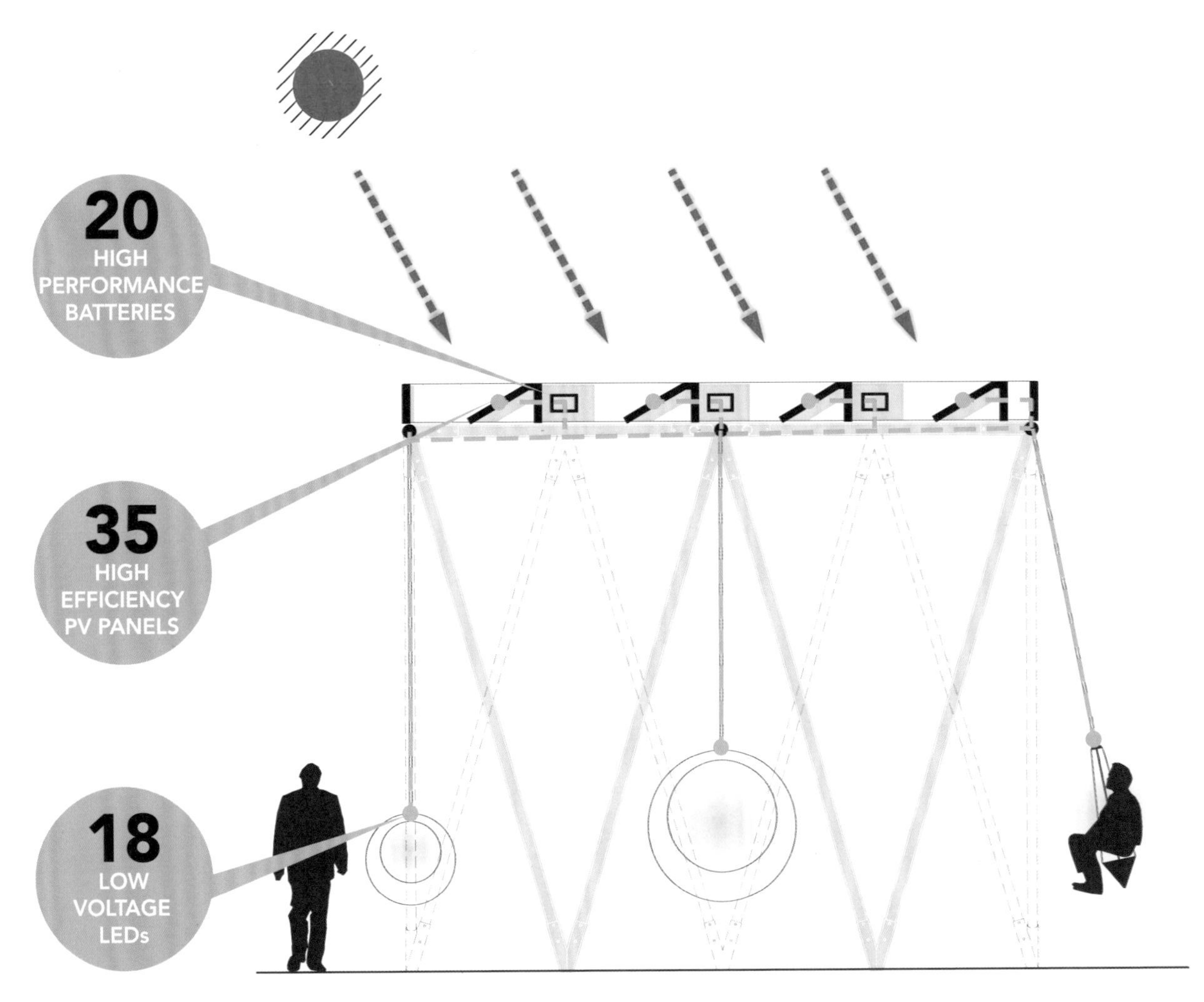

SECTIONAL VIEW

1/2" = 1'-0"

42°

OUR OVERHEAD CANOPY ORIENTS SOLAR PANELS TOWARD THE SOUTH, PERPENDICULAR TO THE SOLAR ANGLE FOR OPTIMAL SOLAR EXPOSURE

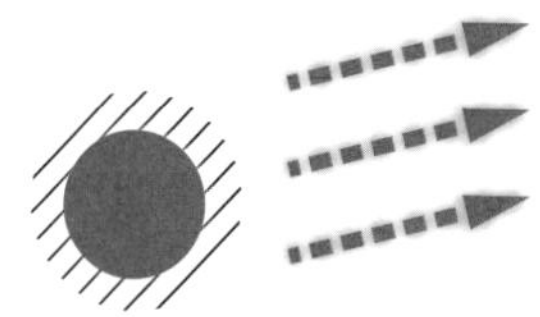

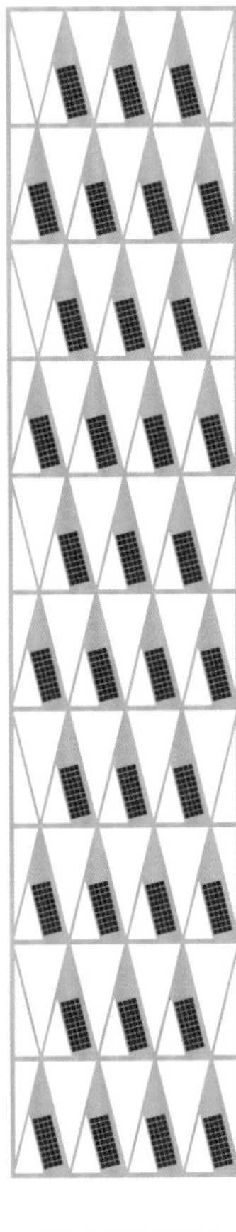

3.84 hrs

THERE ARE 3.84 SUN HOURS PER TYPICAL DAY IN MA

9.6 AMP HOURS

EVERY DAY EACH OF OUR PV PANELS CAN GENERATE 2.5 AMPS AT 12V, WHICH IS ENOUGH ENERGY TO POWER 1.3 – 1.9 SWINGS

PLAN VIEW

1/8" = 1'-0"

336 AMP HOURS

OUR 35 PV PANELS COLLECTIVELY GENERATE 336 AMP-HOURS PER DAY

4.6 kWh / m2 / day

A STANDARD PV SYSTEM IN BOSTON CAN GENERATE AN AVERAGE OF 4.47 KWH / M2 / DAY

3.35 JAN
4.23 FEB
4.76 MAR
4.92 APR
5.34 MAY
5.43 JUN
5.62 JUL
5.65 AUG
5.13 SEP
4.65 OCT
3.13 NOV
2.94 DEC

a person touches the film, their current inhibits the flow of current through the TCF. A lack of current in the TCF triggers the network at the impact location. The touch sensing and gesture recognition provide a wide range of signal and response modes that engage users in a range of modes, from casual play to musical and sonic composition.

Kinesthetic play in public space was further explored through Swing Time, which expands on lessons from the Defensible Dress and Aviary to create a sentient environment for children and adults with 20 interactive and illuminated swings. The swings are each equipped with RGB LED lighting to produce a glowing effect. Each swing contains a microcontroller with an accelerometer to sense when the swing tilts to a certain degree of inclination. When in use, the swings glow with a blue light until the accelerometer detects specific thresholds of inclination, at which the LEDs change color to create a pink "blushing" of the swing, giving feedback to the user. The basic sensor and response mechanisms, coupled with the inviting geometry of the large LDPE swings, create a playful environment. The accelerometer in the swing encourages adults and children to swing higher, let down social norms, and enjoy themselves. Architecture, as manifested in the Swing Time project, seems capable of producing social playfulness and kinetic pleasure.

The emergence of sentient architecture coincides with the proliferation of low-cost sensor technologies and smart devices. The so-called Internet of things (IOT) promises to connect appliances, home entertainment systems, and heating, ventilating, and air conditioning systems, using control devices to personalize and tailor spaces to user preferences. The Nest thermostat knows when we are home and what our preferences are, cooling our living room to our personal thermal specifications just in time for our arrival home. Presumably the sentient IOT future will improve our lives through a more efficient and tailored set of smart spaces.

The Defensible Dress, Aviary, and Swing Time apply material and behavioral thinking with sensor technologies to produce augmented spatial and experiential effects. These effects serve as commentary on social norms or environmental conditions, but also act as inducements to playful interactions and affordances to social behaviors and actions that may not be anticipated. The design research involved in these projects suggests new ways to make architecture that is dense with information, interactive in real time, and constructive of new modes of public space. Sentient and active architectures will provoke new forms of architectural experiences and social behaviors that are only beginning to emerge.

40
Stagecraft and
Architecture
Simon Kim and
Mariana Ibañez

Simon Kim and
Mariana Ibañez

In response to advancing technologies and smart matter that is programmable, intelligent, and responsive, the fabric and tectonics of our buildings and cities are poised to become sentient and dynamic. The built and artificial environments that we inhabit will exhibit agency and motivations that break the subject-object hierarchies hitherto prevalent in architecture. The practice of architecture and urbanism, it is becoming clear, must accept this paradigm shift, learning to engage active matter in ways that still hold to their histories and traditions of culture and meaning.

New forms of life, and new media, have produced an abundance of matter that is now positioned at the core of architecture. Architecture as a discipline is in transformation within a world that is synthetically active and sensate. As such, the architectural discourse on the object, the lineaments of space, and the affective can become expanded into dimensions of retreating and projecting time—buildings that respond in duration and with their own agencies. If the house is a machine in which to live, as Le Corbusier put it, then its machinery should be tested in a life performed at the highest levels: those of dance, dramatic arts, and music. And if architecture, as a future product of the bio-electromechanical age, is to have any merit, it should perform as well as respond under the strongest of human expression in a way to create new forms of meaning.

At the University of Pennsylvania and at Harvard, we have worked to expand the core discipline of architecture into fields of new technologies and performance. Both of these domains have created vital and growing edges from which architecture, within an institutional impetus of research, experimentation, and progress, may project itself forward. By establishing Immersive Kinematics, a research group linked with both the Modular Robotics Lab of Penn Engineering and a broader artist community, we have sought out and established a network of collaborators. These "partner instigators" in the performing arts are paired with institutional design researchers and applied scientists in a tradition stretching from Experiments in Art and Technology back to the Bauhaus.

These partner instigators include the Opera Philadelphia / American Repertoire Council, the Philadelphia Museum of Art, award-winning composer Lembit Beecher, the Dufala Brothers with the Recycled Artist in Residency (RAIR) program, the Institute of Contemporary Art (Philadelphia), and others. Many of these partner instigators have been attracted by mutual curiosity: the American Repertoire Council approached Immersive Kinematics to discuss the future of opera, from which a pilot project was initiated. The directors of Carbon Dance invited us to develop a new work after seeing a TEDx presentation. The RAIR artist-in-residency at Revolution Recovery provided an opportunity to develop an exhibition that expanded on design issues of transformation of life cycles and reclaimed materials.

The products of these investigations have been presented at the Museum of Modern Art (New York), the Annenberg Center's Prince Theater, the Slought Foundation, Traction Company, and the Institute of Contemporary Art (Philadelphia). Each of these investigations was predicated on a paradigm shift that moves the cultural production of design and technology from technocratic fascination toward a new cultural exchange among humans and the environment.

In what is perhaps the most important component of stage and theater, the design of choreography and narrative must find its artful expression in live performance to evoke a reaction from the audience. What is real and what is perceived become separate conditions, with the latter becoming primary. A nonhuman dancer, as produced for *Science per Forms*, for example, has the same charge as its human counterparts, namely to emotionally engage an audience. What this requires is a fundamental shift away from engineering to an artistry for which human actors and performers have trained for years.

Science per Forms

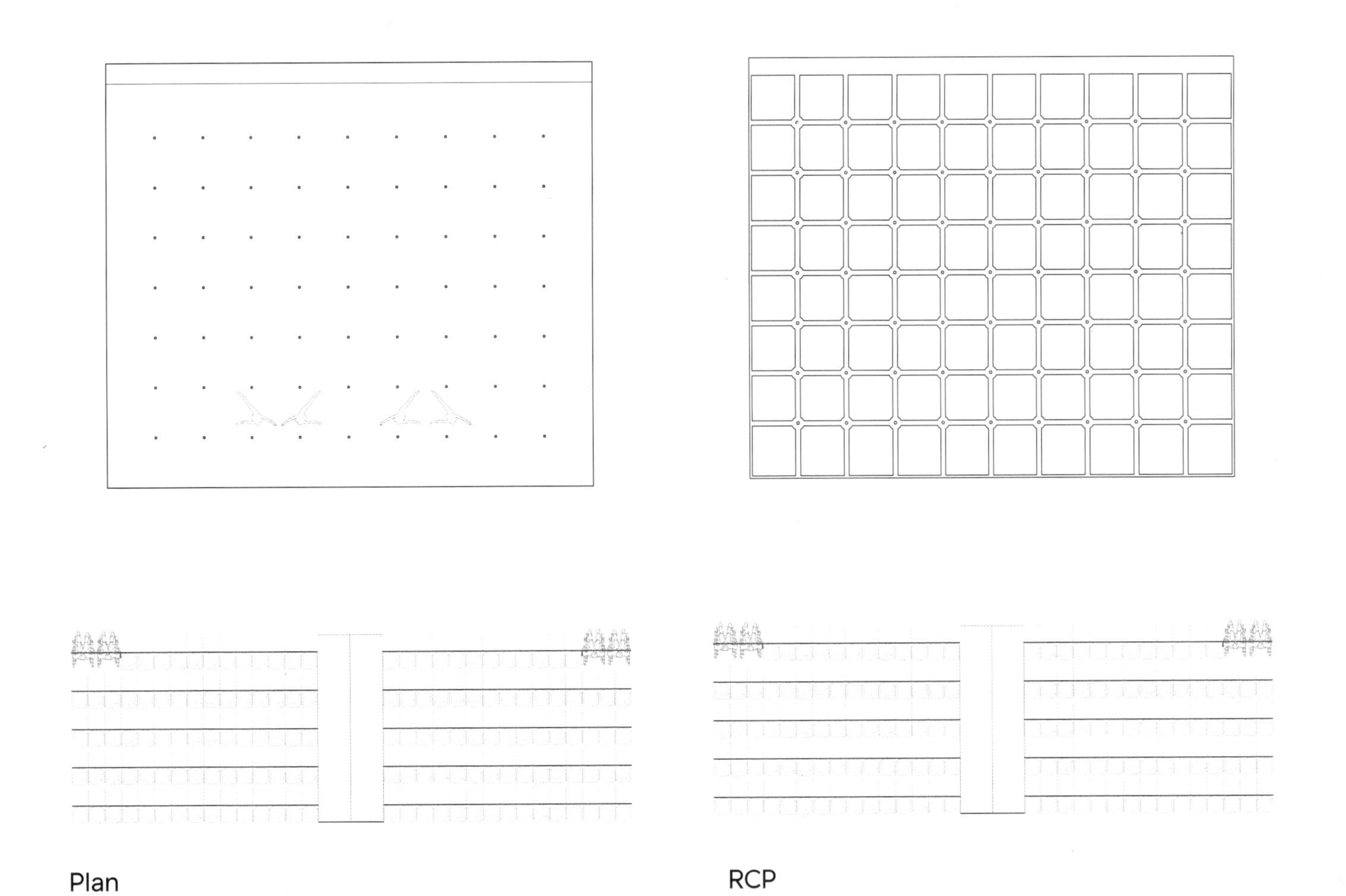

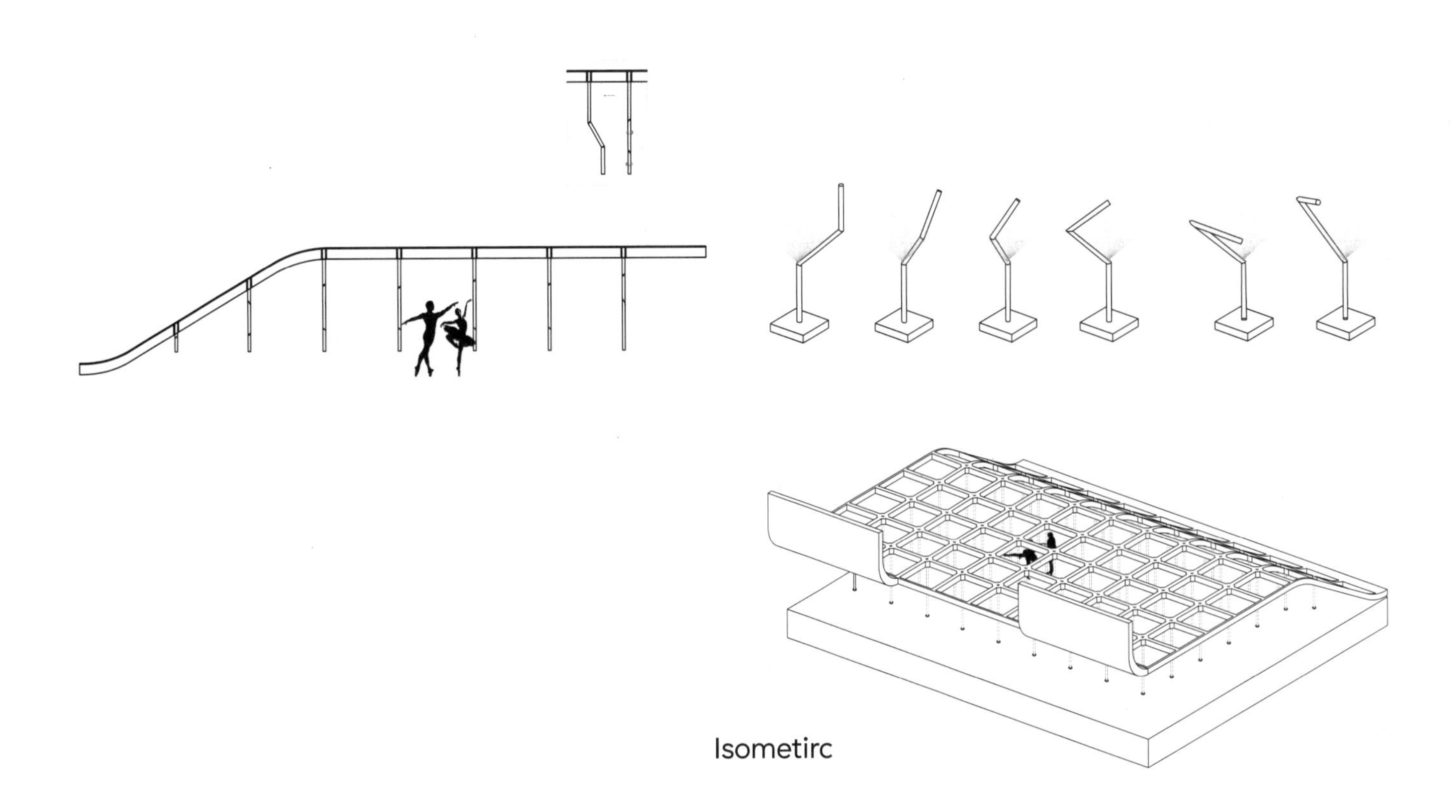

Figure 40.1
Science per Forms, diagrams.

Figure 40.2
Science per Forms.

Figure 40.3
Science per Forms.

Figure 40.4
Science per Forms, interactive projection.

Figure 40.5
Science per Forms.

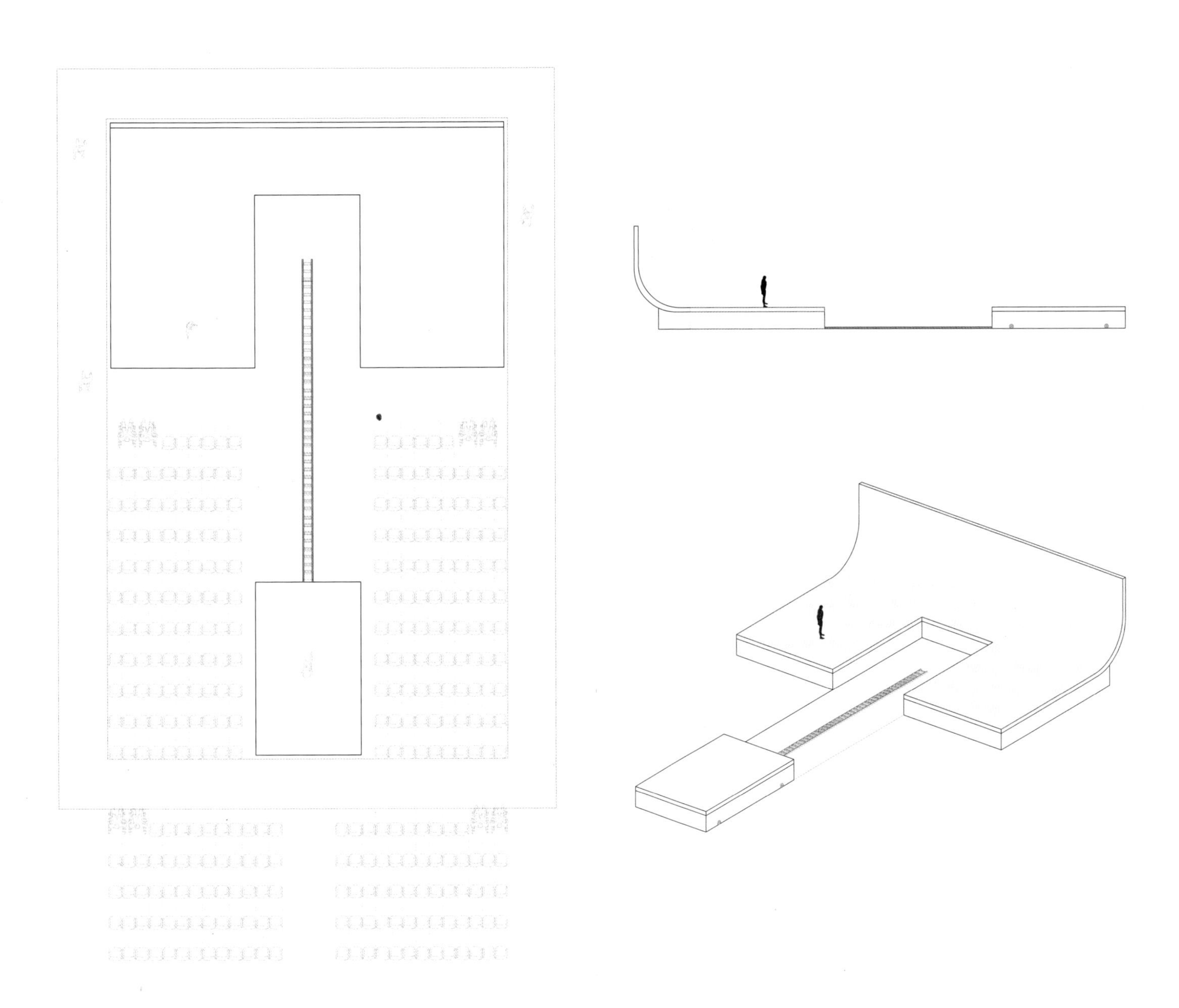

Figure 40.6
Orpheus and Eurydice, diagrams.

Figure 40.7 (top)
Orpheus and Eurydice, “Don't look back! Face recognition.”

Figure 40.8 (bottom left)
Orpheus and Eurydice, Serpent.

Figure 40.9 (bottom right)
Orpheus and Eurydice, Taming the Furies—lute.

For example, the modular arms, developed with Mark Yim's Modular Robotics Lab at the University of Pennsylvania, can be joined together to create longer chains and ranges of motion. The "CKbots" connect structurally via face-matching plates and magnets. These magnets also allow for signaling from a master control unit that permits a form of movement desired for dance and formal gestures. Tethered to a MIDI-based receiver, the modular arms are able to toggle to another mode of interacting with sound. Waves, ripples, and twitches are seen in these movements.

A nonhuman agent that is fully operational, with close tolerance in input data and actuation, must also be trained to be purposeful and expressive in its movement. All of its high fidelity does not matter if it is unable to convey meaning with its movements. Hades' command not to turn around, in *Orpheus and Eurydice*, is shared with the audience as the actors navigate the theater. Their gaze is monitored throughout the performance.

Fundamentally, the successful nonhuman performer, in the context of theater, is one that conveys artful experiences to a human audience. In this role, the nonhuman performer is subject to all the requirements of a human performer. Its movements should show abruptness or languor to indicate a particular emotion. The geometries of its limbs and body should convey intentionality.

What will be of keen interest is to change the parameters of the performer-audience relationship to one where the audience itself is nonhuman. A compelling question is what form of dance a human performer would produce for an artificially sentient audience, as well as what ethos would be produced by synthetic beings that desire theater. The answers to this question will only be answered as new sentience and new intelligence emerge and evolve.

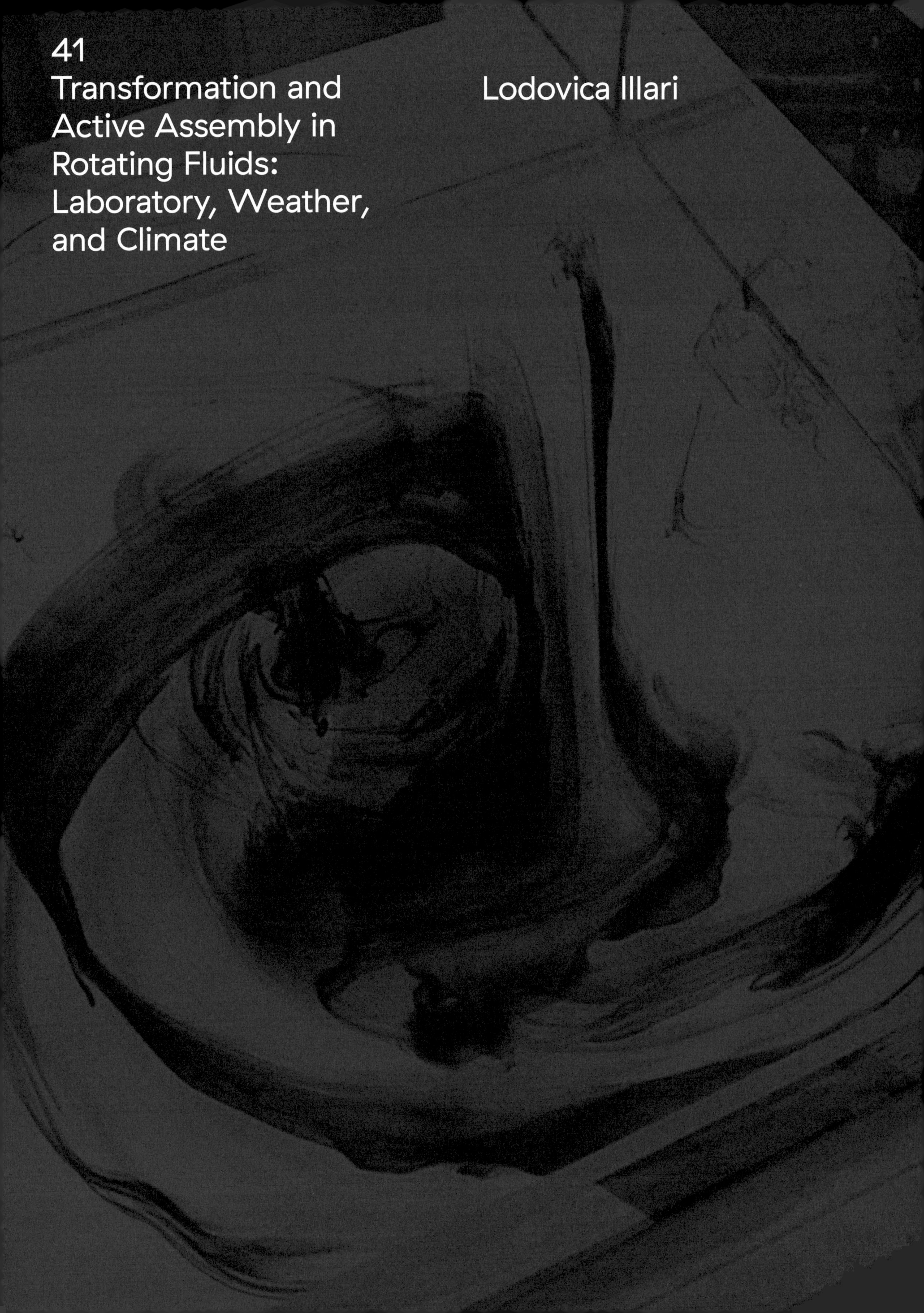

41

Transformation and Active Assembly in Rotating Fluids: Laboratory, Weather, and Climate

Lodovica Illari

41
Transformation and Active Assembly in Rotating Fluids: Laboratory, Weather, and Climate

Lodovica Illari

The general circulation of the atmosphere is extraordinarily complex, comprising many interacting components. A satellite image of clouds over the American continent, for example, as in figure 41.1, shows several different scales of motions embedded within one another—from the planetary scale of several thousand km down to cloud clusters on a scale of tens of kilometers. While the processes at work are very complex, the underlying beauty and order in this figure suggest that organizing principles must be at work.

Where does this order come from? We will see that earth's rotation plays a very important role: it "transforms" fluid behavior by introducing angular momentum constraints. Fluids on earth behave very differently than everyday, nonrotating fluids, because on sufficiently long time scales and broad length scales they "feel" earth's rotation. This branch of science is called geophysical fluid dynamics. It emerged as a branch of applied mathematics in the 1950s as scientists explored the laws of mechanics and thermodynamics that shape the "natural" fluids that make up the atmosphere, ocean, and climate of our planet.

To complement observations and the theoretical (and numerical) analysis of natural fluids, at MIT we have developed an approach to teaching that makes use of simple laboratory experiments of rotating fluids. These illustrate the "transformation" of fluid properties by rotation and the "assembly" of a few ingredients to yield emergent properties that have an uncanny resemblance to meteorological phenomena.[1] In our approach, which we call "Weather in a Tank," the general circulation of the atmosphere is seen to emerge from the "mix" of a few key planetary "ingredients" that are illustrated using simple laboratory experiments, as shown schematically in figure 41.2. A comprehensive guide to the "Weather in a Tank" experiments and how to obtain the apparatus required to carry them out can be found in Illari and Marshall (2006).[2]

1. MATERIAL TRANSFORMATION IN THE DYE-STIR EXPERIMENT

To illustrate the effect of rotation on fluids, we contrast the evolution of fluid flow in two simple, complementary experiments. We place a tank of water on a rotating platform (which we call a "turntable") to which we attach a camera that observes from above, as illustrated

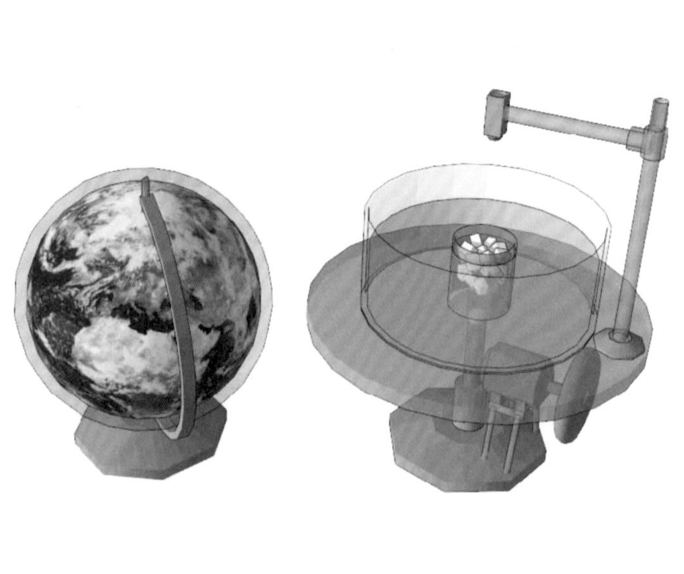

Figure 41.1 (left page, left)
Satellite image of cloud cover from the GOES East satellite on April 14, 2015 (NASA-GSFC Project).

Figure 41.2 (left page, right)
Rotating-fluid experiments explore the interplay of rotation and differential heating/cooling which result in flow patterns that are close analogues of the large-scale circulation of Earth's atmosphere; we call the system "Weather in a Tank."

Figure 41.3 (bottom left)
Left: A non-rotating tank filled with water standing on a stationary turntable. The scene is viewed from above via a camera and simultaneously from the side, making use of a mirror tilted at an angle of 45 degrees. The water is gently agitated by hand and then colored dye (food coloring) is introduced using a pipette to visualize the flow. Right: dye stirred into the non-rotating tank moves every which way, and the two colored clouds of dye mix in a "random" three-dimensional manner.

Figure 41.4 (bottom right)
Left: the same tank is now on a rotating platform and the view is recorded in the rotating frame via a co-rotating camera (as in figure 41.2). In the rotating experiment and before the fluid is agitated, the tank is spun for 10—15 minutes or so into what is known "solid body" rotation. In this state the water moves as a solid body and is at rest when viewed in the rotating frame. Right: on the rotating turntable, dye patches mix through horizontal motion, which sweeps the dye into beautiful interleaving streaks and filaments in the horizontal. The streaks, however, remain columnar in the vertical; see figure 41.5, where the vertical structure is revealed more clearly.

in figures 41.2 and 41.3. When the table is rotating, the camera rotates with it, yielding a view in the rotating frame. Two identical experiments are carried out, except that in one the turntable is kept stationary, and in the other it rotates at a constant speed.

In each experiment we gently agitate the water with our hand and then add some droplets of food coloring to visualize the flow. The water currents created by our manual "agitation" result in the colored dye being stirred into the fluid—hence we call these "dye-stir" experiments. After a few minutes, a striking difference between the nonrotating and rotating experiments is apparent. In the nonrotating tank, the dye disperses in all directions much as we might intuitively expect. But when dye is stirred into the rotating body of water, very different flow patterns are revealed, as can be seen in figure 41.4. Vertical streaks of dye can be seen that are drawn out into exotic horizontal patterns by fluid motion which only varies in the horizontal and not in the vertical: "curtains" of dye wrap around one another into beautiful, highly filamented structures.

The vertical columns are known as "Taylor columns" after G. I. Taylor[3] who discovered them. They are a result of the vertical rigidity parallel to the rotation vector imparted to the fluid by rotation. The water moves around in columns that are aligned parallel to the rotation vector. Since the rotation vector is directed upward, the columns are vertical. Thus, we see that rotating fluids are not really like fluids at all; a remarkable "transformation" has occurred!

A higher-resolution image from a similar experiment reveals the dye curtains even more clearly: figure 41.5 shows the ribbon-like structure of the dye filaments as the dye slowly falls through the water column while being stirred horizontally.

In summary, the fluid is stiffened in one direction by rotation, so that instead of spreading in three dimensions, it moves predominantly in two dimensions as columns parallel to the axis of rotation. In short it is fluid-like in the horizontal, but more like a solid in the vertical. Geophysical fluid dynamics focuses on just such effects and is very different from, for example, the kind of fluid dynamics studied in departments of aeronautics. There fluid phenomena occur at such small scales and are so rapid that rotation has no effect.

Figure 41.6 shows a visual comparison between the dye-stir experiment of figure 41.4 and Jupiter's red spot. Jupiter, like Earth, is a rapidly rotating planet, making a complete rotation in only 9.8 Earth hours though it has a radius that is more than 11 times that of earth. Because of this rapid rotation, Jupiter's atmosphere is under very strong rotational control. The connection with the tank experiment is visually striking. The experiment was performed with a very simple setup, yet patterns of flow resemble those of the circulation on Jupiter. The experiment is not intended to be a simulation of Jupiter's atmosphere but rather an illustration of the importance of rotation on the behavior of its atmosphere.

non-rotating

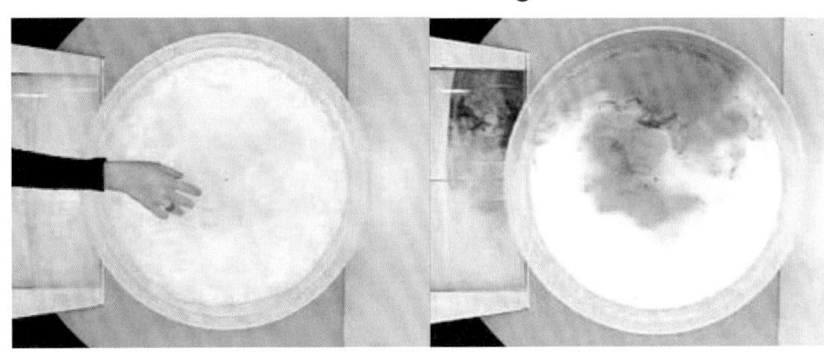

rotating

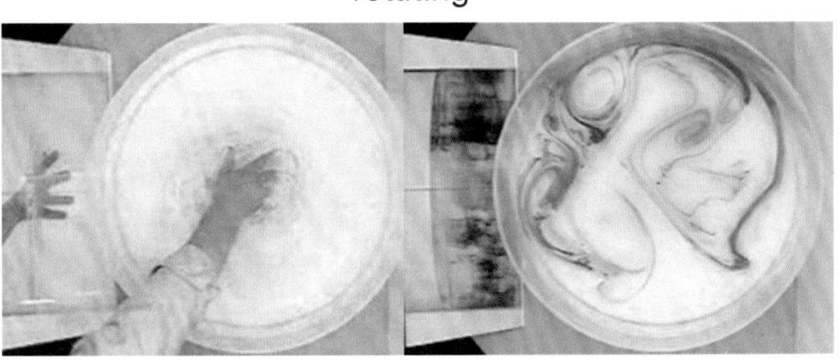

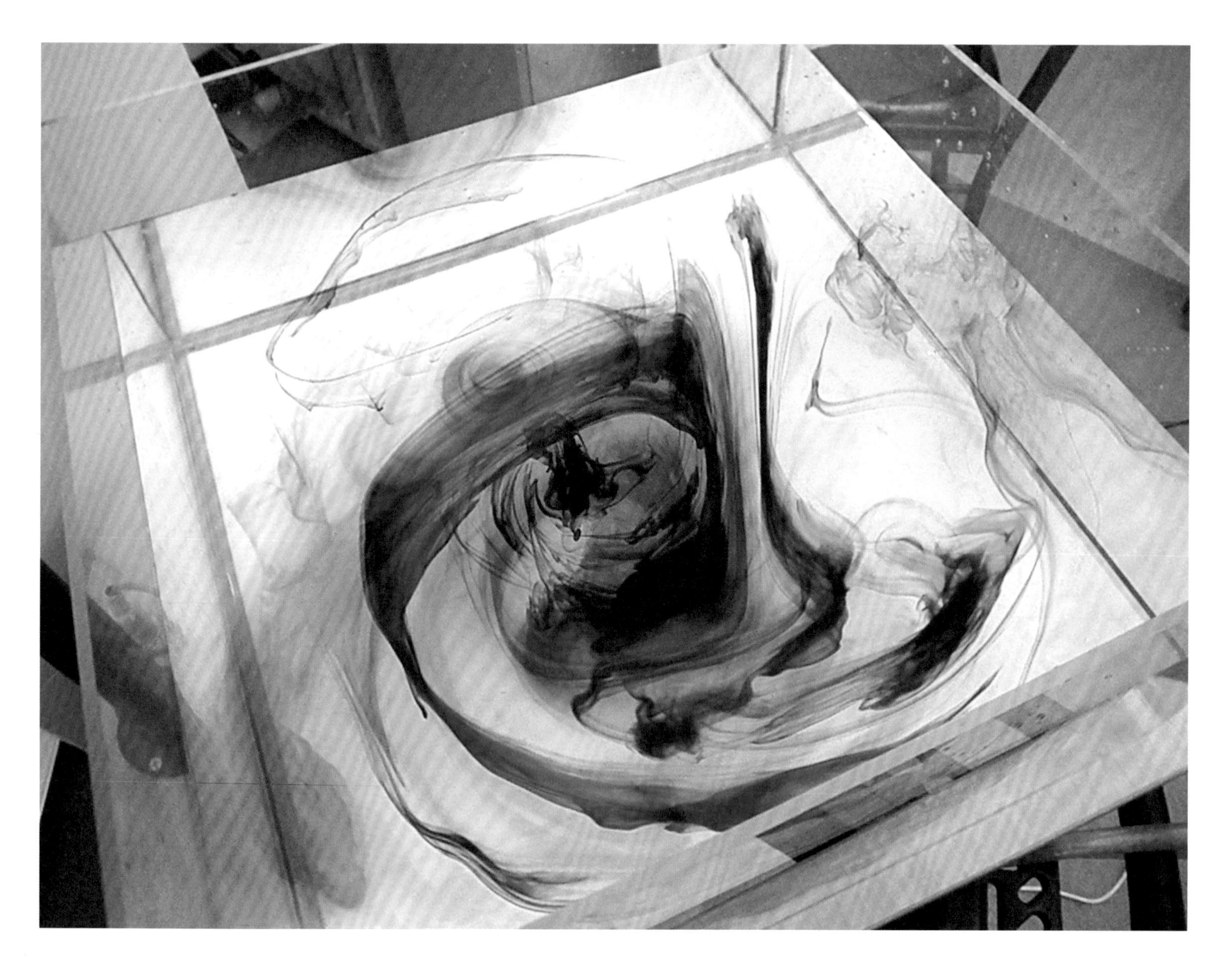

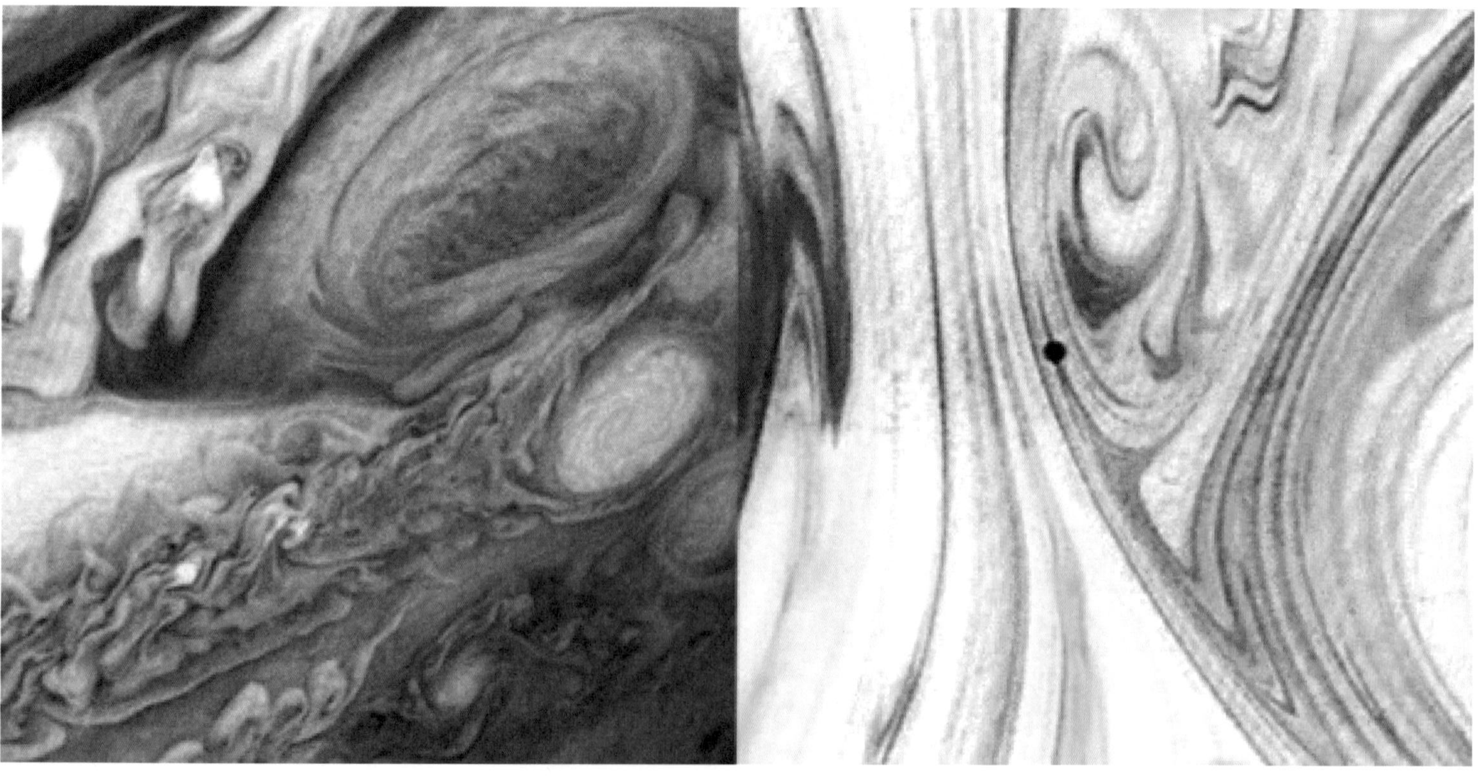

Figure 41.5 (left page, top)
High-resolution image from the dye-stirring experiment in the rotating tank—dye moves in ribbons and curtains.

Figure 41.6 (left page, bottom)
Left: A false-color image of the Great Red Spot of Jupiter, courtesy of NASA. Right: The dye-stir experiment shown in figure 41.4. Rotational constraints are clearly at work in both systems, as is evident from the filamentary interleaving patterns.

Figure 41.7 (right)
Left: Main ingredients for the weather "assembly" experiments in which two key ingredients are brought together: equator-pole temperature difference (ΔT) and earth rotation (Ω). Right: Matrix showing two experiments with ΔT fixed and different Ω, small and large.

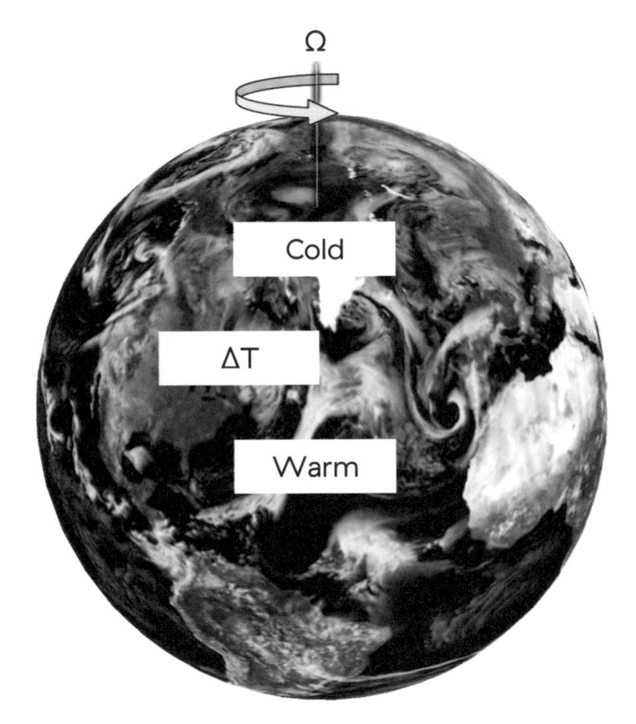

Ω : Earth rotation
ΔT : Temperature Difference Pole – Equator

Two experiments	Two regimes
Ω=large ΔT=large →	Mid-latitude weather systems
Ω=small ΔT=large →	Tropical hadley cell circulation

2. ACTIVE ASSEMBLY

Weather Ingredients

Having transformed the fluid by rotation and inspired by the active assembly theme of this book, we now demonstrate how we can artfully combine a few key ingredients and "assemble" laboratory abstractions that are indicative of the mechanisms and processes that underlie weather patterns.

The first ingredient of importance is rotation, discussed above. The second key ingredient is the equator-pole temperature difference, ΔT: the equator is some 30°C warmer than the poles, and this temperature difference results in motion that attempts to reduce that gradient. But because the earth is rotating, rotation plays a key role in shaping the resulting pattern of wind systems. We now demonstrate the interplay of rotation and temperature gradient in a simple laboratory experiment.

Rotation in the Presence of a Lateral Temperature Gradient

The experimental setup comprises a circular tank of water with a thin-walled stainless steel can at its center containing a mix of ice and water (figure 41.2, right). The melting ice extracts its latent heat of fusion from the surrounding water, cooling it and inducing differential cooling. This is the second ingredient, ΔT, of our active assembly. The only difference between the two experiments is the first ingredient, the rotation rate, Ω, of the turntable on which the circular tank sits.

We carry out the following experiments in turn:

1. slow rotation, Ω = small, less than one revolution per minute (rpm). This is an analogue of the circulation of the tropical atmosphere where the effect of Earth's rotation is felt less than at higher latitudes.

2. fast rotation, Ω = large, order of 6 rpm, an analogue of mid-

latitude weather systems. The two experiments are represented schematically in figure 41.7.

Circulation at Slow Rotation—the Tropical Hadley Circulation

In this experiment the tank rotates very slowly, only once a minute. Nevertheless the rotation of the turntable is still "felt." It imparts a "winding effect" on the fluid, as revealed by the beautiful corkscrew seen in the side view of figure 41.8. At the free surface of the water the currents are moving faster than they are lower down, and in the same direction as the rotation of the tank—see the trajectory of black paper dots in the movie of the experiment.[4] At the bottom of the tank the fluid is moving much more slowly and, in fact, in the opposite direction to the tank—see the permanganate purple streaks in figure 41.8 (top left) as viewed from the rotating camera. The flow at the top is analogous to the upper-level atmospheric westerlies (known as the jet stream), while the flow at the bottom is analogous to the easterly (trade) winds of the low-latitude tropical circulation.

Circulation at fast rotation: mid-latitude weather systems

Following the matrix of experiments outlined in figure 41.7, we can compare the slow-rotation experiment to another with exactly the same laboratory setup but now rotating much more rapidly, at 10 revolutions a minute—Ω is large. This rotation rate yields circulation patterns that are typical of the middle-latitude atmospheric circulation. At these higher rotation rates the axisymmetric circulation of the tropical regime completely disappears. The flow becomes turbulent and more chaotic; tongues of cold and warm fluid intermingle, sliding on top of one another (figure 41.9, left). The analogy to mid-latitude weather systems is very evident. Figure 41.9 (right) shows the 850 mb temperature (at an altitude of some 2 km above the surface) over the northern hemisphere for a typical day in winter. We see tongues of warm air moving northward toward the pole, while at another latitude, and contemporaneously, cold air moves south towards the equator. The effect of this exchange is to equilibrate the pole-equator temperature gradient just as in the rotating tank experiment.

The marked contrast in phenomenology between the low- and high-rotation experiments helps us grasp the importance of rotation in shaping weather regimes on earth. The large-scale flow of the deep tropics is essentially laminar, very different from the highly turbulent weather systems typical of middle latitudes. The underlying cause of this dramatic difference can be readily captured by the appropriate mix of just two ingredients: Ω and ΔT.

Experiments such as these, the so-called "annulus" experiments, were very influential in developing our understanding of the general circulation of the atmosphere and its predictability, as discussed in Lorenz's seminal review on the general circulation of the atmosphere.[5]

3. DISCUSSION AND FUTURE PLANS

In accord with the theme of the "Active Matter" conference, we have seen how rotational constraints result in a remarkable transformation of a fluid's behavior, imparting rigidity parallel to the rotation axis and forcing fluid columns to move in quasi-horizontal planes perpendicular to the rotation vector. Active assembly of weather and climate regimes is achieved by combining this rotational rigidity with a radial temperature gradient, mimicking a cold pole and warm equator. Large-scale patterns of flow emerge that are highly reminiscent of, and have deep connections to, the fundamental properties of atmospheric weather systems and regimes of flow on a differentially heated/cooled rotating earth. In this way we are able to expose underlying principles that go a long way to explaining what might otherwise

Tropical Circulation (Ω=small, ΔT=large)

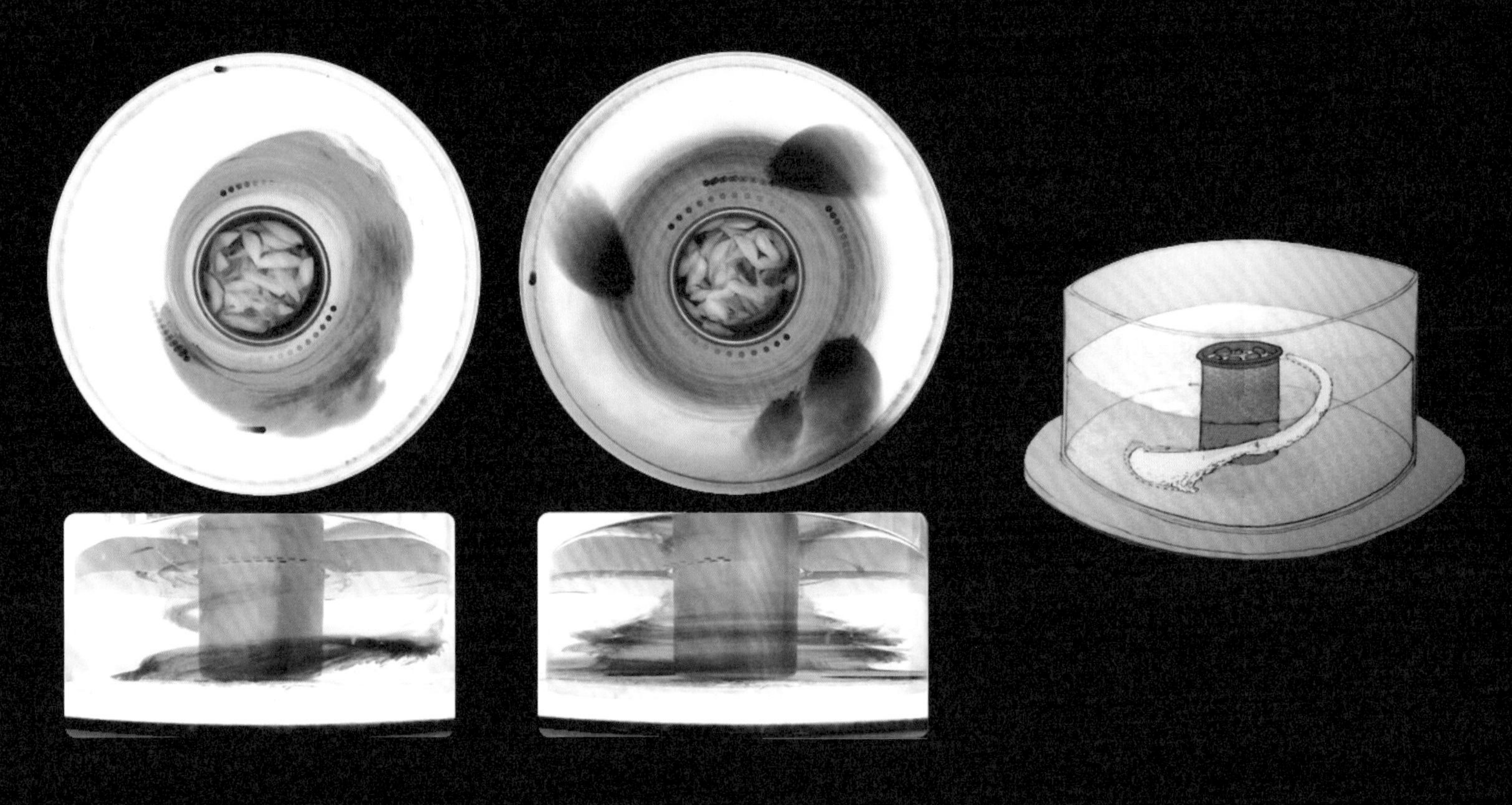

Weather Systems (Ω=large, ΔT=large)

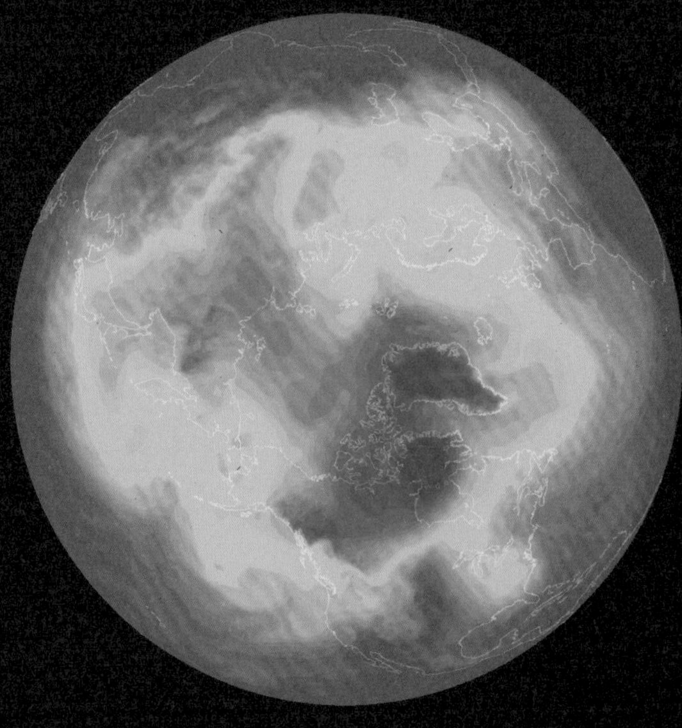

Figure 41.8 (top)
The slow-rotation "tropical circulation" experiment. Left: Top views show the evolving azimuthal circulation. Green dye helps visualize the flow throughout the depth of the water column. Top flow is visualized by tracking black paper dots. Potassium permanganate crystals, dropped into the water, settle on the bottom, creating a purple plume which indicates the direction of bottom flow. Side views show the evolving corkscrew pattern early and later in the experiment. Right: A schematic drawing of the evolving green dye. (For a full movie of the experiment, see http://lab.rotating.co)

Figure 41.9 (bottom)
Weather systems in the laboratory and in nature. Left: View from the co-rotating camera showing turbulent eddies in the high-rotation experiment transferring heat from the warm edge of the tank to the cold center. Right: The 850 mb temperature (at a height of roughly 2 km above the earth's surface) on a typical winter day in the northern hemisphere. We see a burst of cold arctic air traveling south over Canada and the US, while further east a tongue of warm tropical air is carried northward toward Greenland. A video of the high-rotation experiment, together with observations of the evolving 850 mb temperature field, can be found at http://paoc.mit.edu/labguide/circ_exp_fast.html.

seem hopelessly complex phenomena. The approach outlined here motivates the "Weather in a Tank" project that has been extensively used in teaching at MIT and elsewhere.[6]

Finally, to make the "Weather in a Tank" experiments and curriculum available to those who do not have access to rotating tank apparatus, we have developed a growing list of virtual laboratories in which image processing and web technologies are employed in lieu of a real laboratory. Their use in teaching is described by Illari, Marshall, and McKenna (2017).[7] An example of a virtual simulation of the slow-rotation (Hadley circulation) experiment can be found at http://lab.rotating.co/#flyby.

The experiment comes alive just like a PIXAR movie!

4. SUMMARY

We have illustrated through laboratory experiments the remarkable transformation that occurs in the properties of fluids as they are rotated about a vertical axis, as happens to Earth's atmosphere on our rotating planet. Rotating fluids in a very real sense are not fluids at all, because angular momentum constrains them to move as vertical columns made rigid by rotation. Active assembly of weather and climate regimes is achieved by combining this rotational rigidity with a radial temperature gradient, mimicking a cold pole and warm equator. Large-scale patterns of flow emerge that are highly reminiscent of, and have deep connections to, the fundamental properties of atmospheric weather systems and regimes of flow on a differentially heated/cooled rotating earth. The assembly of these ingredients and their use in pedagogy is at the heart of teaching meteorology pioneered in the "Weather in a Tank" project at MIT.

NOTES

1 L. Illari, J. Marshall, P. Bannon, J. Botella, R. Clark, T. Haine, A. Kumar, S. Lee, K. J. Mackin, G. A. McKinley, M. Morgan, R. Najjar, T. Sikora, and A. Tandon, "'Weather in a Tank': Exploiting Laboratory Experiments in the Teaching of Meteorology, Oceanography, and Climate," *Bull. Amer. Meteor. Soc.* 90 (2009): 1619—1632.

2 L. Illari and J. Marshall, "Weather in a Tank: A Laboratory Guide to Rotating Tank Fluid Experiments and Atmospheric Phenomena," 2006, available at http://paoc.mit.edu/labguide.

3 G. I. Taylor, "Experiments with Rotating Fluids," *Proc. Roy. Soc. Lond. A* 100 (1921): 114—121.

4 http://paoc.mit.edu/labweb/lab7/hadley.mpg

5 E. N. Lorenz, *The Nature and Theory of the General Circulation of the Atmosphere* (Geneva: World Meteorological Organization, 1967).

6 http://paoc.mit.edu/labguide/

7 L. Illari, J. Marshall, and W. McKenna, "Virtually-Enhanced Fluid Laboratories for Teaching Meteorology," *Bull. Amer. Meteor. Soc.* doi:10.1175/BAMS-D-16-0075.1, in press.

Aerial Assemblies, a project focused on balloon-filled modules that can assemble in the airspace high above land, construction sites or complex environments. Self-Assembly Lab, MIT + Autodesk Inc.

Fluid Crystallization, Credits:
Self-Assembly Lab, MIT & Arthur Olson

CONCLUSION: ACTIVE MATTER AND BEYOND

Skylar Tibbits

This book embodies the state of the art in active matter—its core principles and phenomena that have converged several threads of research. Some work on molecular interaction and biological functionality; others bring biological principles to synthetic materials; some take a robotic approach to programmability and communication; while yet others take a materials approach to sensing and actuation. Some groups have focused on the assembly, reconfiguration, and disassembly aspects of active matter, while others focus on the transformation, shape change, phase change, or even communication and computing possibilities. Some focus on the fundamentals of the materials, the universal scalability, the design and fabrication tools to enable such advances, or the human interaction, industrial applications, and future construction scenarios. One author may focus on programming bacteria while another describes programming DNA, electronics, wood, textiles, hydrogels, or any other material. We have seen self-organization from atmospheres to bacteria, and from granular matter in sand to building-scale structures. Authors have described the self-reconfiguration of DNA and textiles and that of entire cities. Each body of work is certainly unique in its implementation, forces, or scalar differences, yet all of them also tell us, fundamentally, how to embed information and computation in a physical medium to imbue material agency. The importance of this book is less in the specifics of any one detail than in the global shift we are witnessing in our capability to design with active matter.

There are a number of common threads to highlight. First, how digital information has led to programmable materials and how the ability to program something can lead to a greater collaboration or capability in the medium. With computing we witnessed the shift from initially utilitarian calculations to computing in every aspect our daily lives. Similarly, with materials, we are at a renaissance moment of programmability, where we are seeing new material behaviors, applications, and active transformations. Secondly, we see natural materials programmed genetically, synthesized, grown, fabricated digitally, and harnessed, leading to natural, yet synthetic, novel configurations and properties. We also witness synthetic materials with natural behaviors that only previously existed in the living world—like adaption, repair, growth, and reconfiguration. Now these lifelike material properties are emerging in our products and physical, human-made world. Thirdly, we have demonstrated that local parts can organize themselves and lead to bottom-up principles of manufacturing, from granular jamming to global weather patterns and swarm robot assembly. Finally, we have seen that the fold can be scale-independent, from computational folding algorithms to physical implementations in DNA origami, graphene folding, and self-folding paper. Many of these principles and active material phenomena can now be applied at the smallest and largest of scales.

What comes after programmable and active matter? Will active matter follow a trajectory similar to that of computing, perhaps a new form of material AI? We are currently witnessing greater agency in materials, with logic and simple programs that can be embedded in matter. As electronics attempts to continually scale with Moore's law, chips will likely reach physical limitations in size, and we will likely see more examples of material computation like "reservoir computing" where we can utilize physical material properties for powerful computation, which is far easier to produce in large quantities and perhaps more robust than silicon chips.

However, active matter is arguably useful not only as a platform for computing but also as a radically different style of computation, which is not about capacity, speed, or efficiency per se, but rather about nuance, fluctuation, fluidity, or perhaps creativity, and adaptability—no longer just about pure brute efficiency, but like a more human style of "computation." In this world, the material medium may develop greater agency and a greater capacity to make decisions, more sophisticated embedded programmable behaviors, or nuanced relationships with its environment. Imagine a water computer that can reversibly go from computation mode to fabrication mode simply by lowering the temperature at which "computational water droplets" turn to ice or even to steam and amorphously fill the environment! Or imagine manufacturing scenarios where the parts are not only coming together autonomously, but continually adapting to propose new design solutions for optimal conditions in the environment. In this case the computational medium is more than just a computer; it is a design collaborator and a fluid process for translating information to physical materials and back. Besides giving readers a glimpse at these exciting prospects, processes, and new material capabilities, *Active Matter* shows that newly harnessed materials lead to new material performance and greater material autonomy.

And what about the industrial applications of active matter? Steelcase has described a compelling industry perspective, but we can imagine that the entire manufacturing and product life cycle may once again be completely transformed. We can now see materials that offer unprecedented capabilities like logic, sensing, and actuation without external devices, and even new manufacturing processes where components coalesce into fully functional products. Materials and manufacturing then lead to the design of completely new products that can adapt and transform, with greater performance, intelligence, and unprecedented functional characteristics. With new products we will also have new shipping and distribution processes with smarter packaging that molds itself around the product, adapts in transit to the fluctuating forces and environmental conditions, then releases the product to self-assemble on site. Finally, with new materials, manufacturing processes, products, and shipping, the entire life cycle will be able to start again, fresh, with self-*dis*assembly—the material components will dissolve or self-separate and turn back into fundamental units, to be built once again into functional materials and active products.

Other imaginable applications are even more accessible and yield new research opportunities with their own sets of questions. Think of a stent, for example, a device that surgeons insert into an artery. The stent is meant to expand and in so doing open a collapsed passageway. The stent is traditionally made in a finite number of sizes, each with two states—open or closed. Imagine now the future version of this device. It can be designed and fabricated completely adjusted to the patient's body *and* condition. The new, programmable and active stent can be unique. It might also have new functionality; it might sense temperature, pH, or pressure changes to activate varying degrees of open or closed pathways. It might be customized to sense the environment, to react in a precisely designed way without electromechanical devices or human intervention. Medical applications will likely enter into the future of active matter research, transforming not only products and systems but also our very lives.

Active Matter also highlights incredible advances in textiles and garments. Such work points to a very different relationship with our clothes, our shoes, and the devices that augment our physical comfort. We can foresee clothing that adapts to an athlete's fluctuating body temperature, grows pores for breathability when they start to sweat, closes up when it starts to rain, and changes color to indicate their biometrics. As the athlete runs, their shoes provide the perfect comfort and support; then the moment they step in a puddle the rubber adapts to increase traction, or on grass deep studs grow from the shoe. When they get to work, their garments adjust to the changing environment and add compression to control blood circulation and ease the athlete down from their exercise. At their desk, the chair self-adjusts to provide the perfect individualized lumbar support. These advances may sound like science fiction today, yet, as *Active Matter* lays out, these are the kinds of developments we are collectively working on. Today we have humble lab prototypes that every day show us this future is closer than we think.

One day we may even see truly intelligent materials, and if this happens perhaps we shouldn't be surprised. Isn't active matter somehow closer to human intelligence than computers are to artificial general intelligence? Isn't human intelligence precisely active matter anyway? Not that we completely understand human intelligence today, but we are certainly not made of chips and transistors. What we do know is that our cells, synapses, and biological computation make up our human assembly, evolution, adaptation, decision-making, and physical transformation—*not* actuators, sensors, or traditional computers. In that way, with active matter we may be closer to material intelligence that we previously thought. Perhaps it is only a matter of material design.

Active Matter does not yet address the ethical implications of material agency. What happens, for example, when materials fail or go out of control? Will materials replace humans in assembly or even design when they can self-assemble, reconfigure, and evolve? Who is responsible when active matter fails—is it the designer, the fabricator, the user, or the material itself? With any new technological development all sorts of ethical concerns quickly arise, and it will be important for policymakers, researchers, and end users to collaborate effectively to design safe, functional, and productive relationships with active matter. Certainly we are far from materials taking over or from active matter insurance agencies. Still, it would be naive not to acknowledge the ethical or even the legal implications, just as it is productive to speculate about the relative value we reap and risks we run in developing new materials capable of radically changing our collective lives. We hope that this book sheds the first light on a rapidly emerging field and will help spawn new conversations, research studies, and effective policies around the ethics as well as the implications of active matter in the future.

One of the most important benefits of active matter that is not directly addressed in this book is the potential upside for long-term environmental sustainability, since so much of active matter can engage reversible and renewable processes. Traditionally, if we want to make a product "smart" we tend to introduce devices and systems, new components, or extra power to ramp up the performance. Active matter paves a completely different path to greater functionality and intelligence in our materials and products without the reliance on additional devices. For example, when we want to make smarter buildings we now add complex heating/cooling systems or smart thermostats. When we want to make smart wearables we add electronics, batteries, and sensors—but active matter shows that much of this will not be needed.

Smarter and smarter systems will emerge with less and less—I believe this is the path toward true material elegance in design and functionality: *more with less*. We will use less battery power and nonrenewable energy sources, by using instead more of the passive and abundant energy that is readily available in the environment—temperature, moisture, light, sound, pressure, and many other forms. Active matter uses less of the expensive, complex, failure-prone, and nonrecyclable electronic components that make our current devices "smart."

Ultimately, active matter also means cheaper products can be made dynamic, to do more than their static counterparts—and maybe *even more* than their expensive robotic versions. Active matter means potentially less failure with fewer components and more robustness through adaptability rather than resistance to all forces and variable conditions. Active matter means less assembly time and complexity since the materials will assemble themselves, or more humbly, since smarter materials require fewer physical components and thus reduce the complexity of assembly. All of these unique capabilities elegantly point toward a completely different perspective: an optimistic outlook on the future of active materials that includes improving our relationship with energy and the finite resources of our physical environment.

4D Printed self-folding cube shown transforming from a flat sheet into a 3-dimensional box. Self-Assembly Lab, MIT + Stratasys Ltd. + Autodesk Inc.

FIGURE CREDITS

Figure 0.1: Reprinted with permission of Nokia Corporation.

Figure 0.2: Courtesy MIT Museum.

Figure 0.3: Photo courtesy of Popular Science.

Figure 0.4: Skylar Tibbits & the Center for Bits and Atoms, MIT. Photo: Skylar Tibbits.

Figure 0.5: Photo: Self-Assembly Lab, MIT & E Roon Kang, 2014.

Figure 0.6: Self-Assembly Lab, MIT.

Figure 0.7a: © L. Barry Hetherington, 2015.

Figure 0.7b: Photo credit: Sharon Lacey, courtesy of MIT Center for Art, Science & Technology (CAST)

Figure 0.8: Albert Elias.

Figures 0.9—0.12: Photo courtesy of Steelcase.

Figure 1.1: Photograph © 2014 Studio Tomás Saraceno.

Figures 1.2—1.4: Photograph © 2015 Studio Tomás Saraceno.

Figure 1.5: Courtesy the artist; Tanya Bonakdar Gallery, New York; Pinksummer contemporary art, Genoa; Andersen's Contemporary, Copenhagen; Esther Schipper, Berlin. Photograph © 2016 Studio Tomás Saraceno.

Figures 1.6—1.7: Photograph © 2015 Studio Tomás Saraceno.

Figure 1.8: Courtesy the artist; Tanya Bonakdar Gallery, New York; Andersen's Contemporary, Copenhagen; Pinksummer contemporary art, Genoa; Esther Schipper, Berlin. Photograph © 2010 Studio Tomás Saraceno.

Figure 1.9: Courtesy the artist; Tanya Bonakdar Gallery, New York; Andersen's Contemporary, Copenhagen; Pinksummer contemporary art, Genoa; Esther Schipper, Berlin. Photograph © 2016 Studio Tomás Saraceno.

Figure 1.10: Courtesy the artist; Pinksummer contemporary art, Genoa; Tanya Bonakdar, New York; Andersen's Contemporary, Copenhagen; Esther Schipper, Berlin. Photograph © 2015 Studio Tomás Saraceno.

Figures 2.1, 2.2b—c, 2.3, 2.5—2.6: Zhao Qin, 2016.

Figure 2.2a: Libiakova et al., *Plos One*, 9, no. 8, e104424.

Figure 2.4: Zhao Qin and Markus J. Buehler, 2016.

Figures 3.1—3.2: Image by Fiorenzo Omenetto, 2016.

Figure 3.3: Image by Fiorenzo Omenetto, c. 2012.

Figure 3.4: Image by Fiorenzo Omenetto, c. 2013.

Figure 3.5: Image by Fiorenzo Omenetto, c. 2008.

Figures 4.1—4.2: Bryan Wei, Mingjie Dai, Peng Yin. Harvard University, 2012.

Figures 4.3—4.4: Yonggang Ke, Luvena L. Ong, William M. Shih, and Peng Yin. Harvard University, 2012.

Figures 5.1—5.2: David Muller group, adapted from P. Y. Huang et al., "Imaging Grains and Grain Boundaries in Single-Layer Graphene: An Atomic Patchwork Quilt," *Nature* 469 (2011): 389—392.

Figure 5.3: Jonathan Alden, McEuen group.

Figure 5.4: Peter Rose and Pinshane Huang, McEuen and Muller groups.

Figures 5.5—5.6: McEuen and Muller groups, adapted from M. K. Blees, A. W. Barnard, P. A. Rose, S. P. Roberts, K. L. McGill, P. Y. Huang, A. R. Ruyack, J. W. Kevek, B. Kobrin, D. A. Muller, and P. L. McEuen, "Graphene Kirigami," *Nature* 524 (2015): 204—207.

Figure 6.1: Max Carlson and Michael Short, 2016.

Figure 6.2: R. H. French, R. M. Cannon, L. K. DeNoyer, and Y. M. Chiang, "Full Spectral Calculation of Non-retarded Hamaker Constants for Ceramic Systems from Interband Transition Strengths," *Solid State Ionics* 75 (1995): 13—33.

Figures 6.3, 6.6: Max Carlson, 2016.

Figures 6.4—6.5: Max Carlson and Michael Short, 2016.

Figures 7.1—7.2: Candice Gurbatri and Tetsu Harimoto, 2016.

Figure 7.3: Reprinted with authors' permission from Tal Danino, Octavio Mondragón-Palomino, Lev Tsimring, and Jeff Hasty, "A Synchronized Quorum of Genetic Clocks," *Nature* 463, no. 7279 (2010): 326—330.

Figure 7.4: Reprinted with authors' permission from M. Omar Din, Tal Danino, Arthur Prindle, Matt Skalak, Jangir Selimkhanov, Kaitlin Allen, Ellixis Julio, Eta Atolia, Lev S. Tsimring, and Sangeeta N. Bhatia, "Synchronized Cycles of Bacterial Lysis for In Vivo Delivery," *Nature* 536, no. 7614 (2016): 81—85.

Figures 7.5—7.6: Photo credit: Soonhee Moon.

Figure 8.1, 8.2b: Katia Zolotovsky.

Figure 8.2a: Stefan Schwabe and Emilia Fostreuter.

Figure 8.2c: Reprinted by permission from Macmillan Publishers Ltd. (Nature Publishing Group) (doi:10.1038/nature03461), 2005.

Figures 8.3—8.4: Katia Zolotovsky & Merav Gazit.

Figures 9.1—9.6: Reproduced with permission from John A. Rogers at University of Illinois at Urbana-Champaign and Northwestern University.

Figures 10.1—10.6: Biomimetic 4D Printing. A. Sydney Gladman, Elisabetta A. Matsumoto, L. Mahadevan, and Jennifer A. Lewis: Harvard University and Wyss Institute for Bioinspired Engineering, 2015.

Figures 11.1—11.5: Hyunwoo Yuk, 2015, photos from MIT Soft Active Materials Laboratory.

Figures 12.1—12.7: Images courtesy of Nadia M. Benbernou, Erik D. Demaine, Martin L. Demaine and Anna Lubiw.

Figures 13.1—13.2: The Harvard Microrobotics Lab.

Figures 14.1—14.2: Image courtesy of Daniela Rus, CSAIL, MIT.

Figures 15.1—15.2, 15.8: Self-Assembly Lab, MIT.

Figure 15.3: Self-Assembly Lab, MIT & Christophe Guberan, Erik Demaine, Autodesk Inc.

Figures 15.4—15.6: Self-Assembly Lab, MIT & Christophe Guberan.

Figure 15.7: Self-Assembly Lab, MIT & Carbitex LLC, Autodesk Inc.

Figures 15.9—15.10: Self-Assembly Lab, MIT & Carbitex LLC.

Figure 15.11: Self-Assembly Lab, MIT & Christophe Guberan, Erik Demaine, Autodesk Inc.

Figure 15.12: Christophe Guberan, Carlo Clopath & Self-Assembly Lab, MIT.

Figure 15.13: Christophe Guberan & Self-Assembly Lab, MIT.

Figures 16.1—16.5: Reprinted by permission from Macmillan Publishers Ltd. (Nature Publishing Group) (doi:10.1038/nature18960), 2016.

Figures 17.1—17.5: Ying Liu and Sally Van Gorder.

Figure 18.1: P. T. Brun and Douglas P. Holmes; image courtesy of Douglas P. Holmes, Boston University.

Figure 18.2: Mark Steranka, Matteo Pezzulla, and Douglas P. Holmes; image courtesy of Douglas P. Holmes, Boston University.

Figure 18.3: Abdikhalaq Bade, Matteo Pezzulla, and Douglas P. Holmes; image courtesy of Douglas P. Holmes, Boston University.

Figure 18.4: Henry Hwang, Matteo Pezzulla, and Douglas P. Holmes; image courtesy of Douglas P. Holmes, Boston University.

Figures 19.1, 19.2, 19.4—19.7: Greg Blonder.

Figure 19.3: Greg Blonder and Tucker Toys International (by permission).
Figures 20.1—20.12: Christophe Guberan, 2012.

Figures 21.0—21.7: Self-Assembly Lab, MIT.

Figures 22.1—22.6: Biocouture™. Photo © Gary Wallace.

Figure 23.1: Behnaz Farahi, Pier9/Autodesk, Madworkshop. Photo: Elena Kulikova.

Figures 23.2—23.3: Behnaz Farahi, Pier9/Autodesk, Madworkshop. Photo: Charlie Nordstrom.

Figures 24.1—24.3: Produced by, and in collaboration with, Stratasys, Ltd. Image: Mediated Matter.

Figures 24.4—24.7: Produced by, and in collaboration with, Stratasys, Ltd. Photo: Yoram Reshef, courtesy of Mediated Matter.

Figure 24.8: Produced by, and in collaboration with, Stratasys, Ltd. Image: Mediated Matter.

Figure 24.9: Produced by, and in collaboration with, Stratasys, Ltd. Photo: Yoram Reshef, courtesy of Mediated Matter.

Figures 24.10—24.11: Produced by, and in collaboration with, Stratasys, Ltd. Image: Mediated Matter.

Figures 25.1—25.3: Nervous System (Jesse Louis-Rosenberg and Jessica Rosenkrantz). Photo: Steve Marsel.

Figures 25.4—25.8: Nervous System (Jesse Louis-Rosenberg and Jessica Rosenkrantz).

Figures 26.1—26.11: Photo: Felicia Davis.

Figures 27.1—27.3, 27.5: Jane Scott. Photograph © Cristina Schek, 2015.

Figures 27.4, 27.6—27.7: Jane Scott.

Figures 28.1—28.2: XS Labs. Photo: Shermine Sawalha and Hugues Bruyère.

Figures 28.3—28.7: Photo: Marcelo Coelho.

Figure 28.8: LogicINK. Photo: Skylar Tibbits.

Figures 29.1—29.11: Tangible Media Group, 2017. Image courtesy of Tangible Media Group, MIT Media Lab.

Figure 30.1: Nadya Peek.

Figures 30.2—30.4a: NASA ARC CSL.

Figures 30.4b—30.7: Courtesy of MIT CBA.

Figure 30.8: Amanda Ghassaei, 2016.

Figure 31.1: From Justin Werfel, Kirstin Petersen, and Radhika Nagpal, "Designing Collective Behavior in a Termite-Inspired Robot Construction Team," *Science* 343, no. 6172 (2014): 754—758. Reprinted with permission from AAAS. : Clockwise from top left: Mercury Freedom, CC BY-SA 3.0; Hugo.arg, CC BY-SA 4.0; Harald Süpfle, CC BY-SA 2.5; Werfel, Petersen, and Nagpal 2014 (2 panels); Thomas Schoch, CC BY-SA 3.0.

Figure 31.2: Eliza Grinnel, 2014. Image courtesy of Eliza Grinnel, Harvard School of Engineering and Applied Sciences.

Figure 31.3, 31.6: Juhun Lee, 2016. Images courtesy of Juhun Lee.

Figure 31.4: Max Kanwal, 2014. Image courtesy of Max Kanwal.

Figure 31.5: Devin Carroll, 2014. Image courtesy of Devin Carroll.

Figures 32.1—2: Courtesy of The Living.

Figure 32.3: Photo by Amy Barkow, courtesy of The Living.

Figure 32.4: Photo by Iwan Baan, courtesy of The Living.

Figure 32.5: Photo by Charles Roussel, courtesy of The Living.

Figures 33.1—33.6: Institute for Computational Design, University of Stuttgart.

Figures 34.1, 34.4—34.6: Sabin Design Lab. Photo courtesy of Jenny E. Sabin.

Figure 34.2: Jie Li, Guanquan Liang, Xuelian Zhu, and Shu Yang, University of Pennsylvania. Photo: Felice Macera.

Figure 34.3: Photo courtesy of Randall D. Kamien.

Figure 35.1: Photo: James Weaver.

Figures 35.2—35.4: Photo: Jonathan Grinham.

Figures 35.5—35.14: Photo: John Kennard.

Figure 36.1: Photo: H. M. Jaeger, University of Chicago.

Figures 36.2—36.3: Photo: K. A. Murphy and H. M. Jaeger, University of Chicago.

Figure 36.4: Photo: M. Z. Miskin and H. M. Jaeger, University of Chicago.

Figure 36.5: Photo: L. K. Roth and H. M. Jaeger, University of Chicago.

Figures 37.1—37.6: Andreas Thoma and Petrus Aejmelaeus-Lindström, 2015. Project credit: Gramazio Kohler Research, ETH Zurich, and Self-Assembly Lab, MIT. Photo courtesy of Gramazio Kohler Research, ETH Zurich, and Self-Assembly Lab, MIT.

Figure 38.1: Photographed at University of Southampton, Anechoic Chamber and Psychoacoustics lab; photo: Lotje Sodderland.

Figure 38.2: Photographed at Ricardo Bofill, La Fabrica; photo: Lotje Sodderland.

Figure 38.3: Photographed at LKH-Universitätsklinikum, Thoracic Clinic and Hyperbaric Surgery, Graz; photo: Lotje Sodderland.

Figures 38.4—38.5, 38.7: Lucy McRae and Daniel Gower. Photographed at LaSainte Union Catholic School, UK; photo: Lotje Sodderland.

Figure 38.6: Lucy McRae and Daniel Gower. Photographed at LaSainte Union Catholic School, UK; photo: Julian Love.

Figure 38.8: Lucy McRae. Photographed at Kew Garden Treetop walk; photo: The Helicoptor Girls.

Figures 39.1—39.6: Courtesy of Höweler + Yoon Architecture.

Figure 39.7: Courtesy of Höweler + Yoon Architecture. Photo: John Horner.

Figure 39.8: Courtesy of Höweler + Yoon Architecture.

Figures 39.9—39.11: Courtesy of Höweler + Yoon Architecture.

Figures 40.1—40.9: Provided by Ibañez Kim.

Figure 41.1: GOES East satellite, 2015. (NASA-GSFC Project).

Figures 41.2—41.5: Courtesy of Lodovica Illari, Department of Earth Atmospheric and Planetary Sciences, MIT.

Figure 41.6: Left: courtesy of NASA. Right: courtesy of Lodovica Illari, Department of Earth Atmospheric and Planetary Sciences, MIT.

Figures 41.7—41.9: Courtesy of Lodovica Illari, Department of Earth Atmospheric and Planetary Sciences, MIT.

INDEX

These are the kinds of times when disciplines get recast...

John Main, DARPA Program Manager / Active Matter Summit